Titanium Microalloyed Steel: Fundamentals, Technology, and Products

Xinping Mao
Editor

Titanium Microalloyed Steel: Fundamentals, Technology, and Products

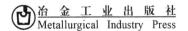

Editor
Xinping Mao
Chinese Academy of Engineering
Wuhan, Hubei, China

ISBN 978-981-13-3331-6   ISBN 978-981-13-3332-3 (eBook)
https://doi.org/10.1007/978-981-13-3332-3

Jointly published with Metallurgical Industry Press, Beijing, China

The print edition is not for sale in China Mainland, people from China Mainland should order the book from Metallurgical Industry Press, Beijing, China.
ISBN of the China Mainland edition: 978-7-5024-7806-3

Library of Congress Control Number: 2018961711

© Metallurgical Industry Press, Beijing and Springer Nature Singapore Pte Ltd. 2019
This work is subject to copyright. All rights are reserved by the Publishers, whether the whole or part of the material is concerned, specifically the rights of translation, reprinting, reuse of illustrations, recitation, broadcasting, reproduction on microfilms or in any other physical way, and transmission or information storage and retrieval, electronic adaptation, computer software, or by similar or dissimilar methodology now known or hereafter developed.
The use of general descriptive names, registered names, trademarks, service marks, etc. in this publication does not imply, even in the absence of a specific statement, that such names are exempt from the relevant protective laws and regulations and therefore free for general use.
The publishers, the authors and the editors are safe to assume that the advice and information in this book are believed to be true and accurate at the date of publication. Neither the publishers nor the authors or the editors give a warranty, express or implied, with respect to the material contained herein or for any errors or omissions that may have been made. The publishers remains neutral with regard to jurisdictional claims in published maps and institutional affiliations.

This Springer imprint is published by the registered company Springer Nature Singapore Pte Ltd.
The registered company address is: 152 Beach Road, #21-01/04 Gateway East, Singapore 189721, Singapore

Foreword

It is gratifying to see that Prof. Xinping Mao's new book "Titanium Microalloyed Steel" is published.

According to the title of this book, it looks like a book on materials. In fact, it gives considerations to both materials and metallurgy, and is a book featuring linking theory with industrial production practice.

Xinping Mao served as the chief project designer in a metallurgy design institute previously. Later, he worked in steel plants as the chief engineer. He accumulated rich practice experience in his career. During his pursuit for Ph.D., he studied in a wide range, paid much attention to theory, and had a good grounding in steel. All these are reflected in this book.

This book theoretically discusses the thermodynamics of titanium in molten steel, followed by the characteristics of solidification process, the characteristics of hot rolling process, the mechanism of titanium as a microalloying element and microalloying technology. Almost the whole production process is covered, which is a significant feature of this book.

Another feature of this book is that it is written based on the process characteristics of electrical furnace-ladle furnace-thin slab casting and direct rolling (TSCR)-continuous rolling, rolling of thin gauge products. The problem of nozzle clogging during tapping of titanium-bearing molten steel is solved, thus the clean titanium-bearing steel can be smoothly produced. By integrating TSCR, microalloying C–Mn steel by single titanium, hot rolling of thin gauge product, thermo mechanical control process (TMCP) technology, and so on, it is able to produce fine grained clean steel with high efficiency, low cost, and high stability. The problem of quality instability is solved. The process exhibits efficient integration and optimal configuration of principles, production technology, and equipment, and has theoretical significance and economic value. In addition, it reflects the process characteristics of producing titanium microalloyed steel by the steel plants and product level in China.

The study and small batch production of titanium-bearing steel were started in the 1960s in China. However, due to several reasons, mass production was not achieved. Since 2000s, because of the efforts of researchers in this field, such as

Prof. Xinping Mao, a great breakthrough was achieved. This book reflects the great progress in the development, production and application of titanium microalloyed steel in China. The products in the form of sheets, especially the hot rolled thin gauge products, are widely used in the fields of containers, automobiles, and engineering machinery, and the annual yield has reached megaton. In addition, the development of titanium microalloyed steel contributes to the development of relevant industry and technology progress. I believe this book is beneficial to the further development, production and application of titanium microalloyed steel.

This book is suitable for the production workers, researchers, designers, teachers, and managers in the fields such as steel metallurgy, materials development and application.

Beijing, China

Ruiyu Yin
Academician of the Chinese
Academy of Engineering

Preface

Titanium microalloying is an important method to improve the performance of steel. Titanium was discovered at the end of the eighteenth century and it was soon used as a microalloying element in steel before the 1920s. Titanium in steel exists mainly in the form of solute dissolved in iron matrix or titanium-bearing secondary phases. Compared with other microalloying elements, such as niobium and vanadium, there are more types of secondary phases in titanium microalloyed steel and its precipitation temperature range is wider, which have more complicated influence on the microstructure and performance of steel and are more difficult to control in the industrial production. Therefore, the behavior of titanium as a microalloying element in steel and the development of titanium microalloyed steel are very attractive topics to do in-depth research. The theoretical research covers the basic principles of chemical metallurgy and physical metallurgy. In addition, it gets involved with the transformation of basic theory into the process technology of actual production.

This book is written based on the authors' research results in the field of titanium microalloying technology over the years. It aims at introducing this field to the students and technicians in need. This book consists of six chapters. Chapter 1 introduces the definition, role, development history, and technical and economic characteristics of titanium microalloyed steel. Chapter 2 systematically discusses the principles of chemical metallurgy of titanium microalloyed steel. It contains the thermodynamics of titanium in molten steel, the reaction between titanium-bearing molten steel and slags and refractories, and the principles of oxide inclusion control technology. Chapters 3 and 4 discuss the physical metallurgy of titanium microalloyed steel. Chapter 3 focuses on the dissolving and precipitation of titanium-bearing phases, and Chap. 4 focuses on the recrystallization and phase transformation. Chapter 5 summarizes the technology of controlling production, structure and performance of titanium microalloyed steel. Chapter 6 introduces the development and application of representative titanium microalloyed steel products.

This book is translated from the Chinese version of "Titanium Microalloyed Steel", which was published by Metallurgical Industry Press in 2016. In this English version, Chap. 6 of the original Chinese version was revised by appending the application of titanium microalloying technology in the development of automobile beam steel, enamel steel and yoke steel.

Wuhan, China Xinping Mao

Acknowledgements

This book systematically summarizes the research achievements in the field of titanium microalloying technology of the authors' team. I am grateful for the careful guidance on our research from Academician Ruiyu Yin, Academician Yong Gan, Academician Shourong Zhang, Academician Guodong Wang, Academician Yuqing Weng and Prof. Naiyuan Tian. I really appreciate that Academician Ruiyu Yin wrote a preface to this book given his busy schedule. I appreciate Prof. Qilong Yong, Prof. Yonglin Kang, Prof. Guangqiang Li, Prof. Xinjun Sun, Prof. Xiangdong Huo, Prof. Gang Zhao, Prof. Liejun Li, Dr. Yizhong Chai, Dr. Jixiang Gao, SE. (senior engineer) Yan Yu, SE. Qilin Chen, SE. Zhenyuan Lin, SE. Chunyan Li, SE. Dayan Zhu, Dr. Jian Zhou, Dr. Liang Su, Dr. Zhaodong Li, Dr. Zhenqiang Wang, Dr. Changjun Wang, Dr. Siqian Bao, Dr. Gengwei Yang, Dr. Jinqiao Xu, Dr. Shuize Wang, Dr. Zhiyong Liu, Dr. Junhui Tao, Dr. Xiaolong Gan, SE. Jiangtao Zhao, and SE. Ming Du, Engineer Wenzhe Tao, Engineer Chang Song, and Engineer Dawei Huang for their fruitful research. I also show my appreciation to Dr. Weihua Sun, Dr. Wanjun Zhu, and SE. Yang Liu for the translation.

Contents

1 **Introduction** 1
   Xinping Mao, Qilong Yong and Xiangdong Huo

2 **Principles of Chemical Metallurgy of Titanium Microalloyed Steel** 35
   Guangqiang Li

3 **Physical Metallurgical Principles of Titanium Microalloyed Steel—Dissolution and Precipitation of Titanium-Bearing Secondary Phases** 71
   Qilong Yong, Xinjun Sun, Zhaodong Li, Zhenqiang Wang and Ke Zhang

4 **Physical Metallurgy of Titanium Microalloyed Steel—Recrystallization and Phase Transformation** 141
   Xinjun Sun, Zhaodong Li, Xiangdong Huo and Zhenqiang Wang

5 **Production, Structure and Properties Control of Titanium Microalloyed Steel** 185
   Jixiang Gao

6 **Design, Development and Application of Titanium Microalloyed Steel** 219
   Qilin Chen and ShuiZe Wang

List of Figures

Fig. 1.1	Crystal structures of Hcp-α-Ti and Bcc-β-Ti.	3
Fig. 1.2	Effects of microalloying elements on the recrystallization-stop temperature of 0.07%C–1.40%Mn–0.25%Si steel.	9
Fig. 1.3	Relationship between the strength increment and the size of the secondary phase according to two different strengthening mechanisms.	12
Fig. 1.4	Relationship between the strength increment and the size of the secondary phase with different volume fractions according to Orowan strengthening mechanism.	13
Fig. 1.5	Effect of secondary phase particles with radius r on suppressing the grain growth.	24
Fig. 1.6	**a** Electron Backscattered Diffraction (EBSD) orientation diagram of high strength steel, **b** Micro-orientation distribution diagram of the ferrite grain boundaries of high strength steel.	25
Fig. 1.7	**a** Morphology of the dislocations in high strength steel, **b** Distribution of precipitates on the dislocations in high strength steel.	26
Fig. 1.8	Particle size distribution of MX phase precipitates.	26
Fig. 2.1	Equilibrium relationship of Ti–O in molten iron at 1873 K.	41
Fig. 2.2	Stable regions for Al–Ti–O inclusions in liquid iron at 1873 K.	44
Fig. 2.3	Effects of the initial oxygen activity and the interval time between Al and Ti addition on the titanium yield and the content of dissolved aluminum ([Al]) [33].	47
Fig. 2.4	Inclusion in Al–Ti–Mg complex deoxidized steel [40].	52
Fig. 2.5	The comparison of inclusion size distribution in steel with different combinations of Al, Ti and Mg deoxidation [40].	53
Fig. 2.6	The heat treatment curve.	55
Fig. 2.7	The evolution of austenite grain at 1200 °C.	57
Fig. 2.8	The change of austenite grain size with time.	58

Fig. 2.9	Microstructure images of complex deoxidized as cast steel through optical microscope (3% nitric acid alcohol corrosion)	59
Fig. 2.10	Images of the acicular ferrite grown on the complex inclusions of MgO–Al$_2$O$_3$–TiO$_x$ and MnS and the EDS of the inclusions in Al–Ti–Mg deoxidized steel.	60
Fig. 2.11	The effect of different Ti/Al ratio in steel on precipitation proportion of MnS on oxide	61
Fig. 2.12	Variation of inclusions' size distribution under different Ti/Al ratio in steel	63
Fig. 2.13	The melting process of Al–Ti deoxidized and Mg treated steel	64
Fig. 2.14	The heat treatment process of Al–Ti complex deoxidized and Mg treated steel	64
Fig. 2.15	The amount and size distribution of inclusions in Al–Ti complex deoxidized and Ca/Mg treated steel	65
Fig. 2.16	The inclusions in the transverse and longitudinal section of C1 and C2 hot rolled plates	65
Fig. 2.17	Thickness of MnS on oxide inclusion surface and amount of inclusion in Al–Ti complex deoxidized and Ca/Mg treated steel	66
Fig. 2.18	The deformation aspect ratio of oxide-MnS complex inclusions in the Al–Ti deoxidized and Ca–Mg treated hot-rolled steel plate	66
Fig. 2.19	EDS maps of the inclusion and SEM images of acicular ferrite at the boundary of various oxides or sulfides on the surface of inclusions	67
Fig. 3.1	The titanium-bearing precipitation in different stages of TSCR process.	72
Fig. 3.2	The solubility products of TiC in austenite.	75
Fig. 3.3	The solubility products of TiN in austenite.	76
Fig. 3.4	The solubility products of TiC in ferrite.	77
Fig. 3.5	The solubility products of TiN in ferrite.	77
Fig. 3.6	The solubility products of TiN in liquid iron	78
Fig. 3.7	The solubility products of TiC and TiN in different iron matrix.	78
Fig. 3.8	The solubility products of TiS in austenite	80
Fig. 3.9	The solubility products of Ti$_2$CS in austenite	80
Fig. 3.10	The solubility products of compounds containing titanium in austenite	81
Fig. 3.11	The solubility products of sulfur containing compounds in austenite	82
Fig. 3.12	The solubility products of various microalloying carbide and nitride in austenite	83

List of Figures

Fig. 3.13	The solubility products of various microalloying carbide and nitride in ferrite	84
Fig. 3.14	The crystal structure of TiC and TiN	89
Fig. 3.15	The crystal structure of TiS. The white atom represents Ti atom, the purple atom represents S atom	90
Fig. 3.16	The crystal structure of Ti_2CS. The black atom represents C atom, the yellow atom represents S atom, the blue atom represents Ti atom	90
Fig. 3.17	Variation of x value of TiC_xN_{1-x} as a function of temperature in a trace Ti-treated steel	101
Fig. 3.18	PTT curves of Ti(C, N) in a trace Ti-treated steel	101
Fig. 3.19	Variation of coefficients of mole fraction of Ti(C, N) versus temperature in Ti microalloyed steels	102
Fig. 3.20	Precipitation-time-temperature curves of Ti(C, N) precipitated at high temperature of Ti microalloyed steels	103
Fig. 3.21	Variation of coefficients of chemical formula Ti(C, N) versus temperature in Ti microalloyed steels during the cooling process after rolling	104
Fig. 3.22	PTT curves of Ti(C, N) precipitated on the dislocation lines during the cooling process after soaking at 1250 °C.	104
Fig. 3.23	PTT curves of Ti(C, N) precipitated at grain boundaries during the cooling process after soaking at 1250 °C	105
Fig. 3.24	Precipitation-time-temperature curves of TiC precipitated on the dislocation lines during the cooling process after soaking at 1250 °C................................	106
Fig. 3.25	PTT curves of TiC precipitated at grain boundaries during the cooling process after soaking at 1250 °C	106
Fig. 3.26	PTT curves of TiC precipitated on dislocation lines during the cooling process after soaking at 1250 °C (with deformation stored energy).................................	107
Fig. 3.27	PTT curves of TiC precipitated on grain boundaries during the cooling process after soaking at 1250 °C (with deformation stored energy).................................	107
Fig. 3.28	PTT curves of TiC nucleated on the dislocation lines from ferrite matrix of different residual Ti content steels (without deformation stored energy)	109
Fig. 3.29	The PTT curves of TiC nucleated on the dislocation lines in different content of Ti microalloyed steels	110
Fig. 3.30	Softening ratio versus time curves of the investigated steels: **a** reference steel, **b** 0.5Mn steel, **c** 1.5Mn steel and **d** 5.0Mn steel. Here P_s and P_f denote the start and finish time of precipitation, respectively	112
Fig. 3.31	Precipitation-time-temperature diagrams of the investigated steels...................................	113

Fig. 3.32	TEM images showing TiC precipitates in 0.5Mn and 1.5Mn steel at 925 °C for various isothermal holding times: **a** 60 s in 0.5Mn, **b** 100 s in 0.5Mn, **c** 200 s in 0.5Mn, **d** 60 s in 1.5Mn, **e** 100 s in 1.5Mn and **f** 200 s in 1.5Mn	114
Fig. 3.33	Calculated solubility curves for TiC in austenite. The dashed line represents the stoichiometry of TiC. The square, circle and triangle denote the locations of the initial compositions ($[Ti]_0$, $[C]_0$) before TiC precipitation of 0.5Mn, 1.5Mn and 5.0Mn steels, respectively	115
Fig. 3.34	Calculated PTT diagrams of the investigated steels	117
Fig. 3.35	Softening ratio versus time curves of the investigated steels: **a** Ti steel and **b** Ti–Mo steel	118
Fig. 3.36	Stress relaxation curves of the investigated steels: **a** Ti steel, and **b** Ti–Mo steel	119
Fig. 3.37	Maximum of $\Delta\sigma$ as a function of testing temperature	120
Fig. 3.38	PTT diagram: **a** two-stage deformation and **b** stress relaxation	120
Fig. 3.39	**a** TEM image showing a (Ti, Mo)C precipitate, **b** selected area electron diffraction (SAD) and **c** EDS	121
Fig. 3.40	HAADF-STEM image showing the cell-like distribution of (Ti, Mo)C carbide on carbon replica film, formed in the specimen deformed at 900 °C and held for 1800 s	122
Fig. 3.41	A carbide particle formed in the specimen deformed at 925 °C and held for 1800 °C: **a** TEM image, **b** HAADF-STEM image and **c** the corresponding SAED. It should be noted that the shape of this precipitate is rectangular pyramid or an octahedron as observed in (**b**) rather than a cube as seen in (**a**). In addition, the two diagonals of this pyramid-like or octahedron-shaped precipitate are on plane {200}, which is identified by SAD in (**c**)	122
Fig. 3.42	**a** HRTEM image showing a carbide particle in the specimen deformed at 925 °C and held for 3000 s and **b** IFFT lattice image of the area marked by a white-line frame on the carbide particle in (**a**)	123
Fig. 3.43	**a–d** TEM images showing carbide particles formed at 925 °C for various holding time in Ti-Mo micro-alloyed steel: **a** 200 s, **b** 600 s, **c** 1800 s and **d** 3000 s. **e–h** TEM images showing carbide particles formed at 925 °C for various holding time in Ti microalloyed steel: **e** 200 s, **f** 600 s, **g** 1800 s and **h** 3000 s	124
Fig. 3.44	Average particle sizes of carbides at 925 °C as functions of isothermal holding time	124

Fig. 3.45	Atomic ratio of Ti/Mo in carbide as a function of isothermal holding time. The figures above Ti/Mo-log(time) curve represent the average particle sizes of the measured precipitates...	125
Fig. 3.46	Morphology of particles in the specimens deformed at different temperature and held for 30 min: **a** 875 °C, **b** 900 °C, **c** 925 °C and **d** 950 °C...	125
Fig. 3.47	Atomic ratio of Ti/Mo in particles in the specimens deformed at different temperatures and held for 30 min...............	126
Fig. 3.48	XRD patterns of the precipitates: **a** Ti steel, and **b** Ti–Mo steel..	126
Fig. 3.49	Quantitative results of PCPA............................	127
Fig. 3.50	Particle size distribution measured by small angle X-ray diffraction: **a** t = 600 s; **b** t = 1800 s; **c** t = 7200 s............	128
Fig. 3.51	Variations of atomic fractions of Ti (**a**), Mo (**b**) and C (**c**) in MC carbide particle with temperature calculated by Thermo-calc with database TCFE6. In (**b**), the measured atomic fraction of Mo in the MC particle were given by assuming that no vacancies are present at C atom positions in the NaCl-type crystal structure, namely perfect NaCl-type structure. The data marked by symbols in the temperature range of 875–950 °C are the present measured data under the condition of 1800 s isothermal holding. The data marked by a quadrangle at 620 °C is from the work of Funakawa et al., whose steel has a similar chemical composition with that of the present studied steel, and was isothermal held at 620 °C for 3600 s after finish rolling and furnace cooled to room temperature...	129
Fig. 3.52	Sub-lattice fraction of Mo in (Ti, Mo)C calculated based on the ideal solution model as a function of temperature......	130
Fig. 3.53	The effect of a small amount of Mo on the Gibbs free energy of precipitation of TiC..................................	132
Fig. 3.54	Comparison of the coarsening rates of various carbides and nitrides in austenite (ideal stoichiometric ratio)..............	135
Fig. 3.55	Comparison of the coarsening rates of various carbides and nitrides in ferrite (ideal stoichiometric ratio)................	136
Fig. 4.1	Relationship of TiN coarsening rate (m) and temperature, titanium content in steel (0.0034% N).....................	143
Fig. 4.2	Influence of Ti/N mass ratio on the austenite grain size during heating in the micro-titanium treatment steel [2].............	144
Fig. 4.3	Stress relaxation experiment of the Ti microalloyed steel......	145
Fig. 4.4	Coarse-grained austenite obtained by thermal simulation test...	145

Fig. 4.5	Stress-strain curves of the high-Ti microalloyed steel deformed with 60% reduction (at true strain of 0.92) **a** at different temperatures and **b** at different strain rates	146
Fig. 4.6	Comparison of stress-strain curves of high-Ti microalloyed steel and carbon steel SPHC	146
Fig. 4.7	**a** Stress relaxation curves and **b** static recrystallization kinetics curves for the studied steel	147
Fig. 4.8	Quenched microstructures of the studied steel at different holding time	147
Fig. 4.9	Austenitic microstructure of the high Ti microalloyed steel after F1 rolling	148
Fig. 4.10	Austenite microstructure of Nb microalloyed steel before and after deformation: **a** coarse austenite grains and **b** incompletely recrystallized grains	148
Fig. 4.11	**a** Stress relaxation curves and **b** recrystallization dynamics curves of studied steel a deformed to different strain and **c** influence of strain on recrystallization time	150
Fig. 4.12	**a** Stress relaxation curves and **b** recrystallization dynamics curves of studied steel (**b**) deformed to different strain and **c** influence of strain on recrystallization time	151
Fig. 4.13	Relationship between the $t_{0.5}$ and strain	152
Fig. 4.14	**a** Stress relaxation curves and **b** recrystallization kinetics curves of steel a at different temperatures and **c** influence of strain on recrystallization time	153
Fig. 4.15	**a** Stress relaxation curves and **b** recrystallization kinetics curves of steel b at different temperatures and **c** influence of strain on recrystallization time	154
Fig. 4.16	Relationship between $t_{0.5}$ and $1/T$ of austenite recrystallization	155
Fig. 4.17	**a** Stress relaxation curves and **b** recrystallization kinetics curves of steel a at different strain rate and **c** influence of strain on recrystallization time	156
Fig. 4.18	**a** Stress relaxation curves and **b** recrystallization kinetics curves of steel b at different strain rate and **c** influence of strain on recrystallization time	157
Fig. 4.19	Relationship between austenite recrystallization $t_{0.5}$ and strain rate	158
Fig. 4.20	Effect of Mo on recrystallization of deformed austenite of Ti microalloyed steel: **a** 0.1%Ti steel, 600 s, **b** 0.1%Ti steel, 1800 s, **c** 0.1%Ti–0.2%Mo steel, 600 s and **d** 0.1%Ti–0.2%Mo steel, 7200 s	159

List of Figures

Fig. 4.21 Optical micrographs showing microstructure transformed from undeformed austenite of the studied steel at different temperatures: **a** 550 °C, **b** 575 °C, **c** 600 °C, **d** 625 °C, **e** 650 °C, **f** 675 °C, **g** 700 °C, **h** 725 °C, **i** 750 °C. 160

Fig. 4.22 The relationship between the isothermal temperature and the hardness transformed from undeformed austenite of the studied steel 161

Fig. 4.23 The relationship between the annealing temperature of Ti-microalloyedsteel (<0.10%C–0.11%Ti) at different times and **a** the ferrite volume fraction and **b** the average microhardness of the ferrite [5]. 162

Fig. 4.24 TEM images showing the TiC preciptates at **a** 675 °C, **b** 700 °C, **c** 725 °C and **d** 750 °C in the ferrite matrix of Ti-microalloyed steel for 1 h of heat treatment [5] 163

Fig. 4.25 Microstructure isothermally transformed from deformed austenite of Ti microalloyed steel (0.065%C–1.8%Mn–0.08% Ti) at different temperatures: **a** 550 °C, **b** 575 °C, **c** 600 °C, **d** 625 °C, **e** 650 °C, **f** 675 °C, **g** 700 °C, **h** 725 °C [6] 164

Fig. 4.26 Hardness of the microstructure isothermally transformed from deformed austenite in microalloyed steels with different Ti contents [6] 165

Fig. 4.27 Microstructure isothermally transformed at different temperatures from undeformed austenite of the 0.044% C–0.5%Mn–0.1%Ti microalloyed steel: **a** 550 °C, **b** 575 °C, **c** 600 °C, **d** 625 °C, **e** 650 °C, **f** 675 °C, **g** 700 °C, **h** 725 °C, **i** 750 °C 165

Fig. 4.28 Microstructure isothermally transformed at different temperatures from undeformed austenite of the 0.044% C–1.0%Mn–0.1%Ti microalloyed steel: **a** 550 °C, **b** 575 °C, **c** 600 °C, **d** 625 °C, **e** 650 °C, **f** 675 °C, **g** 700 °C, **h** 725 °C, **i** 750 °C 166

Fig. 4.29 Influence of Mn content on the hardness of the isothermal transformation microstructure of Ti-microalloyed steel 167

Fig. 4.30 Influence of Mn content on interphase precipitation at 725 °C for isothermal transformation in Ti-microallyed steel 168

Fig. 4.31 Isothermal transformation microstructures of Ti–Mo steel at different temperatures: **a** 550 °C, **b** 575 °C, **c** 600 °C, **d** 625 °C, **e** 650 °C, **f** 675 °C, **g** 700 °C, **h** 725 °C, **i** 750 °C 169

Fig. 4.32 Influence of Mo on the hardness of the microstructure obtained by isothermal transformation of Ti microalloyed steel. (Ti represents the 1.5%Mn–0.1%Ti steel) 169

Fig. 4.33 TEM images showing interphase precipitates at 750 °C in the **a** Ti microalloyed steel and **b** Ti–Mo microalloyed steel 170

Fig. 4.34	Influence of isothermal temperature on the size of interphase precipitates in Ti–Mo microalloyed steel: **a** 750 °C, **b** 725 °C, **c** 700 °C...	170
Fig. 4.35	Mean particle size in the Ti–Mo microalloyed steel and Ti microalloyed steel (1.5%Mn–0.1%Ti)...............	171
Fig. 4.36	**a** HAAADF image showing nano-sized (Ti, Mo)C particles and **b** EDS analysis	171
Fig. 4.37	TEM images showing the crystal structure of (Ti, Mo)C and the orientation relationship with respect to the ferrite matrix: **a** electron diffraction (SAD), **b** bright field and **c** dark field	172
Fig. 4.38	Starting and finishing transformation temperatures of the studied steel at different cooling rates....................	173
Fig. 4.39	Microstructures of the studied steel transformed at different cooling rates: **a** 0.5 °C/s, **b** 1 °C/s, **c** 3 °C/s, **d** 5 °C/s, **e** 10 °C/s, **f** 20 °C/s, **g** 30 °C/s, **h** 50 °C/s.........................	174
Fig. 4.40	The curves of phase transformation fraction with temperature of the studied steel during the simulated coiling.............	175
Fig. 4.41	Microstructure of the studied steel obtained by simulated coiling: **a** 650 °C; **b** 620 °C; **c** 600 °C; **d** 580 °C; **e** 550 °C ...	176
Fig. 4.42	Microhardness of the studied steel after continuous cooling transformation..	177
Fig. 4.43	Microhardness of the studied steel after the simulation of the laminar and coiling cooling processes....................	177
Fig. 4.44	Variation of hardness of the steel plate with annealing temperature by half an hour of isothermal annealing	177
Fig. 4.45	Microstructures quenched at different temperatures after annealing for 30 min: **a** cold rolled, **b** 500 °C, **c** 640 °C, **d** 680 °C, **e** 720 °C, **f** 760 °C, **g** 800 °C, **h** 840 °C	178
Fig. 4.46	Hardness of the steel samples annealed at 630 °C for different time	179
Fig. 4.47	Hardness of the steel samples annealed at 715 °C for different time	179
Fig. 4.48	Microstructure of the samples annealed at 630 °C for different time: **a** 10 h, **b** 25 h.....................................	180
Fig. 4.49	Microstructure of the samples annealed at 715 °C for different time: **a** 10 h, **b** 25 h.....................................	180
Fig. 4.50	TEM images showing the morphology of square particles in steel: **a** cold rolled plate and **b** after 880 °C annealing........	182
Fig. 4.51	The variation of nano-sized TiC precipitates at different stages: **a** cold rolled and **b** after 880 °C annealing................	183
Fig. 5.1	Relationship between soluble aluminum content and soluble oxygen content in molten steel at 1600 °C.................	187

Fig. 5.2	Relationship between soft argon blowing time and oxygen content in molten steel	188
Fig. 5.3	Submerged nozzle adhered by clogs	188
Fig. 5.4	Composition of clogs measured by EPMA	189
Fig. 5.5	Equilibrium phase diagram of $CaO-Al_2O_3$ system	189
Fig. 5.6	Changes in the composition of inclusions in molten steel during LF refining process	190
Fig. 5.7	Morphologies of the typical inclusions in molten steel during refining process	191
Fig. 5.8	Change in nitrogen content during EAF smelting process	193
Fig. 5.9	Influence of nitrogen content in steel scraps on nitrogen content after scraps melting down	193
Fig. 5.10	Schematic diagram of sampling from the cast slab	197
Fig. 5.11	Two dimensional contour maps for the elements in the cross section of the cast slab **a** Carbon, **b** Silicon, **c** Manganese, **d** Phosphorous, **e** Sulfur, **f** Titanium	199
Fig. 5.12	Distribution of density across the cross section of the cast slab	200
Fig. 5.13	Distribution of alumina inclusion content across the cross section of the cast slab	200
Fig. 5.14	Schematic diagram of sampling from the cast slab	200
Fig. 5.15	Comparison of densities of the cast slabs with and without liquid core reduction	201
Fig. 5.16	Relationship between discharge temperature and yield strength	202
Fig. 5.17	Relationship between discharge temperature and tensile strength	203
Fig. 5.18	Relationship between finishing temperature and yield strength	204
Fig. 5.19	Relationship between finishing temperature and tensile strength	204
Fig. 5.20	Relationship between coiling temperature and yield strength	205
Fig. 5.21	Relationship between coiling temperature and tensile strength	205
Fig. 5.22	Microstructure of titanium microalloyed steel plates with different thicknesses: **a** 6.0 mm, **b** 4.0 mm, **c** 1.6 mm	208
Fig. 5.23	Relationship between the size of ferrite grains and the thickness of titanium microalloyed steel (0.0558% Ti)	209
Fig. 5.24	Relationship between yield strength and titanium content	209
Fig. 5.25	Main strengthening mechanisms of steel	210
Fig. 5.26	Relationship between yield strength and titanium content of steel plates with the thickness of 4.0 mm	214

Fig. 5.27	Relationship between strengthening increment and thickness of steel strip containing 0.058% Ti	214
Fig. 5.28	**a** Fe (Mn, Cr, Si, P, S, Cu, P)–C phase diagram, and **b** enlarged diagram of single phase region of ferrite	215
Fig. 6.1	Metallographic microstructure of steel SPA-H (thickness: 1.6 mm)	222
Fig. 6.2	Relationship between precipitation strengthening increment and titanium content	222
Fig. 6.3	Microstructure of typical product (thickness: 1.5 mm)	224
Fig. 6.4	Relationship between grain size and thickness of steel plate	224
Fig. 6.5	Precipitates in the new generation of container steel: **a** TiN precipitated in liquid, **b** TiN precipitated in solid solution, **c** TiC precipitated in austenite and $Ti_4C_2S_2$, **d** TiC precipitated in super-saturated ferrite, **e** TiC formed by inter-phase precipitation	225
Fig. 6.6	Mechanical properties over the coil of the new generation of container steel ZJ550 W	226
Fig. 6.7	Microstructures of different regions of the weld joint of container steel ZJ550 W: **a** weld, **b** weld and transition region, **c** transition region, **d** base metal	229
Fig. 6.8	Samples of the new generation of container steel for cold bending test ($d = 0$)	230
Fig. 6.9	Lightweight containers. **a** 20-foot DV container, **b** 40-foot HC container	230
Fig. 6.10	Orientation map of ferrite grains in steel ZJ700 W (thickness: 2.0 mm)	234
Fig. 6.11	Distribution of precipitates in steel ZJ700 W	235
Fig. 6.12	Comparison of particle size distribution of precipitates in steel ZJ550 W and ZJ700 W	235
Fig. 6.13	Microstructures of different regions of the weld joint of ultra-high strength weathering steel ZJ700 W: **a** weld, **b** transition region between weld and base metal	237
Fig. 6.14	Impact fracture of ultra-high strength weathering steel (Test temperature: −60 °C, thickness: 6 mm)	238
Fig. 6.15	Special 53-foot container	239
Fig. 6.16	Use of thin gauge WJX750-NH to manufacture containers	241
Fig. 6.17	Influence of effective titanium content on the tensile strength of S420MC	243
Fig. 6.18	Influence of intermediate cooling temperature on the microstructure of S700MC: **a** high intermediate cooling temperature, **b** low intermediate cooling temperature	244
Fig. 6.19	Influence of cooling mode and coiling temperature on the strength: **a** tensile strength, **b** yield strength	244

Fig. 6.20	Influence of coiling temperature on the microstructure of S700MC: **a** high temperature coiling, **b** low temperature coiling (thickness: 1.5 mm)................................	245
Fig. 6.21	Particles precipitated during the low temperature coiling	245
Fig. 6.22	Metallographic microstructure of S420MC (thickness: 4 mm)...	246
Fig. 6.23	Tensile strength distribution of S420MC	246
Fig. 6.24	Influence of thickness on grain size	247
Fig. 6.25	Samples of S420MC for cold bending test.................	247
Fig. 6.26	Samples of S700MC for cold bending test: **a** 90 °C, **b** 180 °C...	248
Fig. 6.27	Microstructure of the deformed region	248
Fig. 6.28	Rectangular tube made of S700MC	249
Fig. 6.29	Fatigue curve of S700MC................................	249
Fig. 6.30	Microstructures of different locations of the weld joint of S420MC: **a** base metal, **b** transition zone, **c** weld............	250
Fig. 6.31	Microstructures of different locations of the weld joint of S700MC: **a** base metal, **b** transition zone, **c** weld............	250
Fig. 6.32	Truck parts made of S420MC............................	251
Fig. 6.33	Truck bumper beam made of S700MC....................	252
Fig. 6.34	Bus frame parts made of S700MC	252
Fig. 6.35	Morphology of precipitates in sample 1#.................	256
Fig. 6.36	Morphology of precipitates in sample 2#.................	256
Fig. 6.37	EDS analysis of precipitates in sample 1#	257
Fig. 6.38	EDS analysis of precipitates in sample 2#	257
Fig. 6.39	Microstructures of steel produced by two different heating systems **a** sample 1#, **b** sample 2#.......................	258
Fig. 6.40	Microstructures of QstE600TM produced by two different finishing rolling temperatures: **a** sample 3#, **b** sample 4#......	258
Fig. 6.41	QStE650TM samples for cold bending test................	260
Fig. 6.42	Longitudinal impact energy of different locations of QStE650TM sample.....................................	261
Fig. 6.43	Transverse impact energy of different locations of QStE650TM sample.....................................	261
Fig. 6.44	Fatigue curve of QStE650TM............................	262
Fig. 6.45	CCT curve of steel sample	266
Fig. 6.46	Metallographic microstructure of steel HG785	266
Fig. 6.47	Schematic diagrams of welding assembly and weld bead of HG785	267
Fig. 6.48	Samples for bending test................................	268
Fig. 6.49	Metallographic microstructures of different locations of the weld joint: **a** weld joint, **b** weld (100X), **c** weld (500X), **d** coarse grain zone, **e** fine grain zone, **f** critical heat affected zone	268

Fig. 6.50	Brittle-ductile transition temperature of HG785 sample with the thickness of 8 mm	269
Fig. 6.51	S-N curve of HG785 sample with the thickness of 8 mm	270
Fig. 6.52	Application of welded high strength steel for engineering machinery: **a** crane, **b** truck	271
Fig. 6.53	Microstructure of steel ZM450: **a** hot rolled, **b** heat treated	273
Fig. 6.54	Morphology of micron-scale TiC precipitates: **a** 500X, **b** 200X	274
Fig. 6.55	TiC precipitates obtained by extraction replica technique: **a** micron-scale, **b** nano-scale	274
Fig. 6.56	Comparison of weight loss of ZM450 and HARDOX450 subjected to wear resistance test	275
Fig. 6.57	Floors of the chute of scraper conveyor made by ZM450 plates	276
Fig. 6.58	Schematic diagram of hydro-generator	277
Fig. 6.59	Schematic diagram of rotor yoke	277
Fig. 6.60	Typical metallographic microstructure of magnetic yoke steel	281
Fig. 6.61	Microstructure of WDER750 sample with the thickness of 4 mm	282
Fig. 6.62	Analysis on the precipitates of WDER750 sample with the thickness of 4 mm	283
Fig. 6.63	Magnetic property of different locations of WDER50 strip with the thickness of 4 mm	283
Fig. 6.64	Application of magnetic yoke steel in hydropower projects: **a** Three Gorges Hydroelectric power station, **b** Longtan hydropower station, **c** Xiluodu hydropower station, **d** Xiangjiaba hydropower station	284
Fig. 6.65	Microstructures of different locations of hot rolled enamel steel strip RST360: **a** head, **b** middle, **c** tail	287
Fig. 6.66	Precipitates of different locations of hot rolled enamel steel strip RST360: **a** head, **b** middle, **c** tail	288
Fig. 6.67	Macroscopic surface morphology of enameled layer after accelerated fish scaling test: **a** sample A, **b** sample B, **c** sample C	290
Fig. 6.68	Result of hydrogen permeation test	290
Fig. 6.69	Distribution and density of precipitates of different samples: **a** sample A, **b** sample B, **c** sample C, **d** density	291
Fig. 6.70	**a** Electric water heater, **b** solar water heater, **c** air-source water heater	292
Fig. 6.71	**a** Oil storage tank, **b** chemical reaction tank, **c** large water tank	292
Fig. 6.72	Metallographic microstructure of Q345B (Ti): **a** upper surface, **b** core, **c** lower surface	295

List of Figures

Fig. 6.73	Morphology and composition of precipitates	295
Fig. 6.74	Metallographic microstructure of Q345B of different composition system: **a** 1.5 wt% Mn, **b** 0.5 wt% Mn−0.04 wt% Ti	296
Fig. 6.75	Inclusion grades of Q345B of different composition systems: **a** 1.5 wt% Mn, **b** 0.5 wt% Mn–0.04 wt% Ti	297
Fig. 6.76	Samples for cold bending test.........................	297
Fig. 6.77	Impact energy of Q345B (Ti) at different testing temperatures	298
Fig. 6.78	Metallographic microstructure of the weld joint of Q345B(Ti): **a** weld joint, **b** weld, **c** coarse grain zone, **d** fine grain zone, **e** critical heat affected zone, **f** base metal....	299
Fig. 6.79	Application of titanium microalloyed low alloyed high strength structure steel: **a** rear beam of truck body, **b** excavator arm...	300

Chapter 1
Introduction

Xinping Mao, Qilong Yong and Xiangdong Huo

1.1 Introduction of Titanium Microalloyed Steel

In 1963, the Swedish Noren first proposed the definition of microalloyed steel, namely the Mn-bearing alloy steel or low-alloyed steel with the addition of a small amount of alloying elements. The alloying element has a significant effect on one or several properties of steel, and its amount is smaller than that of traditional alloying element in steel by 1–2 orders of magnitude [1]. This definition has been widely adopted around the world and has been in use up to now. Titanium microalloyed steel is such one kind of microalloyed steel, and titanium is a typical microalloying element. There are other similar elements, such as niobium, vanadium and boron.

As a microalloying element, titanium is present in steel mainly in the form of solute in iron matrix or Ti-bearing precipitate. According to the effect of titanium solute and Ti-bearing precipitate on the recrystallization and phase transformation of austenite, the microstructure of austenite and even ferrite can be significantly refined by properly controlled rolling technology. Therefore, significant grain refinement strengthening effect is achieved. By properly controlling the precipitation behavior of TiC, nano-TiC particles can be obtained, which provides significant precipitation strengthening effect. In addition to the great enhancement of strength and toughness of steel, titanium improves the hardenability of steel and fixes the non-metallic element in steel, leading to its wide use in steel industry [2].

X. Mao (✉)
Baosteel Central Research Institute, Wuhan, China
e-mail: maoxinping@126.com

Q. Yong
Central Iron & Steel Research Institute, Beijing, China
e-mail: yongql@126.com

X. Huo
Jiangsu University, Zhenjiang, China
e-mail: hxdustb@163.com

© Metallurgical Industry Press, Beijing and Springer Nature Singapore Pte Ltd. 2019
X. Mao (ed.), *Titanium Microalloyed Steel: Fundamentals, Technology, and Products*,
https://doi.org/10.1007/978-981-13-3332-3_1

Titanium has been used as a microalloying element since 1920s. Initially, it was mainly used in trace titanium treatment to improve the microstructure and weldability of steel. Since 1960s, with the development of microalloying technology, titanium, as a kind of auxiliary microalloying element, has been widely used in multiple microalloyed steel. During this time, V–Ti and Nb–Ti multiple microalloying technology was developed. Since 1990s, with the development of clean steel smelting technology and thin slab continuous casting and direct rolling (TSCR) technology, single titanium microalloying technology based on TSCR was invented. By addition of 0.04–0.2% titanium into conventional low-alloyed and high strength steel, the maximum yield strength reaches 700 MPa grade. Based on this, Ti–Mo and Ti–V–Mo multiple microalloying technology using titanium as the major microalloying element was further developed, and the maximum yield strength of steel was improved to 900 MPa grade.

With the deep research on the mechanism of effect of titanium in steel over the time, titanium microalloying technology has played a more and more important role in the development of high strength steel. China has the largest reserves of titanium resources in the world. It is of significant economic value and social benefit to make full use of the rich titanium resources and develop titanium micro-alloying technology and titanium micro-alloyed steel.

1.2 Principles of Titanium Microalloying

1.2.1 Characteristics of Titanium

Titanium [2] is a transition metal in the fourth period (the first long period) and the IV subgroup of the periodic table. Its atomic number is 22 and relative atomic mass is 47.867(1) [3]. The structure of its outer electronic layer is $3d^24s^2$. Titanium was discovered by the British chemist R. W. Gregor (1762–1817) when he was studying ilmenite and rutile in 1791. Four years later, the German chemist M. H. Klaproth (1743–1817) also discovered it in the red rutile from Hungary. He advocated the use of nomenclature of uranium and quoted the name of Titan Protoss "Titanic" in the Greek mythology to name this new element as "Titanium" [2].

The titanium studied by Gregor and Klaproth at that time was titanium dioxide powder rather than metallic titanium. It is difficult to prepare metallic titanium because titanium oxide is extremely stable and metallic titanium easily reacts with oxygen, nitrogen, hydrogen, carbon and other non-metallic elements. Metallic titanium was not obtained until 1910 when it was first obtained by the American chemist M. A. Hunter with purity of 99.9% by sodium reduction method (Hunter method) [2].

Solid state titanium is a polymorphic element, and the phase transformation temperature is 882.5 °C. Above 882.5 °C, it is β-Ti with body centered cubic structure. The lattice constant at 900 °C is 0.332 nm. Below 882.5 °C, it is α-Ti with close packed hexagonal structure. The lattice constant at room temperature (20 °C) is

1 Introduction

$a = 0.29506$ nm, $c = 0.46788$ nm, $c/a = 1.5857$, close to the ideal ratio of close packed hexagonal structure, namely 1.633. Accordingly, the distance of nearest atomic neighbor is 0.28939 nm. When the coordination number is 12, the atomic radius is 0.14609 nm, which is larger than that of iron by 14.4%. The theoretical molar volume is 1.0622×10^{-5} m^3/mol. The theoretical and actual densities are 4.506 and 4.506–4.516 g/cm^3, respectively, which are significantly smaller than those of iron (7.875 and 7.870 g/cm^3, respectively). Therefore, titanium is usually classified as light metal. The crystal structures of α-Ti and β-Ti are shown in Fig. 1.1. The specific heats at constant pressure of α-Ti and β-Ti are $C_p = 22.133 + 10.251 \times 10^{-3} T$ J/(K·mol) (298–1155 K) and $C_p = 19.832 + 7.908 \times 10^{-3} T$ J/(K·mol) (1155–1933 K) respectively. The latent heat of the phase transformation α → β is 4142 J/mol.

Titanium is a transition metal with strong atomic binding force. Its sublimation heat is 4.693×10^5 J/mol (25 °C) [4], which is lower than that of tungsten, osmium, tantalum, rhenium, niobium, carbon, iridium, molybdenum, zirconium, hafnium, ruthenium, thorium, boron, rhodium, platinum, vanadium and uranium, but higher than that of any other metallic element. Its melting point is 1660 °C, which is lower than that of tungsten, rhenium, osmium, tantalum, molybdenum, niobium, iridium, ruthenium, hafnium, rhodium, vanadium, chromium, zirconium, platinum, thorium, but higher than that of any other metallic element. Its boiling point is about 3302 °C, which is lower than that of rhenium, tungsten, tantalum, osmium, niobium, molybdenum, hafnium, zirconium, uranium, iridium, thorium, ruthenium, platinum, rhodium, vanadium, lutecium and yttrium, and slightly higher than that of cobalt, nickel, iron and other common metallic elements. The linear expansion coefficient (0–100 °C) of titanium is 8.9×10^{-6}/K [5] or 8.36×10^{-6}/K [2], which is relatively low among the transition metals. It is slightly higher than that of vanadium and much lower than that of iron (12.1×10^{-6}/K). Its average specific heat (0–100 °C) is 0.528 J/(g·K)

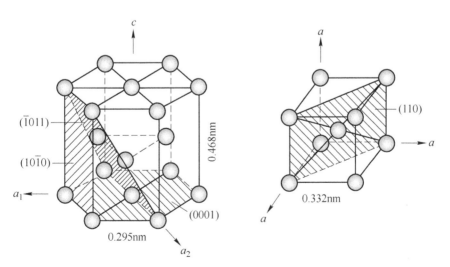

Fig. 1.1 Crystal structures of Hcp-α-Ti and Bcc-β-Ti

[4], higher than that of vanadium (0.498 J/(g·K)) and iron (0.456 J/(g·K)), and much higher than that of niobium (0.268 J/(g·K)).

The Young's modulus and shear modulus of α-Ti at room temperature (20 °C) are 120.2 GPa [4] (115 GPa [2]) and 45.6 GPa [4] (44 GPa [2]), respectively. Its volume compression modulus is 1.084×10^5 MPa [4] and Poisson's ratio is 0.361 [4] (0.33 [2]). On one hand, its elastic modulus is close to that of vanadium and niobium, lower than that of hafnium and tantalum, obviously lower than that of chromium, manganese and iron, significantly lower than that of molybdenum, tungsten and rhenium. On the other hand, its Poisson's ratio is relatively high among the transition metals, only lower than that of gold, niobium, platinum, protactinium, zirconium and vanadium. The elastic stiffness of α-Ti single crystal at room temperature is: $C_{11} = 1.60 \times 10^5$ MPa, $C_{33} = 1.81 \times 10^5$ MPa, $C_{44} = 4.65 \times 10^4$ MPa, $C_{12} = 9.00 \times 10^4$ MPa and $C_{13} = 6.60 \times 10^4$ MPa. The elastic compliance is $S_{11} = 9.69 \times 10^{-12}$ Pa^{-1}, $S_{33} = 6.86 \times 10^{-12}$ Pa^{-1}, $S_{44} = 21.5 \times 10^{-12}$ Pa^{-1}, $S_{12} = -4.71 \times 10^{-12}$ Pa^{-1} and $S_{13} = -1.82 \times 10^{-12}$ Pa^{-1} [5].

The abundance of titanium in the crust is about 0.63%, making it rank the 9th (after oxygen, silicon, aluminum, iron, calcium, sodium, potassium, magnesium). However, as a structural material, its research and application popularity is significantly higher than that of calcium, sodium, potassium, magnesium and so on. Thus, it is called the third metal after iron and aluminum. The main raw materials for producing titanium are ilmenite ($FeTiO_3$), rutile (TiO_2) and titanium slag. The amount of titanium resources valuable for industrial exploitation is about 1.97 billion tons, of which ilmenite (including titanomagnetite) accounts for more than 95%. Ilmenite placers are mainly distributed in South Africa, India, Australia and United State, while ilmenite rock mines are mainly located in Panxi and Chengde Damiao areas of China, Russia, Norway, Canada and United State. Rutile mines are mainly distributed in Brazil and Australia. China is the most abundant country in ilmenite ore resources in the world, having proven reserve of titanium resources of 0.97 billion tons.

The chemical property of titanium is very active and it is added into steel in the form of ferrotitanium. It exists in steel mainly in the form of titanium solute in iron matrix and Ti-bearing precipitate. The role of titanium in steel depends on its form of existence. The effects of titanium solute and Ti-bearing precipitate are quite different. Those effects have both advantages and disadvantages on the performance of steel, and they play different roles in different steels. Therefore, it is necessary to carry out in-depth analysis as well as scientific and reasonable control during the production process.

1 Introduction

1.2.2 Grain Refinement Strengthening of Steel by Titanium

1.2.2.1 Ti(C, N) Inhibits Growth of Austenite Grains

Grain refinement is the only strengthening mechanism to improve both the strength and toughness of steel, which has always attracted much attention. During the grain refinement process, the grain growth must be effectively prohibited, so as to maintain the effect of grain refinement. Among all the methods, pinning the grain boundaries by secondary phase is the most important way to prevent the grain growth. The basic principle of preventing matrix grain coarsening by secondary phase particles in steel was first quantitatively analyzed by Zener [6] and Hillert [7]. They proposed that when the secondary phases are uniformly distributed spherical particles, the criterion for unpinning is:

$$D_C \leq A \frac{d}{f} \quad (1.1)$$

where D_C is the critical average equivalent diameter of grains which can be effectively pinned, d is the average diameter of secondary phase particles, f is the volume fraction of secondary phase particles, and A is the coefficient.

Gladman [8] analyzed the energy change during unpinning and obtained the criterion for unpinning when the secondary phases are uniformly distributed spherical particles.

$$D_C \leq \frac{\pi d}{6f}(\frac{3}{2} - \frac{2}{Z}) \quad (1.2)$$

where $Z = D_M/D_0$ is the factor expressing the non-uniformity of grain sizes. It is the ratio of the diameter of the largest grains (D_M) to the average grain diameter (D_0). The Z value for ideal grains is $\sqrt{2}$. When grains grow normally, the Z value is about 1.7, while it could reach 9 for abnormal grain growth. Based on Eqs. (1.1) or (1.2), the size of critical grains which can be effectively pinned and basically do not grow up is proportional to the average size of secondary phase particles, and inversely proportional to the volume fraction of secondary phase particles. Therefore, in order to prohibit coarsening of matrix grains, there should be enough secondary particles with sufficiently small average size.

It should be noted that there is a critical state for preventing the grain growth by secondary phase [9]. When the size of matrix grains is larger than the critical size, those grains are effectively pinned and almost do not grow. When the size of matrix grains is less than or equal to the critical size, the unpinning occurs and those grains grow abnormally. Therefore, there is a directivity for pinning matrix grains by secondary phase particles. When the volume fraction of secondary phase particles increases continuously or the average size decreases continuously, the uniformity of the size of matrix grains is high. The Z value is about 1.7 and the corresponding coefficient A is 0.1694. Once the grain boundaries are pinned, the pinning will continue to prohibit the grain growth. When the volume fraction of secondary phase particles continues to decrease or the average size increases continuously, the abnormal grains will grow rapidly once the unpinning occurs. The Z value increases to 3 (unpinning after weak pinning) or 9 (unpinning after strong pinning), and the corresponding coefficient A is 4/9 or 2/3 based on Hillert's theory. That is to say, the grains cannot be re-pinned until the grains become large enough. For the case of unpinning after strong pinning, the grain size must grow to the size about 4 times of the original size. Therefore, different requirements of controlling the size and volume fraction of secondary phase are necessary so as to completely control the grain growth under different thermal history conditions.

If the abnormal grain growth is not complete, the phenomenon of mixed grains occurs. Mixed grains cause the non-uniformity of steel performance and seriously damage the plasticity and toughness, so they must be strictly controlled.

In the process of rolling, forging, soaking of heat treatment and rapid heating of welding, it is generally necessary to have secondary phase particles with a sufficient volume fraction and small size to prevent the grain growth. After formation of fine grains through recrystallization or solid state polymorphism phase transformation, it is necessary to have secondary phase particles with a larger volume fraction and a smaller size to prevent the grain growth.

In the production of electrical steel, there must be secondary phase particles with a certain volume fraction to prevent the growth of the initial grains during the soaking process. However, in the rolling process, unpinning needs to occur to activate the abnormal grain growth or the more preferable directional growth, so as to obtain good electromagnetic properties.

TiN or N-rich Ti(C, N) is stable at high temperatures, and their solubility products in iron matrix are very small, so there are still sufficient TiN or N-rich Ti (C, N) particles in iron matrix. The coarsening rate of TiN or N-rich Ti (C, N) particles remains low at high temperatures, so the average size is sufficiently small. The average grain size needs to be controlled at about 200 μm during the high temperature soaking process. Suppose the size of secondary phase particles is controlled at 100 nm, according to Eq. (1.2), if the non-uniformity factor of grain size is 1.7, the volume fraction should be above 0.0085%. While if the factor is 3, the volume faction should be above 0.0218%. When the content of effective titanium in steel is above 0.012% and the nitrogen content is above 0.004%, it is easy for the size and volume fraction of insoluble TiN or N-rich Ti (C, N) at high temperatures to meet the requirements mentioned above, so the growth of matrix grains can be effectively prevented.

A large number of research and actual production show that the effect of TiN or N-rich Ti(C, N) in microalloyed steel to prevent grain growth maintains above 1300 °C. In contrast, the temperatures for Nb(C, N), AlN and V(C, N) are about 1200, 1100 and 1000 °C, respectively. Because the effect of titanium on preventing the grain growth in the soaking of pre-rolling process and the thermal cycles of welding process is irreplaceable, titanium is widely used in microalloyed steel. In order to obtain a sufficient insoluble TiN or N-rich Ti (C, N) at high temperatures and prevent the precipitation of TiN in liquid steel, the content of titanium is generally controlled within the range of 0.01–0.03%. Such steel is called trace titanium treatment steel. Because the size of austenite grains should be controlled during the soaking stage of recrystallization controlled rolling process, trace titanium treatment is necessary. In addition, the steel, which has a high demand to control the grain size in the heat affected zone of welding joint, also requires trace titanium treatment.

TiC in Ti-microalloyed steel precipitates in deformed austenite below 1000 °C by strain induced precipitation and the size is in the range of 10–20 nm. With the rolling process, the temperature decreases continuously, the volume fraction of precipitates increases and the average size decreases. As a result, the pinning effect increases continuously and the non-uniformity of grain size factor is 1.7. Suppose that the size of recrystallized austenite grains needs to be controlled at about 20 μm, if the volume fraction of TiC in steel is in the range of 0.0085–0.017%, then the growth of recrystallized austenite grains can be prevented. It is easy for Ti-microalloyed steel with titanium content of 0.08% or more to meet the requirements of precipitate volume fraction and effective precipitation temperature. In V-microalloyed steel or V–N microalloyed steel, the effective precipitation temperature of V(C, N) is low, so its role on preventing the recrystallized grain growth is small. Therefore, significant grain refinement effect can be obtained when Ti-microalloyed steel is produced by recrystallization controlled rolling process.

1.2.2.2 Titanium Solute and Strain Induced TiC Precipitates Inhibit Recrystallization of Deformed Austenite

Deformation energy is generated and stored in the deformed matrix of steel due to plastic deformation. The stored deformation energy is the driving force of recrystallization of matrix, and it also increases the driving force of subsequent solid state polymorphism phase transformation. The recrystallization of deformed austenite during the hot rolling process, especially dynamic recrystallization, significantly refines the austenite grains. This is recrystallization controlled rolling. The non-recrystallized grains are obviously elongated and flattened, and a large number of intra-granular deformation bands are generated. In the subsequent phase transformation of austenite to ferrite, very fine ferrite grains are produced. This is non-recrystallization controlled rolling. Thus, changing the recrystallization behavior of deformed austenite is critical to achieve good strengthening effect of grain refinement by controlled rolling.

The process of recrystallization involves migration of grain boundaries or subgrain boundaries. If a large number of solute atoms segregate on the grain boundary or sub-grain boundary, the migration of grain boundary requires dissociation from solute atoms or carrying solute atoms. Therefore, the grain boundary migration is hindered and the migration rate is slowed down. This is the solute drag effect on preventing recrystallization. It is obvious that the larger the difference between the size of solute atom and iron atom, the more likely the segregation of solute atoms on the grain boundary. If the diffusion coefficient of solute atoms in iron matrix significantly differs from the self-diffusion coefficient of iron, the grain boundary migration rate is obviously decreased. The atomic size of niobium or boron is quite different to that of iron, and their diffusion coefficient in austenite differs greatly from the self-diffusion coefficient of iron. Hence they have a significant solute drag effect to prevent recrystallization. The atomic size of titanium differs greatly from that of iron, but its diffusion coefficient in austenite is not quite different to the self-diffusion coefficient of iron. Therefore, titanium has a certain solute drag effect, which is not as significant as that of niobium and boron. The atomic size of vanadium, chromium and manganese is very close to that of iron, so the solute drag effect is very small.

In addition, the secondary phase generated by strain induced precipitation on the grain boundary or sub-grain boundary produces a significant pinning effect to inhibit recrystallization. This effect is called Zener pinning effect [6]. A large number of experimental results [10] have shown that, the recrystallization of deformed austenite is significantly delayed sooner after the strain induced precipitation of microalloyed carbonitride in austenite. Because the deformation energy promotes the recrystallization of deformed matrix as well as the strain induced precipitation of secondary phase, these two processes are obviously competitive. If the strain induced precipitation takes place first, then recrystallization is significantly prevented. While if recrystallization occurs first, the precipitation of secondary phase is delayed greatly because of the consumption of deformation energy by recrystallization. As a result, preventing austenite recrystallization by secondary phase is weakened. The effective precipitation temperature of Nb(C, N) or TiC in austenite is above 900 °C, so both of them are able to resist austenite recrystallization by strain induced precipitation. In vanadium micro-alloyed steel with general nitrogen content, V(C, N) does not precipitate in the temperature range of austenite, so it has no effect on the recrystallization of deformed austenite. While in V–N microalloyed steel with high nitrogen content, the effective precipitation temperature of N-rich V(C, N) is around 850 °C, so it has a certain effect to prevent the recrystallization at that temperature.

By summarizing the two effects above, the effects of various microalloying elements to prevent the recrystallization of deformed austenite are obtained and shown in Fig. 1.2 [11].

When the recrystallization of deformed austenite matrix is blocked, the matrix grains are gradually flattened and the grain boundaries are serrated, so the deformation energy is preserved. If the deformation continues, the grain flattening and boundary serration are further enhanced, leading to the accumulation of more deformation energy. The deformation energy stored significantly increases the free energy of austenite. In the subsequent cooling process of ferrite transformation, the deformation energy effectively promotes the formation of ferrite, making the formation temperature of ferrite much higher than the equilibrium temperature A_3, or the amount of ferrite greatly larger than the equilibrium amount [12–14]. Meanwhile, the strain induced precipitation of secondary phase depletes solutes and changes the chemical composition of austenite. Therefore, the free energy of austenite increases and the formation of ferrite is further promoted. In addition, the grain flattening greatly increases the area of austenite grain boundaries, and the large number of deformation bands in the deformed austenite grains is equivalent to increase in the area of

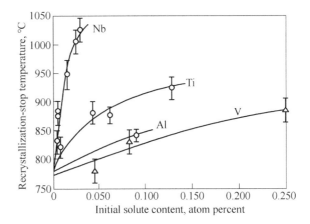

Fig. 1.2 Effects of microalloying elements on the recrystallization-stop temperature of 0.07%C–1.40%Mn–0.25%Si steel

grain boundaries. Moreover, the serration of grain boundaries leads to the formation of a large number of quasi-grain domains on the grain boundaries. Therefore, the nucleation sites of ferrite are not limited to the grain domains of austenite, but are widely distributed on the grain boundaries of deformed austenite. As a result, the heterogeneous nucleation rate of new ferrite is significantly increased, leading to refinement and uniform distribution of ferrite grains.

Figure 1.2 shows that niobium has both the solute drag effect and secondary phase pinning effect. For the competition between recrystallization and precipitation, the solute drag effect is obviously advantageous to strain induced precipitation, so the pinning effect of secondary phase to prevent recrystallization is greatly enhanced. The recrystallization of Nb-microalloyed steel can be completely and easily inhibited during the finishing rolling stage, making Nb-microalloyed steel particularly suitable for non-recrystallization controlled rolling process and even deformation induced ferrite phase transformation process. Vanadium has almost no solute drag effect and the pinning effect of vanadium bearing precipitate is also small, so V-microalloyed steel is particularly suitable for high temperature dynamic recrystallization controlled rolling process. The role of titanium on the recrystallization of deformed austenite is between that of niobium and vanadium, and it has some effect to prevent recrystallization, so Ti-microalloyed steel is suitable for recrystallization controlled rolling, non-recrystallization controlled rolling, or both simultaneously. The combination of recrystallization and non-recrystallization controlled rolling processes generates a very significant grain refinement effect, making it an important direction for the development of microalloyed steel and controlled rolling technology. Ti-microalloyed steel has great advantages in this aspect.

1.2.2.3 TiN Promotes Formation of Intragranular Ferrite

The formation of intragranular ferrite in low carbon steel increases the nucleation rate of ferrite to a certain extent, thereby refining the ferrite grains. The crystal orientations of intragranular ferrite are usually random, while the ferrite nucleated at the grain boundary of austenite and the austenite grains generally follow the K–S orientation relationship, so those ferrite grains have more definite crystal orientations. Therefore, the formation of intragranular ferrite divides the austenite grains, which is favorable for the morphology and distribution of ferrite grains. In recent years, the technology related to intragranular ferrite has attracted wide attention [15].

In fact, the best benefit of intragranular ferrite is that its formation temperature is high and the content of carbon and alloying elements is very small, so it has very high ductility and plasticity. The orientation of intragranular ferrite is thoroughly different to that of ferrite nucleated on the grain boundary of austenite. The intragranular ferrite divides the original austenite grain, which obviously inhibits the formation and directional growth of non-equiaxed ferrite grains. The intragranular ferrite with high toughness completely surrounds the secondary phase particles, thereby significantly reducing or even eliminating their damage to the toughness and fatigue properties of steel.

Research results show that the size of secondary phase must match the size of ferrite nuclei, so as to effectively promote the nucleation of intragranular ferrite. Only those secondary phase particles with the size in the range of 100–1000 nm are capable of promoting the formation of intragranular ferrite. Too small or too large secondary phase particles do not have this effect. In one aspect, it means that the effect of secondary phase to promote the formation of intragranular ferrite is quite limited. Suppose that the volume fraction of secondary phase particles capable of promoting the nucleation of intragranular ferrite is 0.1%, the average size is 500 nm and the austenite grain size is 20 μm, so there are only 1.6 secondary phase particles in each grain. When the austenite grain size is 10 μm, the average number of secondary phase particle per grain is only 0.4. In another aspect, it shows that the secondary phase particles (inclusions) with the size of 100–500 nm and a certain volume fraction are practically harmless. As long as the secondary phase particles in steel are uniformly distributed and their size is generally controlled below 1 μm, those particles with the size of hundreds of nanometers can promote the formation of intragranular ferrite and are surrounded by intragranular ferrite. Therefore, there is no need to pursue that steel does not contain inclusions.

The solubility products of TiN in both steel melt and austenite are very small, so it is difficult to completely suppress the precipitation of TiN at high temperatures. Generally, the size of TiN precipitated in steel melt is in the range of several micrometers to several dozens of micrometers, and the size of TiN precipitated in austenite during the solidification process is in the range of several micrometers to several hundred nanometers. Therefore, it is possible to effectively reduce the actual precipitation temperature of TiN by increasing the solidification cooling rate, so as to limit the size of TiN in the range of several dozens of nanometers to 200 nm. This not only effectively prevents the growth of austenite grains, but also significantly reduces or even eliminates the harmful effect of TiN. During the production of Ti-microalloyed steel by TSCR process, the size of TiN is significantly refined because of the rapid solidification rate. Moreover, TiN particles are completely surrounded by plastic ferrite grains through the intragranular ferrite technology, leading to great improvement of steel performance, especially the plasticity and toughness [16].

1.2.3 Precipitation Strengthening of Steel by Titanium

The secondary phase particles dispersed in steel matrix produce dispersion strengthening. Because the secondary phase is usually produced by precipitation, dispersion strengthening is called precipitation strengthening or aging hardening in high alloy steel.

The secondary phase precipitation strengthening usually decreases the toughness of steel. However, for low-alloyed and high strength steel, the embrittlement vector (increase in the ductile-brittle transition temperature when the strength of steel is increased by 1 MPa) of precipitation strengthening is smaller compared to other reinforcement modes, such as dislocation strengthening and interstitial solid solution strengthening. Therefore, except grain refinement strengthening, the secondary phase precipitation strengthening should be applied with priority to strengthening low alloy high strength steel [5, 17, 18].

Dislocations pass secondary phase particles by cutting through particles and bowing around particles, and the corresponding strengthening mechanisms are cutting mechanism and Orowan mechanism [19]. When the secondary phase is soft or small, the main strengthening mechanism is cutting mechanism. The strength increment is proportional to the size of secondary phase and the square root of volume fraction of secondary phase. When the secondary phase is hard or large, the main mechanism is Orowan mechanism. The strength increment is proportional to the square root of volume fraction of secondary phase and roughly inversely proportional to the size of secondary phase. The competition between the two strengthening mechanisms is shown in Fig. 1.3 [19]. For each specific secondary phase, there is a critical size d_c, below which the cutting mechanism plays the main role and above which the Orowan mechanism is the major mechanism. Maximum strengthening effect is achieved when the size is close to the critical size. Theoretical calculation shows that the critical size d_c of TiC is 2.70 nm.

Research results [18, 19] show that, for most of secondary phases in steel, the main strengthening mechanism is Orowan mechanism. By taking into account the two aspects (1) average spacing of randomly distributed secondary phase particles (2) the dislocation line cannot closely contact the edge of particles when surrounding the particles, leading to the increase in the effective size of particles, the theoretical formula for calculating the strength increment ΔR_P produced by spherical secondary phase particles is derived as follows:

Fig. 1.3 Relationship between the strength increment and the size of the secondary phase according to two different strengthening mechanisms

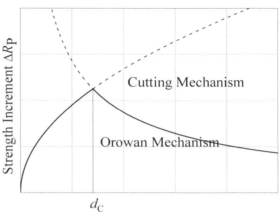

$$\Delta R_\mathrm{P} = \frac{Gb}{\pi K} \frac{1}{(1.18\sqrt{\frac{\pi}{6f}} - 1.2)d} \ln(\frac{1.2d}{2b}) = \frac{Gb}{\pi K} \frac{f^{1/2}}{(0.854 - 1.2f^{1/2})d} \ln(\frac{1.2d}{2b}) \quad (1.3)$$

where G is the shear modulus of matrix, b is the Burgers vector. K is the coefficient, $\frac{1}{K} = \frac{1}{2}(1 + \frac{1}{1-\nu})$, where ν is the Poisson's ratio. f is the volume fraction of secondary phase. d is the size of secondary phase.

Suppose the size of dislocation core is $2b$. The repulsive force exerted on slipping dislocations by phase interface generates regions around secondary phase particles where slipping dislocations cannot enter. This is equivalent to the increase in the size of secondary phase particles by about 20%. Therefore, the size of secondary phase in Eq. (1.3) is multiplied by 1.2.

When the volume fraction of secondary phase is small, ($f^{1/2} \ll 0.854/1.2$), strength increment ΔR_P is derived as follows:

$$\Delta R_\mathrm{P} = \frac{\sqrt{6}Gb}{1.18\pi^{3/2}K} \frac{f^{1/2}}{d} \ln(\frac{1.2d}{2b}) = 0.3728 \frac{Gbf^{1/2}}{K\,d} \ln(\frac{1.2d}{2b}) \quad (1.4)$$

Put material constants (G = 80650 MPa, Poisson's ratio ν = 0.291, b = 0.24824 nm) into the formula above, so

$$\Delta R_\mathrm{P} = 8.995 \times 10^3 \frac{f^{1/2}}{d} \ln(2.417d) \quad (1.5)$$

The units of ΔR_P and d are MPa and nm respectively. Figure 1.4 shows the relationship between the strength increment and the size and volume fraction of secondary phase based on Orowan strengthening mechanism [20].

1 Introduction

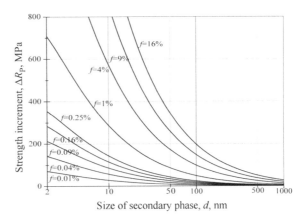

Fig. 1.4 Relationship between the strength increment and the size of the secondary phase with different volume fractions according to Orowan strengthening mechanism

According to Eq. (1.5) and Fig. 1.4, the influence of size of secondary phase is obviously greater than that of volume fraction. For the majority of secondary phases in steel, reduction in the average size significantly improves the strengthening effect. In microalloyed steel, the size of microalloyed carbonitrides is usually controlled within the range of 2–10 nm. Therefore, even the addition of microalloying elements is small and the volume fraction of effective microalloyed carbonitride precipitates is only in range of 0.01–0.1%, the strength increment of dozens of or up to hundreds of megapascal can still be achieved. The volume fraction of cementite in high carbon steel can reach 10%, but if its average size is microns order of magnitude, it can only provide strength increment of dozens of megapascal. If the size is controlled at about 200 nm, the strength increment can reach hundreds of megapascal.

The product of projected area and projected height (i.e. the number of layers of slip planes which can be occupied) on the matrix slip planes of non-spherical secondary phase particles is significantly larger than that of spherical particles with the equal volume, which means the strengthening effect of non-spherical secondary phase particles is significantly larger. Researchers have carried out an in-depth analysis of the strengthening mechanism of disc-shaped niobium carbonitride particles in steel and derived the formula for calculating strength increment [21].

In low-alloyed high strength steel, except the portion of titanium which is in the form of TiN preventing grain growth at high temperatures, the rest of titanium is mainly presented in the form of TiC or C-rich Ti(C, N) in deformed austenite by strain induced precipitation, or precipitated in ferrite during the coiling process. The size of strain induced precipitate TiC in deformed austenite is generally about 10 nm, and the size of TiC precipitated in ferrite can be controlled within the range of 2–5 nm. According to Eq. (1.5) and Fig. 1.4, even if the volume fraction of precipitate is very small, a strong precipitation strengthening effect can still be achieved.

The relative atomic masses of titanium, vanadium and niobium are 47.867, 50.9415 and 92.9064, respectively. The relative atomic mass of titanium is slightly lower than that of vanadium and about half of that of niobium. Therefore, suppose the mass of each alloying element is equal, titanium combines slightly more carbon or nitrogen than vanadium, and much more than niobium. In other words, the mass fraction of TiC or TiN is slightly larger than that of VC or VN, and significantly larger than that of NbC or NbN. In addition, the densities of TiC, TiN, VC, VN, NbC, NbN are 4.944, 5.398, 5.717, 6.097, 7.803, 8.371 g/cm^3, respectively. Titanium carbonitride is lighter than vanadium carbonitride by about 14% and niobium carbonitride by about 56%. Thus, suppose the mass of each carbonitride is equal, the volume fraction of titanium carbonitride is larger than that of vanadium carbonitride by about 14% and niobium carbonitride by 56%. Because the precipitation strengthening effect of micro-alloyed carbonitride is proportional to the square root of volume fraction, the precipitation strengthening effect of titanium carbonitride is greater than that of vanadium carbonitride and significantly larger than that of niobium carbonitride if the mass fraction of each alloying element is equal. A large number of production data show that in niobium micro-alloyed steel with niobium content of 0.02–0.05%, the strength increment due to Nb(C, N) is generally in the range of 50–100 MPa. In V-microalloyed steel with vanadium content of 0.08–0.12%, the strength increment due to V(C, N) is generally in the range of 100–200 MPa. The strength increment in V-microalloyed steel with conventional nitrogen content is close to the lower limit, while that with high nitrogen content is close to the upper limit. Through good production control, the strength increment due to TiC can reach more than 300 MPa in Ti-microalloyed steel with titanium content of 0.08–0.12%.

Obviously, because titanium is liable to form Ti-bearing oxides, sulfides, nitrides, sulfur carbides and so on at high temperatures, the content of titanium capable of forming TiC (effective titanium) significantly fluctuates. This fluctuation leads to the fluctuation of volume fraction of TiC. In addition, due to that the chemical free energy of TiC precipitation fluctuates, the effective precipitation temperature range changes, and thereby affecting the size of precipitates. Therefore, under normal industrial production conditions, the strength increment produced by TiC precipitates fluctuates, making the performance stability of Ti-microalloyed steel significantly lower than that of Nb-microalloyed steel and V-microalloyed steel. This is the key technical problem in the production and application of Ti-microalloyed steel. It is very important to deeply understand the precipitation law of various Ti-bearing phases as well as strictly control the precipitation of various Ti-bearing phases in the actual production process, so as to effectively inhibit the formation of oxide, sulfide, nitride, sulfur carbides and other Ti-bearing phases, and stabilize the volume fraction and effective precipitation temperature of TiC, and thus obtain a stable steel performance.

1.2.4 Effect of Titanium on Toughness of Steel

The material toughness describes the ability of material to absorb energy from deformation to fracturing. When the strength of material is enhanced, sufficient toughness is required to ensure its safe use, so toughness is a very important indicator of material performance. The toughness of low carbon steel is usually characterized by impact toughness or ductile-brittle transition temperature, while fracture toughness is often applied to high carbon steel.

Solute atoms in steel matrix have an important effect on the toughness of steel. They cause lattice distortion of matrix crystal, thereby reducing the toughness. The interstitial atoms lead to serious distortion of matrix lattice, which is very harmful to the toughness. The elements, such as phosphorus, silicon, etc., which result in asymmetric distortion of matrix lattice, also damage the toughness. The effect of solute titanium on the toughness of steel is not significant, and because the amount of solute titanium is small, it has little effect on the toughness of steel.

The grain size of steel matrix has a very important effect on the toughness of steel [9]. Grain refinement is the only way to enhance both the strength and toughness of steel. As described above, the grain coarsening at high temperatures can be controlled by TiN. The growth of recrystallized grains can be prevented by TiC generated by strain induced precipitation during the recrystallization controlled rolling process. Then, the non-recrystallization controlled rolling is used to obtain very fine ferrite grains. Through these procedures, titanium micro-alloyed steel with good toughness is achieved.

According to the fracture characteristics of secondary phase in steel, the secondary phases are generally classified into two types, cleavage type and fracture type [9]. For the cleavage type, the binding force between secondary phase and matrix is weak and the combination is non-coherent. The secondary phase is usually nearly spherical. When external force is exerted, the secondary phase easily breaks away from the matrix along the interface, resulting in micro-cracks slightly larger than the secondary phase particles. For the fracture type, when external force is exerted, the secondary phase is prone to self-fracture, resulting in micro-cracks slightly larger than the short side of secondary phase particles. In addition, for the secondary phase completely coherent or semi-coherent with the matrix, if the size is less than several dozens of nanometers, the binding force between the secondary phase and matrix is strong, and the secondary phase is spherical or nearly spherical. In this situation, the cleavage or self-fracture is not prone to occur. In other words, micro-cracks are not to be generated, so the secondary phase is called non-cracking secondary phase. According to the theory of fracture mechanics, only the micro-cracks with the critical size will expand to cause fracture. By controlling the size of the largest secondary phase instead of the average size, the maximum size of micro-cracks is not to exceed the critical size. This is essential for improving the fracture strength and toughness of material. The critical size of cracks in low strength steel is close to millimeters order of magnitude. As long as the size of the largest inclusion particle is controlled below millimeters order of magnitude, the serious brittle fracture is not to occur. However, the critical size of cracks in ultra-high strength steel is about $10\ \mu m$, so the formation of secondary phase (inclusions) particles with the size of $10\ \mu m$ and above should be strictly controlled.

In addition, the morphology of large secondary phase particles has an important influence on the generation of micro-cracks. For the brittle secondary phase particles with sharp edges and corners, stress significantly concentrates at the sharp edges and corners, which is prone to cause micro-cracks. Significantly elongated film-like, flaky and linear secondary phase particles are very prone to break and generate micro-cracks. The distribution of secondary phase particles plays an important role in the expansion of micro-cracks. When the secondary phase particles are uniformly distributed in the matrix, the interactions of stress fields around particles is small. As a result, even if an individual micro-crack is formed, it is difficult to expand because it is surrounded by iron matrix (the critical size is large). When the secondary phase segregates on the grain boundary of matrix, the grain boundary is weakened obviously. As a consequence, the micro-cracks expand rapidly along the grain boundary and the intergranular fracture is to occur. When the secondary phase particles are distributed in arrays, the stress fields around particles interact, leading to the decrease in the critical size. As a result, the micro-cracks easily expand and inter-connect, and ultimately rapidly expand when the size exceeds the critical size.

Therefore, the laws describing the effect of secondary phase on the toughness of steel depend on the size of secondary phase. The embrittlement vector of strengthening by uniformly distributed fine secondary phase particles in low carbon steel is about 0.26 °C/MPa, which is the least value except that of grain refinement strengthening [9]. In other words, the negative effect of uniformly distributed secondary phase on the toughness of steel is small. Meanwhile, according to the formula describing the strength increment due to secondary phase, the damage of secondary phase to the toughness of steel is proportional to the square root of volume fraction and inversely proportional to the average size. On the other hand, the strengthening effect of non-uniformly distributed large particles (commonly referred to as inclusions) is small, but the damage to the toughness of steel is large, i.e., the embrittlement vector significantly increases. The experimental results [9] show that the damage of large secondary phase particles to the toughness is also roughly proportional to the square root of volume fraction and increases with the average size. Obviously, reducing the volume fraction of large secondary phase particles in steel significantly improves the toughness of steel, while reducing the size leads to a more significant improvement.

TiC particles precipitated in austenite by strain induced precipitation or precipitated in ferrite are small and spherical or disk-like. They are uniformly distributed in the matrix and belong to non-cracking secondary phase. They have a certain adverse effect on the toughness of steel, and this effect is proportional to the strength increment they produce, that is to say, producing a significant strengthening effect at the expense of partial toughness. TiN particles precipitated at high temperatures are large and square, and have obvious damage to the toughness of steel. In addition, TiS and Ti_2CS particles precipitated at high temperatures are also large and have significant damage to the toughness of steel. These coarse precipitates not only have no precipitation strengthening effect, but also cause very serious damage to the toughness of steel, so their volume fraction and size must be strictly controlled to reduce the harmful effects. In terms of thermodynamics, the equilibrium volume fractions of those coarse precipitates are reduced by reducing the content of sulfur and nitrogen in steel. In terms of kinetics, reducing the content of sulfur and nitrogen in steel also reduces the driving force of precipitation and equilibrium precipitation temperature. The actual precipitation temperature is reduced obviously by appropriate rapid cooling. The size of precipitates mainly depends on the actual precipitation temperature. The lower the actual precipitation temperature, the smaller the precipitates. As a result, the size of precipitates can be significantly reduced by this way. At present, with good production control, the precipitation of TiS and Ti$_2$CS particles has completely been inhibited in Ti-microalloyed steel, and the size of TiN particles is below 200 nm, so the damage to the toughness of steel is obviously reduced [16].

1.2.5 Effect of Titanium on Plasticity of Steel

The nature of plastic deformation is the result of large scale slipping of movable dislocations in material. The plasticity of material is divided into two parts, i.e. uniform plasticity and non-uniform plasticity. For structural materials in practical applications, the uniform plasticity is the main concern, because the material has actually failed and cannot be used once the plastic deformation exceeds the uniform deformation stage and enters the necking stage. However, the non-uniform plasticity is also important for the toughness and safety.

The relationship between the stress S and strain ε during the uniform plastic deformation stage can be expressed by the Hollomon equation.

$$S = K\varepsilon^n \tag{1.6}$$

where K is the strain hardening coefficient and n is the strain hardening exponent.

The condition for plastic instability, i.e. the transition from uniform deformation to necking is described as follows.

$$\frac{dS}{d\varepsilon} \leq S \tag{1.7}$$

Thus, the maximum uniform strain ε_B is obtained as follows.

$$\varepsilon_B = n \tag{1.8}$$

Obviously, according to Eq. (1.7), when the increment of strain hardening rate $dS/d\varepsilon$ exceeds the increment of stress S, the uniform plastic deformation continues, otherwise the local necking starts, which leads to fracture eventually. The effect of various micro-defects on the uniform plasticity largely depends on their relative enhancement of the strain hardening rate and the strength of material.

Solute atoms cause the distortion of matrix lattice, adversely affecting the plasticity of material. Interstitial atoms such as carbon and nitrogen seriously reduce the plasticity and lead to yielding. The atoms segregating on the grain boundary such as phosphorus, arsenic, tin and antimony are also very harmful to the plasticity. The substitution atoms with the atomic size and chemical properties similar to that of iron atom, such as nickel, chromium, manganese, have little harm to the plasticity, and even improve the plasticity. Because the size of titanium atom is quite different from that of iron, it has a significant effect to impede the dislocation slipping, which is disadvantageous to the plasticity of steel. However, since the amount of solute titanium is very small, it has little effect on the plasticity of steel.

The matrix grain refinement has a complicated effect on the plasticity. During the deformation at room temperature, the grain boundaries obstruct the dislocations slipping and cause the pilling up of dislocations. The plastic deformation can continue only through the activation of quite a few slip systems and the coordination of plastic deformation among the different grains. Thus, grain refinement, which increases the grain boundaries, is detrimental to the uniform plasticity. This situation is particularly significant for hexagonal metals with less slip systems, but not significant for body centered cubic and face centered cubic metals. The uniform plasticity of ultra-fine grain steel is also very low, because the complete deformation coordination among the grains is difficult to be achieved. On the other hand, grain refinement is advantageous to non-uniform plasticity. A large number of experimental results show that the refinement of austenite grains in steel significantly improves the overall plasticity of steel. When the size of ferrite grains in ferrite pearlitic steel decreases, the uniform plasticity of steel decreases, but the non-uniform plasticity increases. Therefore, the change of overall plasticity depends on the competition between these two types of plasticity. The austenite grains in Ti-microalloyed steel can be refined by recrystallization controlled technology. Suppose that the sizes of ferrite grains are equal, the plasticity of Ti-microalloyed steel is significantly higher than that of Nb-microalloyed steel.

The micron-scale secondary phase particles almost do not improve the strain hardening rate, but obstruct the dislocations and cause the piling up of dislocations, so the plastic deformation resistance increases. Meanwhile, because the elastic modulus of most secondary phases in steel is larger than that of steel matrix, the plastic deformation always occurs in steel matrix. The presence of secondary phase reduces the volume where plastic deformation can occur, so the plasticity decreases. Therefore, the large particles of secondary phase are always to reduce the plasticity of steel. The various oxides, sulfides, nitrides in steel, secondary cementite in hypereutectoid steel, large particles of carbide in high alloy steel and intermetallic compounds, all without exception, are to reduce the plasticity of steel, and the degree of reduction increases with the volume fraction and size of these secondary phases.

The effect of morphology of large secondary phase on the plasticity of steel is very significant. The dislocations can bypass spherical Fe_3C particles by cross slip, but cannot pass lamellar Fe_3C. Therefore, the plasticity of high carbon steel after spheroidization is significantly better than that of steel with pearlite structure formed by slow cooling rate. Similarly, the plasticity of tempered sorbite structure in medium carbon steel subjected to quenching and tempering treatment is better than that of pearlitic ferrite structure obtained by direct cooling.

The secondary phase with size less than 100 nm exhibits a different law. The deformable secondary phase particles do not directly impede the dislocations movement but are cut by the dislocations. Therefore, they do not cause a large increase in the dislocations and have little effect on the work hardening rate. However, these particles increase the flow stress. Hence, the deformable secondary phase particles reduce the uniform plasticity to a certain extent. The non-deformable secondary phase particles continuously produce the dislocation rings according to the Orowan mechanism during the deformation process, resulting in a high work hardening rate. This effect is greater than the increase in the flow stress in a certain strain range, and thus appropriately improves the uniform strain, or at least does not reduce the uniform plasticity.

In addition, the secondary phase destroys the integrity of crystal structure. The formation of stress fields around the secondary phase particles significantly promote the expansion and connection of micro-cracks during the plastic instability of materials. Therefore, the non-uniform plasticity of steel significantly deceases and the overall plasticity decreases.

TiC particles precipitated in austenite by strain induced precipitation or precipitated in ferrite are very small, spherical or disc-like, uniformly distributed in steel matrix and belong to the non-deformable secondary phase. Therefore, they have no adverse effect on the uniform plasticity of steel. The sizes of TiN, TiS and Ti_2CS particles precipitated at high temperatures are large, significantly reducing the uniform and non-uniform plasticity. Through strict control of chemical composition and process parameters, the volume fraction and size of these secondary phases are effectively reduced, so the harm to the plasticity of steel is reduced. Similarly, by reducing the content of sulfur and nitrogen in steel, not only the volume fractions of these coarse precipitates are reduced, but also the equilibrium precipitation temperatures are decreased. Meanwhile, increasing the cooling rate appropriately decreases the actual precipitation temperature. As a result, the size of precipitates is significantly reduced, and the damage to the plasticity of steel is remarkably decreased.

1.2.6 Other Effects of Titanium

When the trace non-metallic elements in steel such as carbon, nitrogen and hydrogen exist in interstitial sites of steel matrix, they usually cause serious damages to some properties of steel. For example, interstitial atoms such as carbon and nitrogen tend to segregate on the dislocation lines and form atmosphere. During the cold working deformation, the atmosphere impedes the dislocation motion. Once unpinned, the yield point elongation occurs. This discontinuous yielding seriously damages the deep drawing performance of steel, resulting in a decrease in the surface quality of cold working steel. For parts with high surface quality requirements such as automotive panels, the existence of solute atoms must be strictly controlled. In stainless steel, the interstitial atoms tend to segregate on the grain boundaries. They react with dissolved chromium and form compounds during the processing or service periods. Because the regions near the grain boundaries are poor in chromium, the intergranular corrosion is prone to occur. In addition, the dissolved hydrogen atoms cause several complex reactions in the processing and service periods, leading to hydrogen embrittlement, hydrogen corrosion, hydrogen induced micro-cracks and delayed fracture.

In order to avoid the harmful effects of trace non-metallic elements, on one hand, it is necessary to strictly control the content of trace non-metallic elements in steel, such as the content of carbon in interstitial free steel (IF steel) often should be below 0.002%. However, the production cost of smelting is inevitably increased significantly. On the other hand, the alloying elements such as titanium and niobium, which have strong metallicity but are not oxidized during the smelting process, react with these non-metallic elements and form stable compounds. As a result, these non-metallic elements are fixed and their harmful effects are eliminated. In order to completely fix the non-metallic elements, it is generally necessary to design the chemical composition according to the ideal chemical composition of compounds, and the content of alloying elements should appropriately exceed the ideal chemical ratio.

The amount of titanium or titanium-niobium added into IF steel is usually beyond the ideal chemical ratio. They react with carbon and nitrogen to form stable carbonitride. As a result, the limitation of carbon content is relaxed and the production cost is significantly decreased [22].

If a proper amount of titanium or titanium-niobium is added into stainless steel, they react with carbon segregating on the grain boundaries to form stable titanium or niobium carbide prior to chromium. Thus, the intergranular corrosion resulted from poor chromium is effectively prevented. This stainless steel is known as stabilized stainless steel.

Addition of titanium or niobium in medium carbon steel leads to the formation of so called "hydrogen trap", which effectively inhibits various hydrogen induced defects and significantly improves fatigue properties, especially the delayed fracture resistance [23, 24].

The position of titanium in the periodic table indicates that it is the strongest carbide and nitride forming element in steel. The combination of titanium and carbon or nitrogen is a very effective way to fix the interstitial elements in steel. Thus, titanium is an important alloying element in stainless steel. Through stabilization treatment, the carbon in steel is combined with titanium to form titanium carbide. Therefore, formation of $Cr_{23}C_6$ by reaction of carbon and chromium near the grain boundaries is inhibited, so the intergranular corrosion resulted from poor chromium is effectively prevented. In order to completely fix carbon, the amount of titanium must be larger than the ideal chemical ratio, that is, the mass ratio of titanium to carbon must be larger than $47.867/12.011 = 3.99$. The ideal chemical ratio of niobium to carbon is $92.9064/12.011 = 7.74$. Thus, suppose equal carbon content, the amount of necessary titanium is much lower and the cost advantage is significant. Therefore, a large amount of stabilized stainless steel is widely micro-alloyed with titanium.

Based on the same principle, titanium is also the most important element to fix the interstitial elements in ultra-deep drawing steel. In ultra-deep drawing steel, the content of carbon and nitrogen is ultra-low, and the amount of titanium and niobium is beyond the ideal chemical ratio, so there are completely no interstitial atoms such as carbon and nitrogen during the cold working deformation process. Due to the absence of yielding phenomenon, the n and r values are high, the "orange peel" phenomenon is not to occur, and the surface quality is excellent. Therefore, this steel is a high-end panel material for automotive and home appliance industries. Similarly, the use of titanium to fix interstitial elements has a significant cost advantage over niobium in terms of the ideal chemical ratio and price of ferroalloy. However, the service strength is low and the dent resistance performance is insufficient. In order to solve the problem, carbon is re-dissolved by several ppm during the annealing procedure of cold rolling process, so that the strength is improved by 30–50 MPa in the soaking procedure of painting. This kind of steel is the baking hardened steel (BH steel). The ratio of titanium to fix nitrogen in BH steel is usually close to the ideal chemical ratio. The ratio of niobium to fix carbon exceeds the ideal chemical ratio, and NbC should be properly re-dissolved during the annealing procedure of cold rolling process [25]. Therefore, titanium is also an important alloying element in BH steel.

In addition, titanium solute in steel significantly improves the hardenability of steel. However, TiN or N-rich Ti (C, N) particles with the size in the range of 200 nm to 1 μm provide sites for heterogeneous nucleation of ferrite, thereby reducing the hardenability. The amount of titanium in low carbon steel is small, so the amount of large TiN or N-rich Ti (C, N) precipitates is effectively controlled and the effect on hardenability is small. Addition of 0.04–0.1% titanium into low hardenability steel with the carbon content of 0.5–0.6% results in the precipitation of a large number of TiN or N-rich Ti(C, N) particles with the size in the range of 200 nm to 1 μm in the austenite region. Thus, the hardenability is significantly reduced.

Studies have shown that the weathering resistance of Ti-microalloyed steel is significantly better than that of carbon manganese steel, Nb-microalloyed and V-microalloyed steel, and close to that of weathering resistance steel [16]. However, the basic principles are not well understood until now.

In summary, Ti-microalloyed steel has unique performance characteristics and great strengthening potential. It is the next important development direction of microalloyed steel and even engineering structural steel. However, the principles of chemical metallurgy and physical metallurgy of Ti-microalloyed steel have not been systematically studied. Moreover, there are still some important technical problems in the industrial production process. Thus, it is necessary to carry out in-depth research under the guidance of basic theory, develop technologies to inhibit the precipitation of TiO, TiS, Ti_2C and control the precipitation behavior of Ti(C, N), so as to achieve the full beneficial effect of titanium and inhibit its harmful behavior, and continuously develop various new Ti-microalloyed steel.

1.3 Titanium Microalloying Technology

1.3.1 Development of Titanium Microalloying Technology

Before 1920s, due to the fact that microalloyed steel had not been widely used, titanium as a micro-alloying element had not attracted much attention. Since then, with the vigorous development of welding technology, it was found that addition of trace titanium in steel significantly improves the weldability of steel, which brought far-reaching impact on the development of Ti-microalloyed steel.

During 1960s and 1970s, important progress on the theory and technology of microalloyed steel had been achieved. The research results on the precipitation strengthening and grain refinement strengthening have provided a theoretical basis for the development of Ti-microalloying steel. Particularly, the international conference "Microalloying 75" [26] established the status of microalloyed steel and determined the development direction, which laid the foundation for the rapid development of microalloyed steel.

In 1980s, the beneficial effect of titanium and vanadium was applied to the development of high temperature recrystallization controlled rolling technology. This application solved the problems, such as the short life span of rolling mill, the long pass interval and the low production efficiency. It was found that it is an ideal way to use recrystallization controlled rolling to produce V–Ti–N series of steel [27]. Thereafter, during the international conference "Microalloying95" held in Pittsburgh, the latest achievements on the development of microalloyed steel during 1975–1995 were summarized. Moreover, a new concept of microalloying technology was put forward, which composes of two types of controlled rolling technology (recrystallization controlled rolling and non-recrystallization controlled rolling) based on high performance steel with austenite regulated. In addition, the concept of thermomechanical control process was proposed by Japanese researcher T. Tanaka.

During late 1990s, the world's major countries in steel production had implemented plans to investigate and develop new generations of iron and steel materials. The microalloying technology became more mature and had attracted wide attention. The great precipitation strengthening effect of nano-carbide was regarded as the important idea on the development of high strength steel or even ultra-high strength steel.

Based on the characteristics of TSCR process, researchers developed a high strength ferritic steel with the yield strength of 700 MPa by single Ti-microalloying technology. This steel has been massively produced and used in industry [28]. Based on the low carbon steel 0.04%C–1.5%Mn, JFE developed a ferritic steel with the tensile strength of 780 MPa by Ti-Mo multiple microalloying technology. The precipitation strengthening effect of nano-carbide particles reached 300 MPa [29], and the steel was named "NANOHITEN" steel [30]. Mittal-Arcelor produced a microalloyed pipeline steel with the yield strength of about 700 MPa by controlled rolling technology [31]. The alloying elements include 0.035–0.05% Ti, 0.08–0.09% Nb and 0.3–0.4% Cr. The microstructure mainly consists of fine ferrite grains. By using vacuum induction furnace melting as well as controlled rolling and cooling technology, researchers at Northeastern University in China developed a kind of Ti-microalloyed high strength steel with the yield strength of more than 700 MPa. The high strength is attributed to the refinement of bainite laths and precipitation strengthening of TiC particles [32].

In order to achieve higher strength, the idea of reinforcing hot rolled ferritic steel by multiple Ti-microalloying technology has been widely accepted and gradually used, such as Ti–Nb, Ti–Mo, Ti–Mo–Nb and Ti–V–Mo. Researchers developed a high strength steel with the yield strength of more than 900 MPa by Ti–V–Mo multiple microalloying technology [33]. Through the combination of multiple Ti-microalloying technology and controlled rolling and cooling technology, the effect of microalloying elements, especially the precipitation strengthening effect of nano-carbide particles, is maximized. This technology will become an important direction of developing Ti-microalloyed steel.

In addition, the interphase precipitation mechanisms attracted wide attention in the 1980s. After 2000, with the development of Ti-microalloyed high strength steel, many researchers [34–37], studied the crystal structure, orientation relationship and strengthening mechanism of nano-carbide produced by interphase precipitation. These research results enriched the theory of physical metallurgy and further promoted the development of micro-alloying technology.

1.3.2 Trace Titanium Microalloying Treatment

Trace titanium microalloying treatment refers to the technical measures to improve the microstructure and weldability of steel by adding 0.01–0.03% titanium into steel. The coarsening of austenite grains at high temperatures are inhibited by TiN particles which are not dissolved, so the microstructure and weldability of steel are improved.

After 1920, with the development of welding technology, it was found that titanium in steel could significantly improve the weldability of steel. Studies have shown that TiN is very stable and does not dissolve at high temperatures during reheating or welding. TiN particles in trace Ti-microalloyed steel prevent the coarsening of austenite grains during the reheating process before rolling. In addition, they effectively inhibit the grain growth in the heat affected zone.

Grain coarsening is a common phenomenon in steel. Preventing grain boundary migration is an effective way to suppress grain coarsening. The area of grain boundaries and the local energy decrease as the grain boundaries and secondary phase particles intersect. When grain boundaries leave secondary phase particles for migration, the local energy rises, which leads to the pinning effect of secondary phase particles on the migration of grain boundaries. Based on the early work of Zener, Gladman derived the maximum particle size r_{crit} that can effectively counteract the driving force of austenite grains coarsening.

$$r_{crit} = \frac{6\bar{R}_0 f}{\pi}(\frac{3}{2} - \frac{2}{Z})^{-1} \tag{1.9}$$

where $\bar{R}_0$ is the average equivalent radius of truncated octahedron grain, i.e. Kelvin tetrakaidecahedron. Z describes the non-uniformity of matrix grain sizes, $Z = 2^{1/2} - 2$. f is the volume fraction of secondary phase particles.

Figure 1.5 [38] presents the radius and volume fraction of secondary phase particles which can prevent the growth of grains with different sizes. Microalloying elements form highly dispersive and small carbonitride particles, which significantly

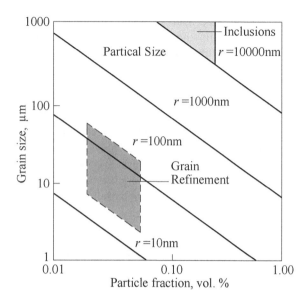

Fig. 1.5 Effect of secondary phase particles with radius r on suppressing the grain growth

improve the resistance to grain coarsening at high temperatures during the austenitizing process. However, at higher temperatures, due to that the secondary phase particles are dissolved or coarsening, their pinning effect fails and the austenite grains grow rapidly [39]. Because TiN is very stable and actually insoluble in austenite, it is able to effectively prevent the grain growth during hot working process. Under the same austenitizing conditions (1230 °C × 10 min), the sizes of austenite grains in three steel (0.055%Ti–0.01%N, 0.075%Ti–0.0102%N and 0.021%Ti–0.0105%N) are 95, 90 and 29 μm respectively. Therefore, the ideal chemical ratio of titanium to nitrogen, which is the most effective in preventing the growth of austenite grains, is close to 2.

1.3.3 Single Titanium Microalloying Technology

Single Ti-microalloying technology refers to the addition of 0.04–0.2% titanium into steel, which significantly improves the overall performance of steel.

The production at the compact strip production (CSP) line of Zhujiang Steel in China shows that the microstructure and strength of low carbon steel strips produced by TSCR technology are significantly better when compared to hot rolled strips produced by traditional cold charge process. The use of clean steel producing technology in Zhujiang Steel significantly reduces the contents of oxygen, sulfur and nitrogen in steel, effectively solving the problem of low recovery rate of ferrotitanium. In the TSCR process, the slab head enters the mill and the tail is still in the soaking furnace, avoiding the influence of temperature difference on the precipitation behavior of Ti

(C, N). The process characteristics of TSCR and the technology of producing clean steel promoted the development of Ti-microalloyed high strength steel. The authors of this book used the single Ti-microalloying technology to develop a ferritic hot-rolled high strength steel with the yield strength of 700 MPa, and massive production was achieved.

For the Ti-microalloyed high strength steel produced by TSCR process, TiN particles precipitated during the soaking stage refine the austenite grains before rolling. Because the ratio of titanium to nitrogen is much higher than the ideal chemical ratio of TiN, TiC particles precipitate in several subsequent stages such as hot rolling, laminar cooling and coiling. The deformation induced precipitation during the hot rolling process decreases the mass fractions of precipitates from interphase precipitation and precipitation in ferrite. The research results show that with the increase in the mass fraction of manganese in steel, the precipitation kinetic curve of TiC is moving to the right and the precipitation process is retarded. Therefore, manganese in Ti-microalloyed steel suppresses the deformation induced precipitation of TiC particles. This suppression causes a large amount of fine TiC particles to precipitate during the subsequent cooling and coiling process, so the resulted precipitation strengthening effect is better. In addition, manganese reduces the $\gamma \rightarrow \alpha$ transformation temperature and increases the carbon content in ferrite after phase transformation, so more TiC particles precipitate in ferrite.

The study of high strength steel microalloyed with single titanium shows that the average size of ferrite grains with high angle grain boundary (greater than 15°) is 3–5 μm, as shown in Fig. 1.6. Figure 1.7 shows that the matrix microstructure contains highly dense dislocations and a large number of nano-particles distributed on the dislocations. Therefore, the effect of precipitation strengthening is significantly improved. Chemical phase analysis shows that the mass fraction of MX phase is 0.0793% (Table 1.1), of which the particles with the size less than 10 nm account for 33.7%, as shown in Fig. 1.8. The strengthening effect provided by the nano-TiC

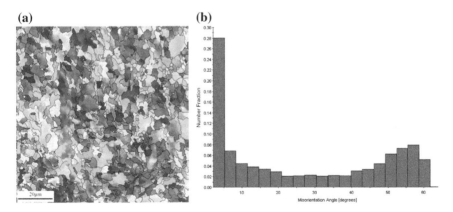

Fig. 1.6 a Electron Backscattered Diffraction (EBSD) orientation diagram of high strength steel, b Micro-orientation distribution diagram of the ferrite grain boundaries of high strength steel

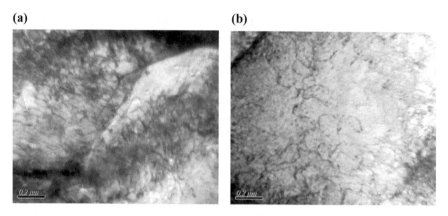

Fig. 1.7 **a** Morphology of the dislocations in high strength steel, **b** Distribution of precipitates on the dislocations in high strength steel

Table 1.1 Mass percent of the elements in MC and M_3C phases relative to the alloy composition

Phase type	Fe	Ti	Cr	Mn	Mo	C	N	Σ
M_3C	0.0500	–	0.0102	0.0027	–	0.0046	–	0.0675
MC	–	0.0589	0.0030	–	0.0009	0.0103	0.0062	0.0793

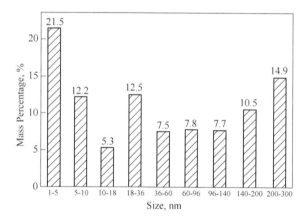

Fig. 1.8 Particle size distribution of MX phase precipitates

precipitate particles reaches 158 MPa. Moreover, the strength is improved by more than 300 MPa through grain refinement. Therefore, grain refinement and precipitation strengthening are the main strengthening mechanisms of Ti-microalloyed high strength steel.

1.3.4 Multiple Titanium Microalloying Technology

Multiple titanium microalloying refers to adding strong carbide forming elements such as niobium, vanadium and molybdenum into steel in addition to titanium microalloying so as to remarkably improve the overall performance of steel.

Multiple microalloying is an important direction of developing microalloying technology. In the aspect of thermodynamics, microalloyed nitride is more stable than carbide. In addition, the stability order of different microalloyed carbides and nitrides is V < Nb < Ti. The solubilities of titanium nitride and vanadium nitride are quite different to that of their carbides, while the solubility difference between niobium nitride and niobium carbide is small. The addition of microalloying elements complicates the precipitation process. However, through strict control of chemical composition and production process, the advantageous effects of different microalloying elements can be effectively achieved.

1.3.4.1 V–Ti Multiple Microalloying Technology

In 1980s, the beneficial effect of titanium and vanadium in the controlled rolling process was confirmed in the development of high temperature recrystallization controlled rolling technology. The addition of titanium and vanadium solved the problems, such as short life span of rolling mills, long pass interval and low production efficiency.

The key points of recrystallization controlled rolling are (1) to obtain fine reheating austenite grains, (2) multiple deformation and recrystallization above the recrystallization temperature, (3) rapid cooling to an intermediate temperature and then air cooling to room temperature.

In order to fully use the advantage of this technology, the composition of desirable steel should have the following metallurgical characteristics: a high grain coarsening temperature, a lower recrystallization temperature during rolling and a low grain coarsening rate after rolling.

It is an ideal way to apply the recrystallization controlled rolling technology to the production of V–Ti–N series of steel. The precipitation of TiN at high temperatures reduces the nitrogen content in austenite, which inhibits the precipitation of VN. In addition, VN is not prone to precipitate in the low temperature rolling process. Therefore, V–Ti–N steel exhibits high grain coarsening temperature, low grain coarsening rate and low recrystallization temperature. Moreover, the residual vanadium in austenite precipitates during the ferrite transformation process, thereby increasing the strength of steel [27].

The addition of titanium into high strength low alloy steel is widely used to control the size of austenite grains. It has been reported that the addition of titanium to steel changes the precipitation behavior of vanadium, resulting in a decrease in the yield strength. The solution temperature of VN is as low as 1094 °C, even if the temperature of slabs decreases to 1000 °C in a short time, vanadium is not to precipitate in the form of VN. However, a small amount of titanium promotes the precipitation of VN in the form of (V, Ti) N. Researchers [40] discovered that (V, Ti) N formed in the casting process is a metastable phase containing the stable phase TiN, and the size of (V, Ti) N particles decreases due to the dissolution of vanadium during the soaking process.

At a given soaking temperature, the yield strength and precipitation strengthening of V–Ti–N steel are lower when compared to V–N steel. The two main factors affecting the precipitation strengthening are volume fraction and average size of precipitate particles. In V–Ti–N steel, V–Ti nitride precipitated in austenite at high temperatures impedes the growth of austenite grains in the rolling process and subsequent process. However, because the particles are large, their contribution to the strength of steel is not great. In contrast, the precipitation of these particles reduces the amount of vanadium and nitrogen dissolved in steel before $\gamma \rightarrow \alpha$ phase transformation, thereby reducing the amount of VN precipitates in ferrite and the corresponding precipitation strengthening effect. With the increase in the final cooling temperature, the precipitation strengthening effect is reduced. At the same final cooling temperature, the precipitation strengthening effect of V–Ti–N steel is significantly lower than that of V–N steel. The addition of titanium in V-bearing steel decreases the yield strength, but slightly refines the grains and improves the impact toughness. The grains in V–Ti–N steel are finer than the grains in V–N steel, but the number of small particles in ferrite matrix of V–Ti–N steel is less than that of V–N steel.

There are two important issues regarding the addition of vanadium and titanium. One issue is whether vanadium reduces the significant role of TiN on inhibiting the grain coarsening. Another issue is whether titanium weakens the precipitation strengthening effect of vanadium.

1.3.4.2 Nb-Ti Multiple Microalloying Technology

In Nb-Ti microalloyed steel, the phase precipitated at high temperatures is TiN. When the ratio of titanium to nitrogen is much higher than the ideal chemical ratio of TiN, the amount of residual nitrogen is small, and niobium carbide and titanium carbide precipitate at low temperatures. When the ratio of titanium to nitrogen in steel is lower than the ideal chemical ratio of TiN, after precipitation of TiN, the residual nitrogen combines with niobium to form NbN, so the high temperature precipitates is Ti–Nb nitride. With the decrease in the temperatures, the proportion of niobium in the precipitates increases, and the low temperature precipitate is a complex of NbC and NbN.

There is a serious microsegregation of substitutional atoms in the casting structure. Because of interdendritic segregation of titanium and niobium, coarse (Ti, Nb) CN is prone to precipitate. Due to rapid solidification and fast cooling rate of TSCR process, the segregation is alleviated and the precipitation occurs at lower temperatures, leading to precipitation of more fine particles.

TiN particles precipitate in liquid or high temperature austenite. They act as the sites for heterogeneous nucleation of Nb (C, N) and TiC in low temperature austenite. If the content of titanium is greater than about 0.011%, the carbonitride precipitated in the interdendritic regions will impede the growth of austenite grains after solidification and during the $\delta \rightarrow \gamma$ transformation process.

In the multiple microalloying technologies using V–Ti and Nb–Ti, the use of titanium is quite conservative. Only the role of TiN on inhibiting the growth of austenite grains is utilized. In addition, the adverse effect of titanium on other microalloying elements also needs to be considered. Therefore, the role of titanium in multiple microalloyed steel is only auxiliary. This is obviously different from the addition of other elements to Ti-microalloyed high strength steel. For example, the addition of molybdenum to Ti-microalloyed steel is intended to further enhance the effect of titanium on steel.

1.3.4.3 Ti–Mo Multiple Microalloying Technology

JFE [41, 42] developed a high strength steel with the strength of 780 MPa by conventional controlled rolling and cooling technology. The composition is 0.04%C–1.5%Mn–0.1%Ti–0.2%Mo, and the microstructure consists of ferrite and nano-carbide. The strength is enhanced by the precipitation strengthening of nano-carbide on the basis of grain refinement strengthening. The volume fraction of nano-carbide particles is increased through blocking the formation of pearlite and cementite at the grain boundaries by molybdenum. The growth of nano-carbide is prevented through decreasing the phase transformation temperature by manganese. The steel has not only high strength and elongation of up to 20%, but also has good machinability. The hole expansion ratio (HER) reaches 120%, several times higher than the steel with the same plasticity. This steel has already been used in automotive parts. In addition, the steel has good strength at high temperatures. When soaked at 600 °C for 25 h, the decrease in the tensile strength is very small.

Chen et al. [34, 35] studied the interphase precipitation of Ti–Mo microalloyed high strength steel during the soaking process. The results show that the isothermal temperatures affect the precipitation behavior of nano-carbide. The interphase precipitation occurs at high temperatures, and the carbide particles in ferrite are dispersive at low temperatures. With the increase in the phase transformation temperature, the carbide particles become denser and finer [36, 37].

Kim et al. [43] developed steel with the basic composition of 0.07%C–1.7%Mn–0.2%Ti–(0.2–0.3%)Mo by thermomechanical control process. The yield strength exceeds 800 MPa, which is attributed to the combination of ferrite grains refinement strengthening and precipitation strengthening. The precipitate strengthening effect is weakened to some extent because of the deformation induced precipitation during the austenite non-recrystallized rolling process. However, the cumulative deformation produces desirable microstructure which provides nucleation sites for $\gamma \rightarrow \alpha$ phase transformation in the subsequent coiling process. As a result, more finer and uniform ferrite grains form in the phase transformation process. The finishing temperature of 880 °C and coiling temperature of 620 °C meet the requirements of strength and toughness of steel.

Compared with the microalloying elements niobium and vanadium, titanium is less used in transformation induced plasticity steel (TRIP steel). Wang [44] added

0.1% Ti and 0.25% Mo into 0.085%C–1.4%Mn–1.5%Si TRIP steel and significantly improved the yield strength and product of strength and plasticity. The matrix microstructure is a typical multi-phase microstructure of TRIP steel, which consists of ferrite, bainite, residual austenite. The bainitic isothermal transformation temperature is the key point of process control, and more volume fraction of high carbon austenite is a key factor to obtain high plasticity. The precipitation of nano-carbide is not affected by bainitic transformation temperature, because it mainly occurs in the cooling process after hot rolling instead of the bainitic isothermal transformation process.

1.3.4.4 Ti–V–Mo Multiple Microalloying Technology

Zhang et al. [33, 45] studied the effect of finishing temperature, cooling rate and coiling temperature on the microstructure of high Ti–V–Mo high strength steel by thermal simulation experiment system. In addition, the multiple precipitation behavior of Ti–V–Mo and effect of coiling temperature on the microstructure and mechanical properties of steel under the actual controlled rolling and cooling conditions was investigated. The ultra-high strength hot-rolled ferrite steel with the yield strength of 900–1000 MPa was developed in the laboratory. In order to improve the strength of the hot rolled steel coiled at low temperatures as well as the uniformity of the microstructure and performance of the whole coil, the tempering heat treatment was used.

Table 1.2 shows the strength increment due to precipitation strengthening σ_p, grain refinement strengthening σ_g and the tensile strength R_m of Ti–Mo microalloyed steel with different compositions but similar rolling process. It shows that σ_p of Ti–V–Mo steel is much higher than σ_p of other Ti–Mo steel, and σ_g of Ti–V–Mo steel is large and reaches 380 MPa. The sum of σ_p and σ_g exceeds 824 MPa, accounting for up to 76.3% of σ_y. It is much larger than that of other Ti–Mo steel, and its R_m is significantly higher than that of other Ti–Mo steel [33, 46–48]. Jha et al. [49] used high Ti–Mo and low V–Nb in the composition design. By the rolling process similar to that of Ti–V–Mo steel, a kind of steel with similar σ_p, σ_g and R_m was developed. This indicates that a large precipitation strengthening increment of greater than 500 MPa could be achieved by the addition of high content of microalloying elements with high solubility in steel such as vanadium and molybdenum as well as appropriate rolling process and soaking at the coiling temperature. The high strength is maintained and the plasticity is not significantly decreased. However, because some amount of titanium, vanadium and molybdenum are not fully dissolved in Ti–V–Mo steel, the maximum precipitation strengthening effect of microalloying elements is not fully exploited. Therefore, a portion of these elements is wasted. However, the solubilities of vanadium and molybdenum in ferrite are much larger than that of titanium. If the addition of vanadium and molybdenum is small, the desirable volume fraction of precipitates cannot be achieved, thereby affecting the precipitation strengthening effect. This problem needs further study.

Table 1.2 Strength increment of Ti–Mo steel with different compositions

Composition	Process in the lab		σ_p(MPa)	σ_g(MPa)	R_m(MPa)	Ref.
	Finishing rolling temperature (°C)	Coiling temperature (°C)				
0.04%C–0.092%Ti–0.19%Mo	900	620	300	312	820	[46]
0.075%C–0.17%Ti–0.275%Mo	880	620	276	318	951	[46]
0.059%C–0.23%Ti–0.19%Mo	900	620	<200	365	769	[47]
0.096%C–0.25%Ti–0.45%Mo–0.031%Nb–0.074%V	900±10	600	330–430	420–450	1020–1170	[49]
0.09%C–0.093%Ti–0.26%Mo–0.14%V	780	600	310	361	955	[33]
0.10%C–0.10%Ti–0.12%Mo	850–930	620	160	285	627	[48]
0.16%C–0.20%Ti–0.44%Mo–0.41%V	870	600	444–487	380	1134	[45]

1.3.5 Economic Characteristics of Microalloying Elements in Steel

In recent years, the iron and steel industry has entered difficult period of meager or even no profit. Reducing the production cost becomes an important development direction. The cost of microalloying in the production of microalloyed steel requires strict control.

The price of ferroalloy is closely related to the abundance of the element in nature. Only the element having rich mineral deposit and capable of economic exploitation and smelting has a wide range of development prospects. The abundances of niobium, vanadium and titanium in the crust are 20, 120 and 5600 ppm, respectively, which indicates titanium has the largest reserve in nature, followed by vanadium, and niobium has the least. Therefore, the prices of ferroalloys are significantly different. The current prices of 66% niobium iron, 50% vanadium iron and 35% titanium iron are about 30,000, 9000, and 1300 $ per ton, respectively. The costs of adding 0.1% of the alloys are 44, 18 and 4 $ per ton steel, respectively. Taking into account that the burning of titanium in the smelting process is slightly larger and the current rate of recovery is about 80%, the cost of adding titanium alloy increases slightly, about 5 $ per ton steel. Obviously, as a microalloying element, titanium has a very significant cost advantage.

References

1. Noren T M. Special report on columbium as a microalloying element in steel and its effect on welding technology [R]. Washington: Ship Structure Committee, 1963.
2. Leyens C, Peters M (eds). Titanium and titanium alloys [M]. Weinheim: Willey-VCH, 2003.
3. Loss R D. Atomic weights of the elements 2001 [J]. Pure Appl. Chem., 2003, 75: 1107–1122.
4. Brandes E A. Smithells metals reference book [M]. 6th edition. London: Butterworth & Co. Ltd., 1983.
5. Yong Q L, Ma M T, Wu B R. Physical-mechanical metallurgy of microalloyed steel [M]. Machinery Industry Press, Beijing, 1989.
6. Zener C, Smith C S. Grains, phases, and interfaces: an interpretation of microstructure [J]. Trans AIME, 1948, 175:47.
7. Hillert M. On the theory of normal and abnormal grain growth [J]. Acta Metal., 1965,13: 227–238.
8. Gladman T. The theory of precipitate particles on grain growth in metals [C]. Proc. Roy. Soc., 1966, 294A: 298–309.
9. Pickering F B. Physical metallurgy and the design of steels [M]. London: Applied Sci. Pub., 1978.
10. Balance J B. The hot deformation of austenite [C]. New York: TMS-AIME, 1976.
11. Cuddy L J. The effect of microalloy concentration on the recrystallization of austenite during hot deformation [C]. In: DeArdo A J, Ratz G A (eds), Thermomechanical processing of microalloyed austenite, Warrendale: TMS-AIME, 1984: 129–140.
12. Dong H, Sun X J, Liu Q Y, et al. Deformation induced ferrite transformation-phenomena and theory. Iron and Steel, 2003, 38(10), 56–67.
13. Dong H, Sun X J. Deformation induced ferrite transformation in low carbon steels [J]. Current Opinion in Solid State and Materials Science, 2005, 9: 269–276.

14. Sun X J, Dong H, Liu Q Y, et al. On post-dynamic austenite-to-ferrite transformation in a low carbon steel [C]. Proceedings of the 3rd International Conference on Advanced Structural Steels.Gyeongju (Korea), 2006: 105–110.
15. Hajeri K F, Garcia C I, Hua M J, et al. Particle-stimulated nucleation of ferrite in heavy steel sections [J]. ISIJ Inter. 2006, 46(8): 1233–1240.
16. Mao X P. Microalloying technology on thin slab casting and direct rolling process [M]. Metallurgy Industry Press, Beijing, 2008.
17. Gladman T. The physical metallurgy of microalloyed steels [M]. London: The Institute of Materials, 1997.
18. Cahn R W. Physical metallurgy [M]. Netherlands: North-Holland, 1970.
19. Yong Q L. Theoretical analysis on the mechanism of precipitation strengthening of microalloyedcarbonitride in ferrite. Chinese Science Bulletin, 1989, 34(19): 707–709.
20. Yong Q L, Sun X J, Yang G W, et al. Solution and precipitation of secondary phase in steels: phenomenon, theory and practice.
21. Yong Q L, Zhen L, Sun Z B, Precipitation and precipitation strengthening of niobium carbide in ferrite in microalloyed steel. Acta Metallurgica Sinica, 1984, 20(1): 9–16.
22. Takechi H. Metallurgical aspects on interstitial free sheet steel from industrial viewpoints [J]. ISIJ Inter. 1994, 34(1): 1–8.
23. Hui W J, Dong H, Weng Y Q, et al. Effect of vanadium microalloying on delayed fracture resistance of high strength steel [J]. Heat Treatment of Metals, 2002, 27(1):10–12.
24. Hui W J, Dong H, Weng Y Q, et al. Effect of titanium on delayed fracture resistance of high strength steel, Journal of Iron and Steel Research, 2002, 14(1):30–33.
25. Baker L J, Daniel S R, Parker J D. Metallurgy and processing of ultralow carbon bake hardening steels [J]. Mater. Sci. Tech., 2002, 18(4): 355.
26. Gladman T, Dulieu D, Mcivor, I D. Structure-property relationships in high-strength microalloyed steels [C]. In: Proc. of Symp. On Microalloying 75, Union Carbide Corp., New York, 1976: 32–55.
27. Zhen Y Z, Fitzsimons G, Fix R M, et al. Recrystallization controlled rolling and air cooling of V-Ti-N microalloyed steel [J]. Iron Steel Vanadium Titanium, 1985(3): 12–19.
28. Mao X P, Huo X D, Sun X J, et al. Strengthening mechanisms of a new 700MPa hot rolled Ti-microalloyed steel produced by compact strip production [J]. Journal of Materials Processing Technology, 2010, 210:1660–1669.
29. Funakawa Y, Shiozaki T, Tomita K, et al. Development of high strength hot-rolled sheet steel consisting of ferrite and nanometer-sized carbides [J]. ISIJ Int., 2004, 44:1945–1951.
30. Seto K, Funakawa Y, Kaneko S. Hot rolled high strength steels for suspension and chassis parts "NANOHITEN" and "BHT® Steel" [J]. JFE Technical Report, 2007(10): 19–25.
31. Shanmugam S, Ramisetti N K, Misra R D, et al. Microstructure and high strength–toughness combination of a new 700MPaNb-microalloyed pipeline steel [J]. Materials Science and Engineering, 2008, 478 A: 26–37.
32. Yi H L, Du L X, Wang G D, et al. Development of a hot-rolled low carbon steel with high yield strength [J]. ISIJ International, 2006, 46 (5): 754–758.
33. Zhang K, Li Z D, Sun X J, et al. Development of Ti–V–Mo complex microalloyed hot-rolled 900MPa-grade high-strength steel [J]. Acta Metall. Sin.(Engl. Lett.), 2015, 28(5), 641–648.
34. Chen C Y, Yen H W, Kao F H, et al. Precipitation hardening of high-strength low-steels by nanometer-sized carbides [J]. Materials Science and Engineering A, 2009, 499: 162–166.
35. Wang T P, Kao F H, Wang S H, et al. Isothermal treatment influence on nanometer-size carbide precipitation of titanium-bearing low carbon steel [J]. Materials Letters, 2011, 65: 396–399.
36. Yen H W, Huang C Y, Yang J R. Characterization of interphase precipitated nanometer-sized carbides in a Ti-Mo-bearing steel [J]. ScriptaMaterialia, 2009, 61: 616–619.
37. Yen H W, Chen P Y, Huang C Y, et al. Interphase precipitation of nanometer-sized carbides in a titanium-molybdenum-bearing low-carbon steel [J]. ActaMaterialia, 2011, 59: 6264–6274.
38. Deardo A J. Metallurgical basis for thermomechanical processing of microalloyed steels [J]. Ironmak.Steelmak., 2001, 28(2): 138–144.

39. Cuddy L J. Microstructure developed during thermomechanical treatment of HSLA steels [J]. Metall. Trans. A, 1981, 12A(7): 1313–1320.
40. Zhang J, Baker T N. Effect of equalization time on the austenite grain size of simulated thin slab direct charged (TSDC) vanadium microalloyed steels [J]. ISIJ International, 2003, 43(12): 2015–2022.
41. Funakawa Y, Seto K. Coarsening behavior of nanometer-sized carbide in hot rolled high strength sheet steel [J]. Mater. Sci. Forum, 2007, 539–543: 4813-4818.
42. Funakawa Y. Mechanical properties of ultra fine particle dispersion strengthened ferritic steel [J]. Mater. Sci. Forum, 2012, 706–709: 2096–2100.
43. Kim Y W, Song S W, Seo S J, et al. Development of Ti and Mo micro-alloyed hot-rolled high strength sheet steel by controlling thermomechanical controlled processing schedule [J]. Mater. Sci. Eng. A, 2013, 565: 430–438.
44. Wang C J. Research on the control and mechanical behavior of metastable austenite and precipitates in multi-phase structure steel [D], PhD thesis, Central Iron and Steel Research Institute, Beijing, 2013.
45. Zhang K, Yong Q L, Sun X J et al. Effect of coiling temperature on microstructure and mechanical properties of Ti-V-Mo complex microalloyed ultra-high strength steel [J]. Acta Metallurgica Sinica, 2016, 52(5): 529–537.
46. Kim Y W, Kim J H, Hong S, et al. Effects of rolling temperature on the microstructure and mechanical properties of Ti-Mo microalloyed hot-rolled high strength steel [J]. Materials Science and Engineering: A, 2014, 605: 244–252.
47. Shen Y F, Wang C M, Sun X. A micro-alloyedferritic steel strengthened by nanoscale precipitates [J]. Materials Science and Engineering: A, 2011, 528: 8150–8156.
48. Jha G, Das S, Lodh A, et al. Development of hot rolled steel sheet with 600MPa UTS for automotive wheel application [J]. Materials Science and Engineering: A, 2012, 552: 457–463.
49. Jha G, Das S, Sinha S, et al. Design and development of precipitate strengthened advanced high strength steel for automotive application [J]. Materials Science and Engineering: A, 2013, 561: 394–402.

Chapter 2
Principles of Chemical Metallurgy of Titanium Microalloyed Steel

Guangqiang Li

The main roles of the titanium in the titanium microalloyed steel are the grain refinement strengthening and the precipitation strengthening. The smelting of titanium microalloyed steel should satisfy that the most of the titanium dissolves in the molten steel, and precipitates in the form of carbide or carbon nitride after the subsequent solidification, rolling and heat treatment processes. Due to the fact that the affinity between titanium and oxygen is less than that of aluminum and oxygen, but stronger than that of silicon or manganese and oxygen, the addition of titanium can lead to the formation of large amounts of Ti-containing oxides in case of the molten steel is not properly deoxidized during the smelting process. The high content of nitrogen will lead to the formation of titanium nitride inclusions in the molten steel. The titanium nitrides or the titanium oxides will affect the composition and properties of the mold fluxes, which easy to cause the clogging of submerged entry nozzle, and disturbs the smooth production of the continuous casting. The active chemical properties of titanium lead to the difficulty on accurate control of titanium content in steels, and also cause the fluctuation of the mechanical properties of steel plates. Therefore, the widely application of Ti microalloying technology has been limited. During the refining process in pyrometallurgy of titanium microalloyed steel, it is need not only to reduce the content of oxygen, sulfur, phosphorus, nitrogen, hydrogen and other impurities as that of the low carbon manganese steel, but also to deal with the relationship between the titanium and aluminum, or that between the titanium and the refining slag or the refractory materials, the aim is to avoid the numerous formation of larger size Ti-containing inclusions, ensuring the cleanliness of steel and increase the yield of titanium. As the general principle of the smelting process for the clean steel has been introduced by many books and journals, to improve the cleanliness of steel and provide the thermodynamic basis for the yield of titanium, this chapter focuses on the thermodynamics of the deoxidation of molten steel with titanium and

G. Li (✉)
Wuhan University of Science and Technology, Wuhan, China
e-mail: liguangqiang@wust.edu.cn

aluminium, the interaction of titanium and aluminum in the molten steel and the effect of the titanium and aluminum content on the composition of inclusions in the titanium microalloyed steel. In addition, the control of oxide inclusions in titanium microalloyed steel and its application in oxide metallurgy are also introduced. With the development of the control level in the metallurgical process, the Ti microalloying technology reflects a good adaptability in the thin slab continuous casting and rolling (TSCR) process, so, the Ti microalloying technology has broad application prospects.

2.1 Fe–Ti Binary Phase Diagram and Fe–Ti Alloy

The smelting of titanium microalloyed steel requires the understanding of the Fe–Ti binary phase diagram. According to the Fe–Ti binary phase diagram (mass fraction of titanium within 30%) calculated by the Thermo-Calc thermodynamic software combining with the TCFE3 thermodynamic data base [1], it is found that the iron and titanium are completely mutually dissolvable in the liquid status. The maximum solubility of titanium in the austenitic iron is 0.69%, and the solubility of Ti at the boundaries of austenite and ferrite is 1.24%, whereas the maximum solubility of titanium in ferrite is 8.4%. As the temperature decreases, the solubility of titanium in ferrite decreases sharply. For example, the solubility of titanium in iron at 600 and 400 °C are 0.53 and 0.15%, respectively.

Titanium is added into the steel in the form of Fe–Ti alloy during the smelting of the titanium microalloyed steel. The China National Standard [2] issued in 2012 stipulated 15 grades of Fe–Ti alloys, of which the titanium content is 30–80%, and the titanium content interval is 10%. Table 2.1 shows the allowable range of titanium content and the impurity content of various grades of Fe–Ti alloys. It should be noted that Fe–Ti alloys contain higher aluminum content which is due to the alloy is produced by the aluminothermic reduction method. Therefore, the selection of the Fe–Ti alloy should consider the requirement of Al limit with individual grade of steel during the steelmaking process. Besides the impurities listed in Table 2.1, the Fe–Ti alloy also contains a small amount of calcium and oxygen. In the national standard, the size of Fe–Ti alloy is divided into 5 classes of lump, granule and powder [3]. Joanne L. Murray published the fully composition range binary phase diagram of the Ti–Fe alloy [3]. By the phase diagram, the melting point of the Ti–Fe alloy with 40% of Ti is about 1317 °C, and the melting point of the Ti–Fe alloy containing Ti of 50–80% is lower than 1320 °C. The melting point (eutectic point) of Ti–Fe alloy containing 68% titanium is the lowest of 1085 °C.

The schematic representation of the melting, dissolution, and reaction after the addition of ferroalloy as a deoxidizer to the molten steel mainly consists of four stages [4]: (1) With the intermediate formation of a steel shell due to the local cooling, the alloy melts or dissolves with absorbing heat from the surrounding molten steel (stage I). Melting or dissolving depends on the melting temperature of the ferroalloy; (2) nucleation of deoxidation products (oxide inclusions) in the vicinity of a deoxidizer

Table 2.1 The grades and chemical composition of Ti–Fe alloy (GB/T 3282—2012)

Grades	Chemical composition (mass fraction) (%)							
	Ti	C	Si	P	S	Al	Mn	Cu
		Less than or equal						
FeTi30-A	25.0–35.0	0.10	4.5	0.05	0.03	8.0	2.5	0.10
FeTi30-B	25.0–35.0	0.20	5.0	0.07	0.04	8.5	2.5	0.20
FeTi40-A	>35.0–45.0	0.10	3.5	0.05	0.03	9.0	2.5	0.20
FeTi40-B	>35.0–45.0	0.20	4.0	0.08	0.04	9.5	3.0	0.40
FeTi50-A	>45.0–55.0	0.10	3.5	0.05	0.03	8.0	2.5	0.20
FeTi50-B	>45.0–55.0	0.20	4.0	0.08	0.04	9.5	3.0	0.40
FeTi60-A	>55.0–65.0	0.10	3.0	0.04	0.03	7.0	1.0	0.20
FeTi60-B	>55.0–65.0	0.20	4.0	0.06	0.04	8.0	1.5	0.20
FeTi60-C	>55.0–65.0	0.30	5.0	0.08	0.04	8.5	2.0	0.20
FeTi70-A	>65.0–75.0	0.10	0.50	0.04	0.03	3.0	1.0	0.20
FeTi70-B	>65.0–75.0	0.20	3.5	0.06	0.04	6.0	1.0	0.20
FeTi70-C	>65.0–75.0	0.40	4.0	0.08	0.04	8.0	1.0	0.20
FeTi80-A	>75.0	0.10	0.50	0.04	0.03	3.0	1.0	0.20
FeTi80-B	>75.0	0.20	3.5	0.06	0.04	6.0	1.0	0.20
FeTi80-C	>75.0	0.40	4.0	0.08	0.04	7.0	1.0	0.20

depending on the local supersaturation (stage II); (3) the growth and agglomeration of the inclusion particles in the molten steel (stage III); (4) finally the removal of these inclusions by various mechanisms (stage IV).

The early process of melting and dissolution of pure titanium and Fe–35%Ti in the molten iron with total oxygen content of 100–140 ppm was studied [4]. The dissolution of pure titanium in molten steel is relatively slow. Generally, the titanium is added into molten steel after Al deoxidation in industrial steel production, and the steel contains 0.03–0.06% of aluminum with the dissolved oxygen of 4–6 ppm. It is difficult to generate titanium oxide inclusions in this condition unless the dissolved titanium met the previously formed alumina inclusions, then the alumina can possibly be reduced by titanium to form the Ti–Al–O inclusions. The mass fraction ratio of the Ti/Al surrounding the alumina inclusions can be easily greater than 10, so, the reduction of alumina by titanium can occur based on the thermodynamic calculation. Due to the lower melting point of Fe–70%Ti than that of pure titanium, the Fe–70%Ti dissolves much faster than pure titanium in molten steel. Once the Fe–Ti alloy is dissolved in the molten iron, the oxide inclusions can be formed in the surrounding of dissolved Fe–Ti alloy or the titanium in case of the oxygen concentration is high enough to satisfy the thermodynamic conditions. The Fe–70%Ti contains Al, and the Ti–Al–O inclusions are generally formed during the steelmaking process. In addition, Ca is also an impurity element in the Fe–70%Ti alloy, thus, the Al–Ti–Ca–O inclusions can also be formed. The Fe–35%Ti alloy is produced by the aluminothermic reduction process and contains higher total oxygen, and part of the total oxygen is alumina inclusions. The early dissolution of Fe–35%Ti in the molten steel results in the formation of an oxide shell at the interface between the liquid iron and the ferroalloy. The thickness of the shell increases with the time extension. The oxide layer is derived from the oxidation of Ti and Al in Fe–35%Ti alloy and the agglomeration of alumina inclusions in the alloy itself. Subsequently, with the dissolution of the Fe–Ti alloy, it is observed that the size of these alumina inclusions decreases with time due to the reduction of the alumina inclusions from Fe–35%Ti alloys by the high concentration of titanium to form Al–Ti–O inclusions. The statistical analysis of the data for industrial applications of Fe–35%Ti alloys also comes to the same conclusion [5].

2.2 Thermodynamics of Titanium in Liquid Iron

2.2.1 Equilibrium between Titanium and Oxygen in Liquid Iron

Thermochemic data is very important to the thermodynamic calculation of steelmaking reactions, which are based on the dilute solution theory established by C. Wagner. In 1988, the Japanese Society for the Promotion of Science published the "Steelmaking Data Sourcebook" in English [6]. The thermochemic data for steelmaking process is optimized and the recommended values are given. In 2010, Hino from Tohoku University, Japan and Ito from Waseda University, Japan published the English version handbook of "Thermodynamic data for steelmaking" [7], in which the thermodynamic data after the literature [6] are included. But these new data are not evaluated and optimized, readers can choose to use them appropriately. The equilibrium constants and activity interaction coefficients (where [ ] means the elements are in dissolved state in liquid iron and the concentration is in mass percent, the same as below) in the Fe–[Ti]–[O] system have been given in this book as shown in Table 2.2.

The deoxidation reaction of Ti equilibrated with Ti_3O_5 in the molten iron is as follows:

$$Ti_3O_5(s) = 3[Ti] + 5[O] \tag{2.1}$$

$$\log K = \log(a_{Ti}^3 a_O^5) = 3\log a_{Ti} + 5\log a_O = \log K' + 3\log f_{Ti} + 5\log f_O \tag{2.2}$$

$$\log K' = \log\bigl([Ti]^3[O]^5\bigr) \tag{2.3}$$

The deoxidation reaction of Ti equilibrated with Ti_2O_3 in the liquid iron is as follows:

$$Ti_2O_3(s) = 2[Ti] + 3[O] \tag{2.4}$$

$$\log K = \log(a_{Ti}^2 a_O^3) = 2\log a_{Ti} + 3\log a_O = \log K' + 2\log f_{Ti} + 3\log f_O \tag{2.5}$$

Table 2.2 Comparison of equilibrium constants and activity interaction coefficients in Fe–[Ti]–[O] system [6, 8–11]

References	Thermodynamic data at 1873 K			Oxide in equilibrium	Method
	e_O^{Ti}	$\log K$	$\log K - T(K)$		
[6]	−1.12	−16.1	−30,349/T + 10.39	Ti_3O_5	0.013 < [Ti] < 0.25
[8]		−5.81		Ti_2O_3	Based on the standard free energy of formation
[9]		−6.06		Ti_2O_3	Based on the standard free energy of formation
[10]	−1.0	−19.26	−91,034/T + 29.34	Ti_3O_5	Estimated based on various data, [Ti] < 0.4
	−0.42	−11.69	−55,751/T + 18.08	Ti_2O_3	Estimated based on various data, 0.4 < [Ti] < 8.8
[11]	−0.34	−16.86		'Ti_3O_5'	Based on the experimental data with 'Ti_3O_5' crucible and other data, 0.0004 < [Ti] < 0.36
	−0.34	−10.17		Ti_2O_3	Based on the experimental data with Ti_2O_3 crucible and other data, 0.5 < [Ti] < 6.2

$$\log K' = \log\left([Ti]^2[O]^3\right) \quad (2.6)$$

The activity interaction coefficient between titanium and oxygen in the liquid iron is shown in Table 2.3.

Cha et al. [11] has studied the relationship between temperature and the deoxidation equilibrium of Ti equilibrated with 'Ti_3O_5' or Ti_2O_3 in liquid iron [13]. The single quotes from 'Ti_3O_5' and the later 'TiO' are used to denote the nonstoichiometric compounds, i.e., the atomic ratio of Ti/O is not strictly consistent with that in the chemical formula. For Eq. (2.1), the corresponding thermodynamic data is given as Eq. (2.7).

$$\log K_{\cdot Ti_3O_5\cdot} = -\frac{68280}{T} + 19.95, (1823\,\text{K} < T < 1923\,\text{K})$$

$$\left\{\begin{array}{lll} -17.50, & 0.01 < [Ti] < 0.28, & 1823\,\text{K} \\ -16.52, & 0.006 < [Ti] < 0.40, & 1873\,\text{K} \\ -15.56, & 0.0045 < [Ti] < 0.52, & 1923\,\text{K} \end{array}\right\} \quad (2.7)$$

For Eq. (2.4), the corresponding thermodynamic data is given as Eq. (2.8).

$$\log K_{Ti_2O_3} = -\frac{42940}{T} + 12.94, (1823\,\text{K} < T < 1923\,\text{K})$$

$$\left\{\begin{array}{lll} -10.61, & 0.28 < [Ti] < 4.89, & 1823\,\text{K} \\ -9.99, & 0.40 < [Ti] < 6.22, & 1873\,\text{K} \\ -9.39, & 0.52 < [Ti] < 2.79, & 1923\,\text{K} \end{array}\right\} \quad (2.8)$$

The activity interaction coefficient between titanium and oxygen in the molten iron is shown in Eqs. (2.9) and (2.10).

$$e_O^{Ti} = \frac{1701}{T} + 0.0344, \quad (1823\,\text{K} < T < 1923\,\text{K}) \quad (2.9)$$

$$e_{Ti}^{Ti} = \frac{212}{T} - 0.0640, \quad (1823\,\text{K} < T < 1923\,\text{K}) \quad (2.10)$$

Figure 2.1 shows the equilibrium relationship of Ti–O in liquid iron at 1873 K. The equilibrium relationship of Ti–O system in the liquid iron at 1873 K according to the thermodynamic data of Tables 2.2 and 2.3 are also included in this figure. It should be worth noting that the deoxidation products associated with the titanium content in Fig. 2.1 by Cha's work are 'Ti$_3$O$_5$', 'Ti$_2$O$_3$', or 'TiO', which are determined by the electronic backscatter patterns for the deoxidation products observed in the quenched steel in which Ti was in equilibrium with them.

Table 2.3 The activity interaction coefficient between titanium and oxygen in the molten iron (1873 K) [6, 10–12]

References	e_O^{Ti}	e_{Ti}^O	e_{Ti}^{Ti}	e_O^O	r_O^{Ti}
[6]	−1.12	−3.36	0.042	−0.17	–
[10]	−1.0	0	0.013	0	–
	−0.42	0	0.013	0	0.026
[11]	−0.34	−1.0261	0.042	−0.17	–
[12]	−0.6	−1.8	0.013	−0.20	0.031

2.2.2 Activity Interaction Coefficient among Aluminum, Silicon and Titanium in Liquid Iron

Three different values of e_{Ti}^{Si} in liquid iron at 1873 K were given in literatures [6, 14, 15] as 2.1, 1.43 and −0.0256, respectively. The e_{Ti}^{Si} from literature [14] was obtained using the method of equilibrium between Fe–Si–Ti alloy and liquid Ag, and the literature [15] employed the method of equilibrium between Fe–Si–Ti alloy with TiN crucible by controlling N_2 partial pressure. Obviously, the results are quite different. The equilibrium constant log $K = -6.059$ [9], $e_O^{Si} = -0.066$ [6] in Eq. (2.4), and the values of e_{Ti}^{Si} are 2.1 and 1.43 from the literatures [6, 14]. One can find other activity interaction coefficients in Table 2.3, then the equilibrium relationship of Ti–O in liquid iron containing Si at 1873 K can be obtained.

K. Morita et al. also measured the activity coefficients of Al to Ti at 1873 K in liquid iron by using the same method as that in literature [14], the $e_{Ti}^{Al} = 0.024$ and $e_{Al}^{Ti} = 0.016$ [16] were obtained. In case of $a_{Al_2O_3} = 1$, the logarithm of the deoxidation equilibrium constant of Al was as follow [6]:

$$\log K = \log(a_{Al}^2 a_O^3 / a_{Al_2O_3}) = -13.591 \tag{2.11}$$

For reaction $Ti_2O_3(s) = 2[Ti] + 3[O]$, using $e_O^{Ti}, e_{Ti}^O, e_{Ti}^{Ti}, e_O^O$ in Table 2.3 [6], the equilibrium relationship of the O–Al in liquid iron containing 0.5 and 1% Ti was calculated [7].

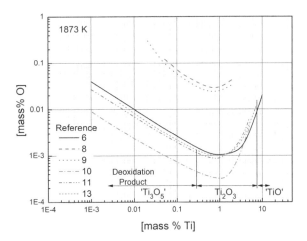

Fig. 2.1 Equilibrium relationship of Ti–O in molten iron at 1873 K

2.2.3 Equilibrium of Titanium–Aluminum–Oxygen in Liquid Iron

The microalloyed steel is a kind of steel with special quality, which requires strict control of the content of impurities, reducing of the number of non-metallic inclusions, and adjusting the morphology and distribution of sulfides. The smelting of microalloyed steel is similar to that of low carbon steel. The difference is that the deoxidation and desulphurization during the smelting of steel need to pay more attention. The order of adding alloy materials should be studied to increase the yield of the alloy elements. The refining process of microalloyed steel is indispensable. According to the requirements of different composition and grades of steel, the proper combination of refining conditions should be applied, especially to prevent the secondary oxidation of molten steel and various defects during the continuous casting process. The main function of titanium in Ti-microalloyed steel is to improve the strength of steel by precipitation strengthening for the formation of carbide and nitride. Meanwhile, small size titanium nitrides and Ti-containing complex oxides can also play a significant role in grain refinement, which can improve the strength, toughness and weldability of steel, too. In order to improve the yield of titanium, and fully play the role of titanium precipitation strengthening, on the one hand, the titanium should be used for the formation of small amount of titanium-containing oxides inclusions during the deoxidation of Ti-microalloyed steel, so as to play the role in the oxide metallurgy, and on the other, the excessive oxidation of titanium should also be avoided.

In general, the titanium is a kind of precious rare metals, so the consumption of titanium for deoxidation should be avoided no matter Ti-microalloyed steel, IF steel, or stainless steel. The deoxidation is often conducted by aluminum or Si–Mn–Al complex deoxidant, and sometimes combines with the RH vacuum carbon deoxidation. Because of the relative cheaper aluminum and its good deoxidizing effect, the dissolved oxygen in the molten steel can be reduced to several ppm in case of the suitable content of dissolved aluminum and oxygen are in equilibrium with alumina in the liquid steel.

The molten steel contains numerous solid inclusion particles, which will lead to nozzle clogging during continuous casting. The nozzle clogging often occurs in the production of Al-killed steel and Si-killed steel. The presence of titanium in the molten Al-killed steel will increase the possibility of nozzle clogging [17, 18]. Kimura [17] clarified that the reason of the frequent occurrence of nozzle clogging was that the dissolved Ti in the molten steel reacted with the refractory of immersed nozzle, leading to the change of its surface quality, and resulting in the bonding of alumina inclusions and the inner wall of the submerged nozzle refractory. Kawashima et al. [18] proposed that the frequent nozzle clogging is caused by the following reasons:

(1) The formation of liquid Al–Ti–O oxides caused by reoxidation can be the binder for alumina inclusions during the casting process;

(2) Ti can improve the wettability between nozzle refractory and molten steel, resulting in the acceleration of the change of surface quality of nozzle materia caused by the reduction of silica in nozzle refractory by aluminum in the molten steel;
(3) The reduction of oxygen activity in the melt contributes to the decomposition of alumina inclusions, resulting in a large number of small size alumina inclusions.

In fact, the nozzle clogging is attributed to the inclusions that formed by the reaction between the locally supersaturated titanium which is thermodynamically instable and the transition state complex inclusions from the initial deoxidation products, instead of the Al_2O_3 from the initial aluminum deoxidation products that is thermodynamically stable alumina [19].

Some laboratory investigations indicate that the spherical titanium oxide are formed by titanium deoxidation. Then the secondary adding aluminum after the Ti-deoxidation of steel will cause the previously formed titanium oxide inclusions being reduced by aluminum to generate the alumina inclusions, which further aggregate to the alumina clusters. The experimental results confirmed that Ti–Al–O complex oxides and the double-phase (composed of alumina and titania particles) oxide inclusions exist in the Ti–Al deoxidized steel.

The thermodynamics of the Ti–Al–O system in the molten steel has not been perfectly described. This is because: (1) there are a variety of Ti oxides, and the mutual solid solubility among them cannot be determined in a wide range of concentration; (2) the possible complex oxide in the Ti–Al–O system are still not fully known, and the reports in the existing literatures are not consistent. Ruby-Meyer et al. [20] calculated the Ti–Al–O equilibrium phase diagram at 1793 K, predicting that there may be a liquid region between the stable regions of Ti_2O_3 and Al_2O_3. Jung et al. [21] predicted that the solid Ti_3O_5 exists by using the Fe–Ti–Al–O equilibrium phase diagram at 1873 K calculated with FactSage, which didn't exist in the phase diagram calculated by Ruby-Meyer et al., whereas, no liquid phase was found in the phase diagram calculated by Jung et al.

Based on the thermodynamic data in Table 2.4 and the interaction coefficients in Table 2.5, the stable region of inclusions in Al–Ti–O system in liquid iron can be calculated, and the change of the concentration of aluminum and titanium dissolved in the molten steel in these regions is revealed, as shown in Fig. 2.2.

The solid solubility between Al_2O_3 and Ti_2O_3 at 1873 K is 4.5 mol% Ti_2O_3 dissolving in Al_2O_3, and 4.5 mol% Al_2O_3 dissolving in Ti_2O_3 [26]. Because of the same crystal structure of Al_2O_3 and Ti_2O_3, and the same valence of Al and Ti, the

Table 2.4 Reactions and standard free energy change for Fe–Al–Ti–O system

Reactions	ΔG^0 (J/mol)	References
$Al_2O_3(s) = 2[Al] + 3[O]$	$867300 - 222.5T$	[22]
$Ti_2O_3(s) = 2[Ti] + 3[O]$	$822100 - 247.8T$	[13]
$Ti_3O_5(s) = 3[Ti] + 5[O]$	$1307000 - 381.8T$	[13]
$Al_2TiO_5(s) = 2[Al] + [Ti] + 5[O]$	$1435000 - 400.5T$	[6, 13, 23]

Table 2.5 The activity interaction coefficients among Al, Ti and O in liquid iron [13, 22]

$e_i^j (j \rightarrow)$	Al	Ti	O
Al	$\frac{80.5}{T}$ [6]	0.0040 [24]	$-\frac{9720}{T} + 3.21$
Ti	0.0037 [24]	$\frac{212.2}{T} - 0.064$	$2.9925 e_O^{Ti} - 0.00864745$ [25]
O	$-\frac{5750}{T} + 1.90$	$-\frac{701}{T} + 0.0344$	$-\frac{1750}{T} + 0.76$ [6]

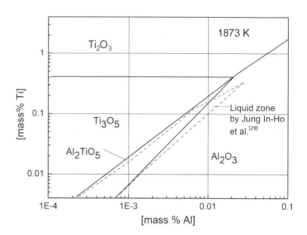

Fig. 2.2 Stable regions for Al–Ti–O inclusions in liquid iron at 1873 K

activity coefficient of Al_2O_3 in Ti_2O_3 solid solution and the activity coefficient of Ti_2O_3 in Al_2O_3 solid solution is similar, thus the activity ratio of Al_2O_3 to Ti_2O_3 is of approximately 1. The solid solubility of Ti_3O_5 in Ti_2O_3 has also been reported but was not well determined [27]. Assuming that Ti_3O_5 is a simple compound, the free energy of formation of Ti_3O_5 is also reported [13], in which the Ti_3O_5 is coexistent with the Ti_2O_3. Although the literature [20] mentioned the existence of liquid region, no enough thermodynamic data supports this. For those steels containing lower dissolved aluminium, the Ti_3O_5 or Ti_2O_3 will be formed when the Ti content is high or partially supersaturated with dissolved titanium (such as the moment of adding Fe–Ti alloy).

Matsuura et al. [28] found that Al deoxidation is inclined to be completed within 2 min after the aluminum addition into molten steel, resulting in spherical alumina inclusions. While aluminum and titanium are both added to the molten steel at the same time, alumina is early generated. The titanium oxide is later formed on the surface of alumina, which is almost converted to alumina in the end. But the alumina contains 20 mol% Ti, depending on the content of titanium. Even if all the experiments are carried out in the range of alumina stable, some titanium oxide will form, which is attributed to the local concentration of aluminum decreases during the alumina particle growth, and the melt composition moving to Al_2TiO_5 or titanium oxide stable region. Titanium oxide will be reduced by the aluminum in molten steel, resulting in the change of morphology of inclusions from spherical shape to polygon. The shape change of these inclusions from transition reaction promotes the formation of inclusion clusters, which can cause the subsequent nozzle clogging.

The liquid phase region in Al–Ti–O exists or not is still a suspending problem. The main difficulty is the lack of equilibrium thermodynamic data for the Al–Ti–O (Al_2O_3–Ti_2O_3–TiO_2) system, especially under low oxygen partial pressure. Jung In-Ho et al. [29] have evaluated and optimized the existing thermodynamic data and phase diagrams, including Al_2O_3–TiO_2, Al_2O_3–Ti_2O_3 and Al_2O_3–Ti_2O_3–TiO_2 systems for liquid-state slag and all solid phases, the temperature is from 298 K to above the liquid temperature under the pressure of 1 bar. The liquid oxide phases were studied by using the modified quasi chemical model. A liquid phase was predicted to be existed in the Al_2O_3–Ti_2O_3–TiO_2 system under the conditions of secondary steel refining, as shown in Fig. 2.2, in the dashed line surrounded area.

Because of the various valence of titanium, the stoichiometric ratio of titanium oxide varies with the oxygen partial pressure of the system. The oxygen partial pressure in equilibrium with 'Fe_tO' in liquid at 1600 °C is 5.0×10^{-9} bar. The equilibrium oxygen partial pressure in conventional aluminium deoxidized liquid iron at 1600 °C is 5.0×10^{-15} bar (dissolved aluminum concentration of 0.05%). Jung In-Ho et al. found that Al_2TiO_5 was not existed in the Al–Ti deoxidized liquid iron under the oxygen partial pressure of 10^{-15} bar in the Al_2O_3–Ti_2O_3–TiO_2 system. Based on the calculation, the Al_2TiO_5 is stable at a higher oxygen partial pressure (e.g., 10^{-12} bar or more) in the temperature range above 1264–1271 K, whereas below 1264–1271 K, (depending on the oxygen partial pressure) Al_2TiO_5 will be decomposed into Al_2O_3 and TiO_2. But a liquid region exists in Fig. 2.2 at 1873 K. The liquid region expands with the decrease of the content of dissolved aluminum and titanium in the molten steel, which is due to that the increase of the oxygen concentration in the liquid iron, and the liquid oxide partly dissolves FeO with a maximum of 10 mol%. The liquid phase disappears in case of [mass% Ti]>0.3 and [mass% Al]>0.03 when the equilibrium oxygen partial pressure in molten steel is below 10^{-15} bar.

Basu et al. [30] and Park et al. [31] found that the nozzle clogging of Ti-containing steel is often caused by the Ti–Al–O inclusions with alumina as nucleus generated by reoxidation. Jung In-Ho inferred that this Al–Ti–O phase which caused the nozzle clogging is most probably a liquid oxide phase saturated with solid Al_2O_3 core.

Van Ende et al. [32] investigated the formation and evolution of inclusions in steel during Al deoxidation and titanium alloying in the RH refining processing (1600–1650 °C) in an induction furnace under Ar atmosphere. The induction furnace used in the experiment can obtain intermediate samples without breaking the atmosphere in the furnace, and can real-time measure the dissolved oxygen by a zirconia probe.

In their Al/Ti deoxidation experiment, the initial dissolved oxygen in the molten steel is 817 ppm. After adding 0.2% Al to the molten steel, the oxygen activity (the same as dissolved oxygen) is quickly reduced to 1.5 ppm. Three minutes later after aluminum is added, 525 ppm of titanium is added to the steel. Then the oxygen activity in the steel is almost the same, and the dissolved aluminum concentration decreases gradually with the dissolved titanium loss of only 45 ppm. The loss of dissolved aluminum is oxidized by the residual oxygen in argon atmosphere, and there is no slag coverage on the surface of the molten steel. The experimental results show that the efficiency of titanium alloying is very high in case of sufficient deoxidation of aluminum, and only pure alumina inclusions are observed in the experimental steel samples.

After the molten steel treated by Al-deoxidation and Ti-alloying, the secondly deoxidation by the addition of aluminum into above-mentioned molten steel was conducted. They found that the concentration of dissolved aluminum in the steel decreased gradually with time, and the concentration of dissolved titanium increased gradually. The titanium oxide in the titanium-containing oxide inclusion was subsequently reduced by aluminum, but the complete reduction of Ti–Al–O inclusions would take a long time. When the content of titanium and aluminium in steel is high, the content of the dissolved titanium is also high. The main reduction reactions of titanium oxides by aluminum are show as follows:

$$Ti_3O_5(s) + 10/3[Al] = 5/3\,Al_2O_3(s) + 3[Ti] \tag{2.12}$$

$$Ti_2O_3(s) + 2[Al] = Al_2O_3(s) + 2[Ti] \tag{2.13}$$

They got the conclusions as following:

(1) Only alumina inclusions are observed in the molten steel when strong Al deoxidation prior to Ti alloying was performed. While with partially deoxidized steel prior to Ti addition ($a_{[O]}$ = 140–280 ppm), Ti deoxidation took place. In that case, large Al_2O_3 clusters and small Ti–Al oxide particles randomly distributed in the steel were found. After the second Al addition, only Al_2O_3 clusters and few small Ti–Al–O inclusions remained due to the reduction of Ti oxides by [Al].

(2) The extent of [Ti] oxidation increased with the increasing of $a_{[O]}$ before Ti alloying. No Ti deoxidation occurred with low $a_{[O]}$ prior to Ti alloying. On the other hand, increasing the Ti addition to 1500 ppm had no significant effect on Ti deoxidation.

(3) The reduction of Ti oxides through the secondary Al addition was confirmed by inclusion observations and the chemical composition analysis for the steel samples. Ti recovery from Ti–Al–O inclusions occurred rapidly after the secondary Al addition and was enhanced by adding Al immediately after Ti alloying.

(4) The combination of experimental and calculated results shows that, in order to prevent Ti oxidation, it is essential to maintain [Al] to a sufficiently high value. Approximately 200 ppm [Al] is necessary prior to Ti addition to minimize Ti loss during the alloying process.

Wang et al. [33] investigated the activity of oxygen ($a_{[O]}$) before deoxidation, the interval time between Al and Ti additions and the [Al] content after RH refining on the yield of titanium. The studied IF steel grade requires the total aluminium of 0.03% and the titanium of 0.07%. This IF steel was produced by the combination process of

BOF(210 tons)-RH-CC, in which the tapping of the BOF was operated without full killed deoxidation, and the RH vacuum regular circulation method was used for the decarburization. After the decarburization, deoxidation by aluminum was conducted based on the dissolved oxygen measurement in the molten steel. Then the FeTi70 (70 mass% titanium) was applied for the alloying after the aluminum was added into the molten steel for 2–5 min. Finally, the RH vacuum cycling refining was ended after a high degree of vacuum was maintained for 4–6 min.

Figure 2.3 illustrates the effect of the initial activity of oxygen and the interval time between Al and Ti addition on the titanium yield and the dissolved aluminum content ([Al]s). The results show that the titanium yield is higher in case of the low initial oxygen activity with the same content of aluminum, while the titanium was added after the aluminum deoxidation for 3–5 min. Actually, the relationship between the dissolved aluminum and titanium can be predicted according to the Reactions (2.12) and (2.13) and the corresponding thermodynamic data. In general, log[Ti] has linear relationship with log[Al], and the slope is in good agreement with the ratio of the stoichiometric coefficients of Al and Ti in those reactions [34].

Wang et al. [33] analyzed the inclusion phases and the removal effects comparatively before and after ferro-titanium alloying. When controlling the activity of oxygen ($a_{[O]}$) before deoxidation below 350 ppm, the interval time between Al and Ti addition above 3 min can guarantee the titanium yield above 85%. With $a_{[O]}$ above 350 ppm, the interval time between Al and Ti addition should be extended more than 5 min. Extending the interval time between Al and Ti addition can improve the titanium yield at the same $a_{[O]}$ and [Al]$_s$. During RH refining process, Al_2O_3 inclusions with the equivalent diameter above 200 μm can float up and be removed within 5 min, but the removal time of Al–Ti–O inclusions with the same size is 1–2 min more than that of pure Al_2O_3. Al–Ti–O complex inclusions form around Al_2O_3 after the titanium alloy being added into the melt, and the titanium yield decreases.

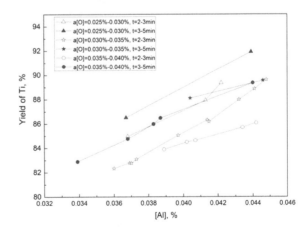

Fig. 2.3 Effects of the initial oxygen activity and the interval time between Al and Ti addition on the titanium yield and the content of dissolved aluminum ([Al]) [33]

2.3 Reaction among Titanium-Bearing Molten Steel and Molten Slag or Refractories

2.3.1 Reaction between Titanium-Bearing Molten Steel and Molten Slag

During the steel refining process, the reaction between slag and steel is very important to control the chemical composition of molten steel and the inclusions in the steel. The presence of multi-valence titanium oxide makes the reaction between steel and slag and the composition of inclusions complicated. The literatures about titania-containing slag mainly involve the phase diagrams and equilibrium experiments of the slag-metal phases or slag-gas phases. These experiments were explored by changing the different composition and different experimental conditions of the slag (mainly different oxygen potential) for the related studies. However, it was difficult to identify the consistency of the results from different experiments. In addition, these experiments only partially covered the concerns of the steel makers for the range of oxygen activity and alloy elements. Ruby-Meyer et al. [35] analyzed the precipitation of non-metallic inclusions in Ti-deoxidized steels with the multiphase equilibrium code CEQCSI which is based on the IRSID slag model. The slag model has been extended to systems containing titanium oxides, and the applied to slag-metal reactions in the Ti-containing steel refining, using data available in the literatures, and validated with industrial results obtained on various steel grades.

In the IRSID slag model, the titanium oxide content of the slag is expressed as % TiO_2, although the calculation takes into account Ti^{3+} and Ti^{4+}; The steel characteristics are expressed with the Ti and Al activities, to unify on the data from carbon and stainless steels. For carbon steels, the activities are practically equal to the contents in weight %, whereas they are substantially different in the case of stainless steels ($a_{Al} = 3$–4 times %[Al], and $a_{Ti} = 7$–8 times %[Ti] for the grades considered). The model calculations for slag contains 5 and 10% TiO_2, 10% MgO, %CaO/%$Al_2O_3 = 1.4$ at 1500 and 1600 °C, agree with the data points from industrial experiments reasonably. The increase of temperature, the decrease of the aluminum content and the increase of TiO_2 content in slag all lead to the increase of the Ti equilibrium distribution ratio between slag and metal.

2.3.2 Reaction between Titanium-Bearing Molten Steel and Refractory

Ruby-Meyer et al. [35] also studied the reaction between Al–Ti deoxidized steel and refractory in which MgO is the main component. The experiment was conducted to study changes of oxide inclusions and the chemical composition of Al–Ti deoxidized steel by the reduction effect of the MgO based refractories. Multiple heats of

1 kg molten steel containing 0.12% Mn and 0.03% Si were melted in the sintered MgO crucibles. Argon blowing was kept for 20 min to 1 h in the sealed induction furnace at 1580 °C. The crucible material consists of 95% MgO, 3.5% SiO_2, and 1.5% CaO. After adjusting the initial Al and Ti content in steel, the intermediate metal samples were drawn every 2–5 min with quartz tubes and rapidly quenched. Then the composition and the inclusions in the samples were analyzed respectively. The inclusions observed in the liquid metal are about 1 to several microns, and the inclusions formed during the solidification of the sample are much smaller. In the first sample, the quantity of inclusions is much larger than that in the later ones.

The change of chemical composition and inclusion characteristics of steel depends on the initial content of aluminum and titanium. For two extreme cases, the results are as follows:

(1) For the initial content of 0.03%Al and 0.04%Ti, only the concentration of aluminum decreases persistently; while the content of aluminum reaches 0.006% titanium began to decline. The inclusions contain alumina and spinel, and the proportion of spinel increases during the reaction proceeding.
(2) For the initial 0.006%Al and 0.08%Ti, Both the content of aluminum and titanium decreases during the experiment. At the same time, two inclusions were observed: spinel, a phase rich in Al_2O_3 and TiO_2.

In two experiments, the content of silicon in steel increases, and the content of magnesium increases slightly, while the magnesium is concentrated in spinel inclusions. The material balance shows that the reoxidation of molten steel is mostly due to the silica in the crucible. The gas phase also causes very slight reoxidation.

If trace amount of magnesium is added to the molten steel, the calculation shows that spinel $(Mg, Mn)O–Al_2O_3$ which coexists with Al_2O_3 or liquid oxides is generated while oxygen reaches saturation. When supersaturated oxygen is supplied to the oxygen saturated molten steel, the aluminum content begins to decrease. And more and more Mg is immobilized in the liquid oxide, which leads to the reduce of spinel content and possibly even disappears. On the contrary, in the Al_2O_3 stable region, the spinel phase always exists.

2.4 Control of Oxide Inclusions in Titanium Microalloyed Steel and Application of Oxide Metallurgy

Titanium in the titanium microalloyed steel not only improves the strength of the steel in form of the TiC, TiN precipitation strengthening, but also enhances its toughness and welding performance. Appropriate size and quantity of titanium oxide inclusions are benefit for the austenitic grain refinement in hot rolling process and promote the nucleation of acicular ferrite in the transformation of the austenitic and acicular ferrite. Many researches indicated that compared with the carbon/titanium nitride, titanium oxide is more stable, which cannot be dissolved under high temperature. In addition, fine oxides can pin austenitic to refine the grains. On the other hand it can induce the nucleation of acicular ferrite, which divides the grain into several subgrains to significantly improve the performance of the welding heat affected zone. In the early 1960s, it was found that spherical inclusions in size of several microns to tens of nanometers existed in weld metals [36, 37]. Harrison and Farrar [38] found that this kind of spherical oxide inclusions can benefit for the growth of ferrite, and increase the toughness of the weld. In 1990, Japanese scholars have proposed the concept of "oxide metallurgy" based on its profitable function, which improves the toughness and weldability of the steel through controlling the generation of tiny, dispersive and high melting point oxide inclusions with the acicular ferrite grown in the steel [39].

2.4.1 Effect of Aluminum–Titanium–Magnesium Complex Deoxidation on Inclusions and Microstructure of Steel

Number of oxide inclusions formed by Ti deoxidation is very large. While oxide inclusions formed by Mg deoxidation grow slowly. To obtain the large quantity of fine oxide inclusions, Ti and Mg are sequentially added in the molten steel to take the advantages of these two deoxidizer. However, there are various types of inclusions and complex reactions in the system of Ti–Mg–Al–O. The change of Al and Ti content affects the immediate type, quantity and size of complex inclusions. Song et al. [40] studied the effects of Al–Ti–Mg complex deoxidization on inclusions and their effects on the microstructure of steel. The chemical composition of the Al–Ti–Mg deoxidization experimental steel is showed in Table 2.6.

Sample 1 with the Al–Ti complex deoxidation, aluminum content is as high as 0.210%. It can be seen in the previous Fig. 2.11 that Al_2O_3, as is the stable deoxidation products, but titanium oxide inclusions has not been observed in the scanning electron microscopy (SEM). In order to get more titanium oxides, the amount of Al has been reduced or not adding it. Sample 3 employed titanium deoxidation directly without aluminum addition. The titanium content is four times that of total aluminum (industrial pure iron raw material contains a small amount of aluminum) in the steel, and extensive Ti_3O_5 inclusions were found by scanning electron microscope. The composition of inclusions is inconsistent with the thermodynamic equilibrium calculation due to the following reasons: (1) the analysis of aluminum and titanium content is applied by the total aluminum and titanium; (2) the existence of local supersaturation when deoxidizer initially added; (3) inclusions composition variation with time; (4) heat preservation time is not long enough for the small experimental crucibles. Some SEM images of representative inclusions in these samples are shown in Fig. 2.4.

Figure 2.4a is the TiN wrapped Al_2O_3 complex inclusions observed in the specimen 1. Atomic percentages of the inclusions that obtained from X-ray energy spectrum (EDS), were presented in numbers for analyzed elements in the figure. The other figures are the same. Figure 2.4b is titanium oxide in specimen 3. The SEM

Table 2.6 The chemical composition of the Al–Ti–Mg deoxidization experimental steel

No.	Types of deoxidizer and initial addition/%					Composition of steels after Al–Ti–Mg deoxidation/%						
	Mn	Fe-Si	Al	Ni-Mg	Ti-Fe	Mn	Si	Al	Ti	Mg	O	N
1	0.65		0.670		0.30		0.010	0.210	0.034	0.0017	0.0056	0.0012
2	0.65	0.30	0.056			0.36	0.013	0.045		0.0017	0.0071	0.0012
3	0.60	0.29			0.60	0.35	0.012	0.025	0.100	0.0017	0.0067	0.0062
4	0.56	0.23	0.056		0.30	0.32	0.010	0.045	0.057	0.0017	0.0096	0.0051
5	0.57	0.24	0.056	0.06	0.30	0.35	0.011	0.032	0.063	0.0021	0.0093	0.0032
6	1.16	0.23	0.056	0.06		0.33	0.010	0.032		0.0018	0.0066	0.0051
7				0.06	0.30	0.46	0.008	0.027	0.050	0.0020	0.0094	0.0060

images of specimen 1 and specimen 3 were compared and found that only the inclusion of TiN and TiN with Al_2O_3 exists in the Al deoxidized sample. In Ti deoxidized steel (0.025% aluminum,and other drag-in elements with ferroalloy), large quantity of independent titanium oxide inclusions was observed, EDS shows that atoms ratio of Ti and O respectively occupy 33.30, 64.75%. The inclusion is approaching Ti_3O_5. Different deoxidized products are corresponding with its stable zone when employed Ti–Al complex deoxidization. There are also TiN inclusions in the sample. Figure 2.4c is a complex inclusion of TiN with $Al_2O_3 \cdot MgO$. Figure 2.4d is a complex inclusion of the titanium oxide and $Al_2O_3 \cdot MgO$ in specimen 5. This sample is treated by Ti–Mg after aluminum deoxidization. The Ti–Mg complex deoxidized steel has not only individual inclusions of TiN, Ti_3O_5 and MgO, but also massive complex inclusions of Ti_3O_5, $Al_2O_3 \cdot MgO$ and TiN with $Al_2O_3 \cdot MgO$, in both complex inclusions the core are $Al_2O_3 \cdot MgO$. Hence, it is indicated that formation of vast tiny

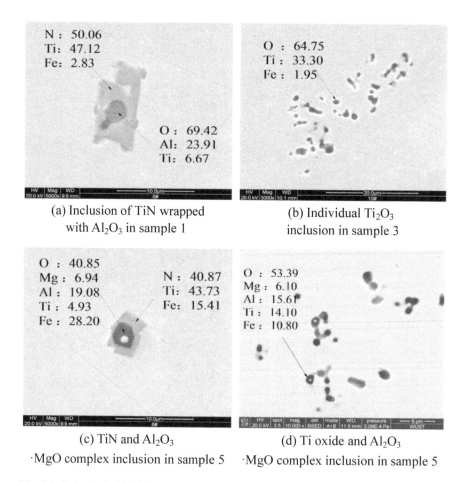

Fig. 2.4 Inclusion in Al–Ti–Mg complex deoxidized steel [40]

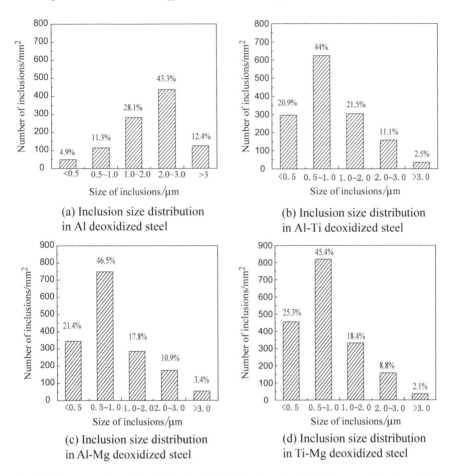

Fig. 2.5 The comparison of inclusion size distribution in steel with different combinations of Al, Ti and Mg deoxidation [40]

$Al_2O_3 \cdot MgO$ has provided conditions for the dispersing oxide, nitride inclusion in Al killed steel with Mg treatment. The comparison of size distribution in steel samples with different combinations of aluminum, titanium and magnesium deoxidation is shown in Fig. 2.5.

As shown in Fig. 2.5a is Al individual deoxidized steel, with the maximum quantity of larger size inclusions due to the formation of alumina inclusion clusters. Figure 2.5b and c are respectively Al–Ti, Al–Mg deoxidized steel, the inclusions size distribution has obvious changes. The amount of inclusions under 1 micron have greatly increased, and there are the maximum tiny inclusion and minimum medium inclusion in Ti–Mg killed steel because the quantity of Ti deoxidized products and low growth rate of Mg deoxidizing products as shown in Fig. 2.5d. Figure 2.5c shows that the inclusions size of below 1 μm in sample 6 are 67.9% of the total inclusions, of which 0.5–1 μm is 46.5% and more is $Al_2O_3 \cdot MgO$. The inclusion of 3 μm is also present, but only 3.4% of the total. Figure 2.5d shows that the inclusions with the size less than 1 μm in the sample 7 is 70.7%, which close to the sample 6, but the total inclusion number is more than it. It was compared and found that the amount of oxide inclusions in Ti–Mg killed steel are more than that in Al–Mg killed steel, and the inclusion particle size distribution is more concentrated, most of the inclusion size is below 1 microns. The main reason is that the relatively large number of titanium deoxidation products, and slowly growth rate of magnesium deoxidation products with tiny and dispersed particles. Thus, Ti–Mg complex function could integrate these two advantages [41, 42]. So titanium deoxidation can produce a large number of dispersed fine oxide inclusions, on the basis of the Ti deoxidation following with Ni–Mg alloy magnesium for further processing, which can generate more the complex inclusions that conducive to formation of the intragranular ferrite. In this experiment, unit area of the total amount of inclusions is analyzed, and the total number of inclusions is approximately 1000–2000/mm^2, which is correspond to the data from Ohto and Suito [43].

2.4.2 Effect of Titanium-Magnesium Complex Deoxidation on Austenite Grain Coarsening

In the process of steelmaking, one of the most interesting problems is removing or reducing various non-metallic inclusions. However, non-metallic inclusion is a double-edged sword. In another way, if the size of the residual inclusions could be controlled under a certain critical size, it is beneficial for the microstructure and mechanical performances of steels, such as preventing grains growing up, improving the strength and toughness of steel and so on. It is difficult and uneconomical to produce high purity steel with inclusions in low content. In the process of steelmaking, it is worthwhile to try to get inclusions, like very fine oxide particles, to improve the property of steel. High melting point oxide particles got by titanium, magnesium deoxidation can be used as second phase particle to organize austenite grow up during reheating the steel. It can be also used as heterogeneous nucleation particles when cooling the steel, it can induce acicular ferrite structure inside the austenite, the initial austenite will be split into multiple acicular ferrite grains, and then, the steel structure will be refined.

Therefore, the steel with titanium and magnesium deoxidized was kept at 1200 °C by a laser confocal high temperature scanning microscope, by which the effect of titanium and magnesium oxide inclusions on austenite grain coarsening was studied. The experimental objects are samples a and b. Sample a was deoxidized with manganese, silicon, aluminum, sample b was deoxidized with manganese, silicon, aluminum, titanium and magnesium. The sample was cut into a small column with 8 mm diameter and 4 mm height. And then, the sample was polished and put into the crucible in the laser microscope. Finally, the sample was heat treated, as shown in Fig. 2.6, and the change of grain size with time was observed. The composition of samples could be seen in Table 2.7.

The austenitic grain growth is not only affected by time, but also related with temperature. With the temperature increasing, the atomic diffusion velocity increasing exponentially, and the austenite grows up rapidly. The alloy like titanium, zirconium, vanadium, aluminum and so on in steel will prevent the austenitic grain growth strongly, and increase the austenite coarsening temperature. There are two situations when grains grow up: (1) equally distributed grain size, (2) abnormal grain grow up. Under isothermal condition, the grains grow up gradually, the total area of grain boundary decreases gradually, and the energy decreases. Assuming that the instantaneous rate of grain growth is proportional to the grain boundary energy in unit volume, the kinetics equations of grain growth can be established. The Arrhenious equation of grain growth can be seen in Eq. (2.14) [44]:

$$D^{\frac{1}{n}} - D_0^{\frac{1}{n}} = Kt \cdot \exp(-\frac{Q}{RT}) \quad (2.14)$$

where D_0 is initial grain diameter (μm), D is grain diameter (μm), Q is activation energy for grain growth (J/mol), t is time (s), T is temperature in Kelvin, R is gas constant (8.314 J/(K·mol)), n is grain growth index, when $T > 1273$ K, n is 2, K is a proportional constant.

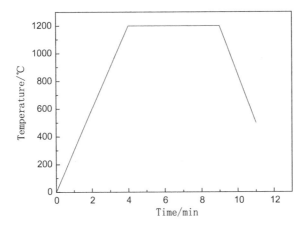

Fig. 2.6 The heat treatment curve

Table 2.7 The composition of samples used in laser confocal high temperature scanning microscope observation (%)

Sample	Mn	Si	Al	Ti	Mg	O	N
a	0.732	0.102	0.045	0	0.0020	0.0071	0.0012
b	0.855	0.171	0.024	0.236	0.0029	0.0040	0.0038

It can be seen from Eq. (2.14), the heating temperature plays an important role in austenitic grain growth. When temperature is constant, the higher the temperature is, the more quickly the austenitic grain grows. And the change of austenitic grain size with heating time can be shown as Eq. (2.15) [45].

$$D = K't^n \tag{2.15}$$

where K' is a proportional constant.

The existence of inclusions and the second phase particles in steel will prevent grain boundary migration in heat preservation stage. When diffusion resistance of dispersive distributed inclusions and the drive force of grain growth reach balance, the grain will stop growing. And Zener analyzed and proposed the equation which is for the dispersed second phase particles pinning austenitic grain growth [46]:

$$D = \frac{4}{3} \times \frac{r}{V_f} \tag{2.16}$$

where D is grain diameter (μm), r is second phase particles radius (μm), and V_f is the volume fraction of second phase.

It can be seen from Eq. (2.16), the smaller the inclusion particle size is, the bigger the volume fraction is. In another way, the more the amount of fine dispersed of inclusion in steel is, the smaller the grain diameter is and the more prominent the inhibition effect to austenitic grain growth is. This experiment mainly studied the influence of adding Ti–Mg on austenitic grain growth in heat preservation at 1200 °C. As mentioned above (Fig. 2.5d), adding Ti–Mg to steel, which can increase the amount and volume fraction of the second phase particles, compared to deoxidized with Mn and Si, the number of the second phase particles which can pin austenite grain boundary increases significantly. The evolution of austenite grain of sample a and sample b at 1200 °C was shown in Fig. 2.7.

The austenite grain size in the picture was measured, the change of austenite grain size with time was shown in Fig. 2.8. The change of sample a is more obviously. As time goes on, the austenite grain of sample a grows up gradually, small grains were swallowed by big grains. When it's 250 s, the average austenite grain size of sample a is 23 μm, after 270 s, there is only one whole grain in the field of view, and the average size of the austenite grain is 68.4 μm, which is more than three times as much as that in 250 s. In contrast, the austenite grain size of sample b is smaller. When it's 250 s, the austenite grain size is only 8.8 μm, and it is 15.5 μm at 520 s, increased 6.7 μm.

Many researchers suggest that dispersive distributed second phase particles have good effects in precipitation strengthening and grain refining. The second phase particles which were formed by Ti–Mg deoxidizing can pin the austenite grain and prevent the grain grow up during heat preservation at 1200 °C. It can be seen from Eq. (2.16), the greater volume fraction and smaller average diameter of second phase particles in steel can get smaller austenite grain size. In the sample a, the average diameter of inclusion is 1.039 μm. The pictures got from laser microscope show that

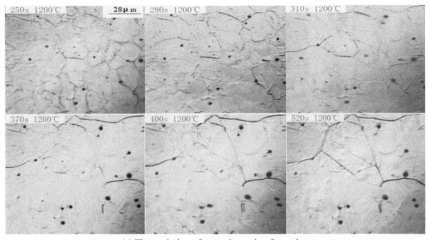

(a) The evolution of austenite grain of sample a

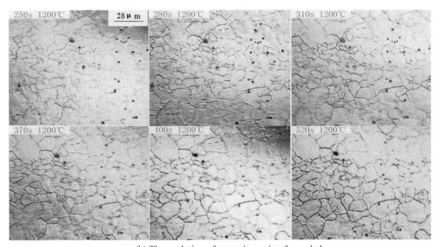

(b) The evolution of austenite grain of sample b

Fig. 2.7 The evolution of austenite grain at 1200 °C

the austenite grain of sample a is homogeneous but coarsened during heat preservation at 1200 °C for 5 min, and the grain inhomogeneity coefficient is 1; as for sample b, the average diameter of inclusion is 0.673 μm, and the grain inhomogeneity coefficient is 1.3. The volume fraction of sample a and b were 0.00304, 0.00415. It can be calculated from Eq. (2.16), the critical grain size of sample a is 89 μm, and the critical grain size of sample b is 62 μm. After holding for 5 min at 1200 °C, the grain size of sample a grew up to 68.4 μm, close to but still less than the critical grain size D_c, The grain will continue growing up with the time extension or the temperature increasing, the grain size of sample b grows up to 15.5 μm, which is far less than

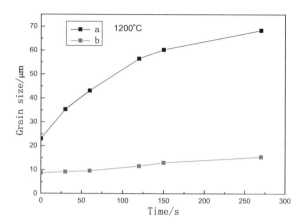

Fig. 2.8 The change of austenite grain size with time

the critical grain size, and the pinning effect of inclusions and second phase particles to the austenite grain is obvious. Therefore we can conclude that Ti–Mg complex deoxidation for steel has good effects in refining austenite grain.

Figure 2.9 is the images of as cast microstructure through optical microscope of some specimens in Table 2.6 corroded by 3% nitric acid and alcohol.

Figure 2.9a shows the microstructure magnified 25 times of sample 2 in Tables 2.6 (the same below) are mainly pearlite and polygonal ferrite, both of which are coarse grains. Figure 2.9b shows the microstructure magnified 100 times of sample 3. It can be seen that the ferrite has been refined, showing a typical acicular ferrite distribution, and the grain is small. Figure 2.9c is the microstructure magnified 100 times of sample 5. The microstructure of this sample is finer, which are mainly pearlite and fine acicular ferrite. Figure 2.9d shows the microstructure magnified 100 times of sample 7, in which the microstructure are mainly pearlite and ferrite but the ferrite is much more than that of sample 3 and grain size is finer than that of sample 2.

The microstructure in Fig. 2.9a is coarse and the ferrite and cementite are in polygonal shape. From the microstructure comparison of Fig. 2.9a–d, the addition of Ti can obviously refine the microstructure in steel because the fine Ti_3O_5, TiN or other small complex oxide inclusions can be formed by the addition of Ti, which can promote the nucleation of acicular ferrite and thus refine the microstructure of steel. Hence, lots of acicular ferrites are formed and the acicular grains are uniform and fine, interlocking each other. The microstructure of Fig. 2.9b, c are almost similar, but the distribution of ferrite in (d) is not as dispersed as (b), which can be contributed the no addition of Mn in (d). The element Mn can increase the relative amount of pearlite, so the reduction of Mn content in steel will increase the amount of ferrite and bring it together. The microstructure in (c) is the finest. On the one hand, it is due to the addition of Ti, on the other hand, because of the Mg treatment after the Ti deoxidation, the number of inclusions favorable to the nucleation of acicular ferrite increases, which benefit to the formation of acicular ferrite and make the microstructure finer. Figure 2.10 is image of the acicular ferrite grown on the complex inclusions of $MgO–Al_2O_3–TiO_x$ and MnS and the EDS of the inclusions. The number after the element symbol in

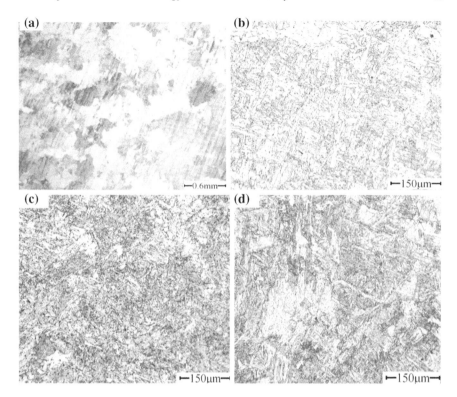

Fig. 2.9 Microstructure images of complex deoxidized as cast steel through optical microscope (3% nitric acid alcohol corrosion)

the images of inclusions is the atom percentage of the corresponding element. It can be seen that the complex inclusions of $MgO–Al_2O_3–TiO_x$ and MnS are suitable for nucleation of acicular ferrite.

The carbon content in the steel mentioned above is very low (0.003%), and the oxygen content in the steel is high. It is prone to form oxide inclusions containing titanium when adding ferrotitanium into steel. However, the medium and high carbon steel is not so. The medium and high carbon steel are fully deoxidized by aluminum, so the oxygen content in the molten steel is low, and the titanium oxide inclusions has small numbers. Therefore, the effect of oxide metallurgy is difficult to perform. Li [47] studied the influence of Al–Ti deoxidation on the precipitation behavior of MnS and microstructure in non-quenched and tempered steels. The composition of the steel sample at the end of the deoxidation experiment is shown in Table 2.8.

The results for typical morphology and composition of inclusions in the samples at the end of experiment show that the inclusions in the steel are independent precipitated MnS and MnS precipitated with oxide as core when using titanium instead of aluminium as deoxidizer for non-quenched and tempered steel (number 3 in Table 2.8). In the two case (number 3 and 2 in Table 2.8), the size and morphology

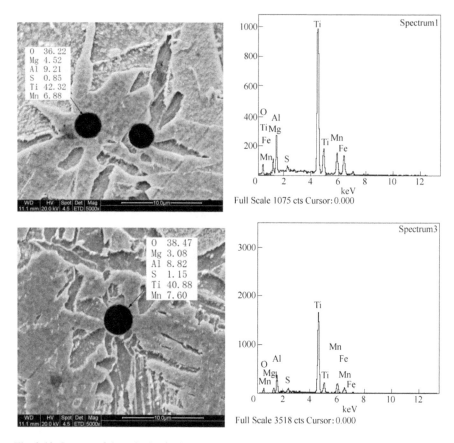

Fig. 2.10 Images of the acicular ferrite grown on the complex inclusions of MgO–Al$_2$O$_3$–TiO$_x$ and MnS and the EDS of the inclusions in Al–Ti–Mg deoxidized steel

of inclusions vary greatly. Using titanium as deoxidizer, the inclusions in the steel are spherical with smaller size, and the precipitation rate of MnS on oxide is also different.

In the final sample, independent precipitated MnS inclusion was observed, and MnS inclusion with oxide as core was also observed. However, the ratio of MnS precipitated on oxides is different due to the different Ti/Al ratio in molten steel. According to the photos of 1000 times amplified by SEM and EDS, the proportion of MnS with oxide as core in different samples was calculated, and the relationship between the proportion of MnS with oxide as core and Ti/Al ratio in steel liquid, is shown in Fig. 2.11. It can be seen from Fig. 2.11 that with the increase of Ti/Al ratio in steel, the ratio of MnS precipitated on oxide increases, and the ratio of MnS precipitated independently decreases. When the Ti/Al ratio is higher than 6.5, the MnS precipitated on the oxide is more than 95%. Therefore, in non-quenched and

Table 2.8 Chemical Composition of each sample at the end of the deoxidation experiments

Number	C	S	Mn	V	Al	Ti	O	N
1	0.34	0.037	1.26	0.09	0.043	0.016	0.0050	0.0068
2	0.37	0.032	1.11	0.08	0.046	–	0.0045	0.0052
3	0.35	0.041	1.10	0.09	0.011	0.070	0.0035	0.0062
4	0.35	0.033	1.35	0.08	0.029	0.170	0.0050	0.0061
5	0.36	0.047	1.12	0.09	0.021	0.080	0.0047	0.0067
6	0.40	0.041	1.13	0.10	0.020	0.110	0.0031	0.0071
7	0.38	0.039	1.08	0.08	0.011	0.037	0.0035	0.0045
8	0.33	0.04	1.14	0.07	0.011	0.040	0.0062	0.0051
9	0.36	0.043	1.41	0.13	0.011	0.018	0.0087	0.0062
10	0.42	0.038	1.09	0.12	0.011	0.062	0.0058	0.0048
11	0.38	0.041	0.94	0.11	0.010	0.009	0.0057	0.0059

tempered steel, the increase of Ti/Al ratio in steel is beneficial to the precipitation of MnS on oxides.

Figure 2.12 shows the variation of the inclusions' size distribution with the Ti/Al mass ratio in the steel. It can be seen from the figure, with the increase in the mass ratio of Ti/Al in steel, the proportion of inclusions small than 1 μm in steel was increased, and when Ti/Al increased from 0 to 0.9, the proportion of inclusions small than 1 μm in steel had a large increase; when Ti/Al is higher than 1.63, the frequency was relatively stable at 70–85%; with the increase of Ti/Al in steel, the proportion of inclusions in 1–3 μm had a decreasing trend, and when Ti/Al was more than 1.63, the frequency is relatively stable, and mostly below 20%; with the increase of Ti/Al

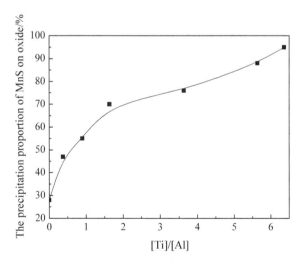

Fig. 2.11 The effect of different Ti/Al ratio in steel on precipitation proportion of MnS on oxide

in steel, the proportion of inclusions larger than 3 μm in steel is decreased, and when Ti/Al is higher than 1.63, the frequency basically less than 5%.

Because the content of Mn and S in non-quenched and tempered steel is high, lots of inclusions of MnS would be formed during solidification of molten steel. The result of SEM and EDS in the final samples also shows that most of the inclusions in the steel are MnS inclusions (alone or with oxide as core). The inclusions' size distribution statistics in Fig. 2.12 can be considered as the size distribution rule of MnS. Therefore, it can be concluded that, with the increase of Ti/Al in steel, the proportion of MnS less than 1 μm has an increasing trend, and the proportion of MnS in 1–3 μm, and the proportion of MnS larger than 3 μm, all decrease. When Ti/Al is higher than 1.63, the proportion of MnS in each diameter tends to be stable.

After using titanium to treat separately, a lot of acicular ferrites are formed in the steel, and the ferrite structures bite each other, and the steel structure is obviously refined. A large number of crossed acicular ferrite structure is formed after the Al–Ti–Mg complex treatment, and the steel structure is finer than that of the steel with titanium treating only.

Ti microalloying is carried out on the basis of aluminum deoxidation. Aluminum deoxidized steel is usually treated with Ca. Calcium treatment can modify alumina and sulfide, and reduce their harm for nozzle clogging. In the process of steel refining, magnesia refractory material is generally contacted. Some steels utilize oxide metallurgy technology to improve microstructure, refine grain, improve strength and toughness, and may also add small amounts of magnesium. For the control of quantity and composition of inclusions in Ti deoxidized steel, a very important factor is the Ti/Al ratio in steel, as previously observed, and the particle size distribution of inclusions in steel is affected deeply by the Ti/Al ratio. On the other hand, the aluminum content is also an important factor, and according to Fig. 2.2. The different concentrations of titanium and aluminum in steel will lead to different inclusions. Experimental results show that the number and size of inclusions will change. Finally, the inclusions' size distribution and interface state of complex oxide and the complex of oxide and manganese sulfide are different, which deeply influence the results of oxide metallurgy. In view of this, Zheng Wan and Wu Zhenhua studied the influence of aluminum content on inclusions' properties and steel microstructure in Al–Ti deoxidized steel treated with calcium or magnesium [48, 49].

The experiments were carried out in a 25 kg vacuum induction furnace with crucibles of MgO. Taking Al–Ti complex deoxidized steel treated with Mg for example, the composition of experimental steel is shown in Table 2.9, and the melting process of steel is shown in Fig. 2.13.

The ingot was about 15 kg, and it was reheated and rolled into steel plates of 10 mm thick and 100 mm long, according to the conditions and procedures shown in Fig. 2.14.

Figure 2.15 compared the amount and size distribution of inclusions in Al–Ti complex deoxidized and Mg/Ca treated steels. In general, there were more inclusions in Al–Ti complex deoxidized steel with lower aluminum content, and the inclusions less than 1 μm found in low Al content and Mg treated steel were most.

Fig. 2.12 Variation of inclusions' size distribution under different Ti/Al ratio in steel

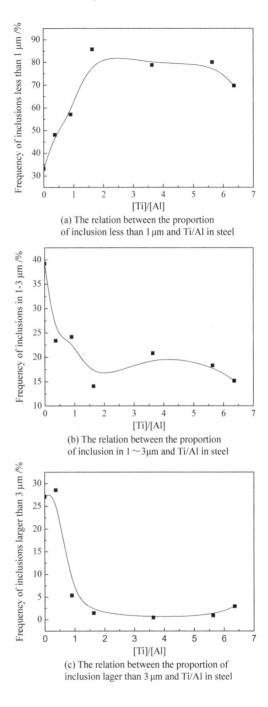

(a) The relation between the proportion of inclusion less than 1 μm and Ti/Al in steel

(b) The relation between the proportion of inclusion in 1∼3μm and Ti/Al in steel

(c) The relation between the proportion of inclusion lager than 3 μm and Ti/Al in steel

Table 2.9 The chemical compositions of Ti-Al deoxidized and Ca or Mg treated steel (%)

Sample	C	Si	Mn	S	T. O	T. N	Al$_s$	Ti	Ca	Mg
C1	0.03	0.21	1.89	0.0048	0.0072	0.0020	0.0055	0.018	0.0010	–
C2	0.03	0.25	1.87	0.0032	0.0034	0.0020	0.0260	0.022	0.0011	–
M1	0.06	0.17	1.78	0.0045	0.0074	0.0020	0.0056	0.011	–	0.0049
M2	0.03	0.18	1.86	0.0034	0.0045	0.0020	0.0340	0.021	–	0.0050

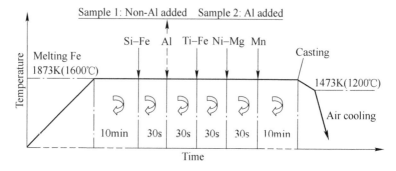

Fig. 2.13 The melting process of Al–Ti deoxidized and Mg treated steel

This benefits from the fact that there are many Ti–Mg complex oxides and they are difficult to grow.

The inclusions in these experimental steels were mostly the complex ones of oxide and MnS, and MnS precipitated in the outer layer of oxide. When the steel is hot rolled, the out layer MnS is easy to deform, and the larger the MnS thickness is, the more serious the deformation is. As shown in Fig. 2.16, the inclusions' deformation along the steel cross section is small, but the deformation along the rolling direction is large, and the inclusions' deformation of the low aluminum sample is small.

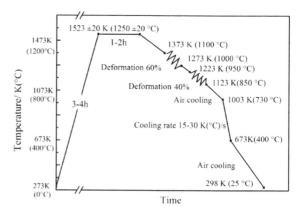

Fig. 2.14 The heat treatment process of Al–Ti complex deoxidized and Mg treated steel

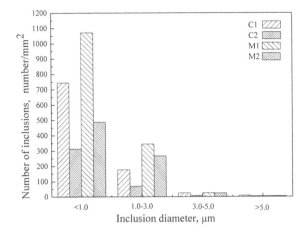

Fig. 2.15 The amount and size distribution of inclusions in Al–Ti complex deoxidized and Ca/Mg treated steel

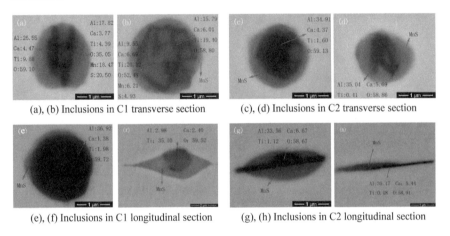

Fig. 2.16 The inclusions in the transverse and longitudinal section of C1 and C2 hot rolled plates

Figure 2.17 compared the thickness of MnS on oxide inclusion surface and amount of inclusion in Al–Ti complex deoxidized and Ca/Mg treated steel. As shown in the figure, the inclusions in low aluminum samples had large amount, especially a large number of small inclusions, and the thickness of the inclusions surface of MnS is small, which indicate that a large number of fine oxide precipitates as the core of MnS dispersed MnS.

The deformation ratio of inclusions can be measured by their aspect ratio. The aspect ratio is defined as the ratio of the length to width of the deformed inclusion. Figure 2.18 compares the aspect ratio of oxide-MnS complex inclusions in Al–Ti deoxidized and Ca–Mg treated hot-rolled steel plate. It shows that the deformation aspect ratio of inclusions in the transverse and longitudinal section of the hot-rolled

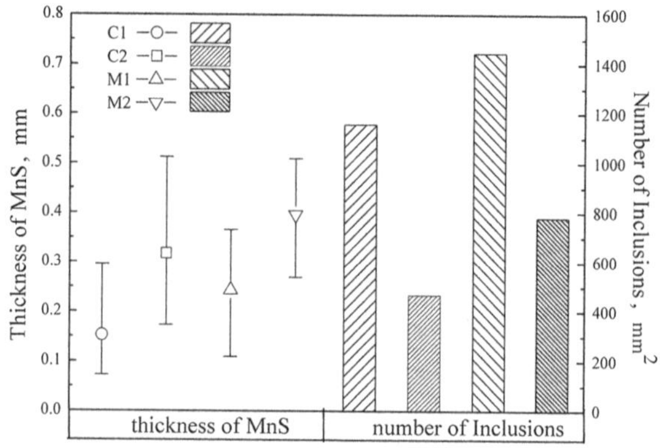

Fig. 2.17 Thickness of MnS on oxide inclusion surface and amount of inclusion in Al–Ti complex deoxidized and Ca/Mg treated steel

Fig. 2.18 The deformation aspect ratio of oxide-MnS complex inclusions in the Al–Ti deoxidized and Ca–Mg treated hot-rolled steel plate

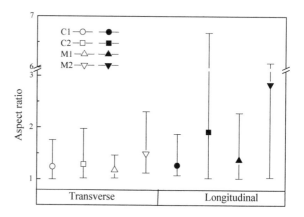

steel plate with low aluminum is relatively lower, which means the deformation degree of inclusions in low aluminum steel is smaller mostly owing to the MnS dispersed by lots of small oxide inclusions and its thinner surface thickness in complex inclusions.

Actually, the complex oxide surface will form the multiphase interface between the oxide and sulfides if Al, Ti, Mg and S content in the steel is appropriate and the deoxidation operation is proper. These multiphase interfaces are beneficial to the formation of acicular ferrite, the increase of the steel strength, and the improvement of steel toughness and welding performance. Figure 2.19 is a SEM image of the acicular ferrite at the borders of MnS, TiO_x–MgO, and Al_2O_3–MgO and the EDS maps of element distribution in inclusion. Acicular ferrite is formed and grown near the boundary of various oxides or sulfides on the surface of inclusions.

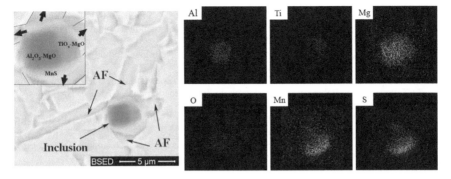

Fig. 2.19 EDS maps of the inclusion and SEM images of acicular ferrite at the boundary of various oxides or sulfides on the surface of inclusions

References

1. http://www.calphad.com/iron-titanium.html.
2. China Iron & Steel Association. Ferrotitanium, GB/T 3282-2012 [S]. Beijing: Standards Press of China, 2013. (in Chinese).
3. Joanne L Murray. The Fe–Ti (Iron-Titanium) system [J]. Bulletin of Alloy Phase Diagrams, 1981, 2(3): 320–334.
4. Manish Marotrao Pande, Muxing Guo, Bart Blanpain. Inclusion formation and interfacial reactions between FeTi alloys and liquid steel at an early stage [J]. ISIJ International, 2013, 53(4): 629–638.
5. Pande M M, Guo M, Devisscher S,et al. Influence of ferroalloy impurities and ferroalloy addition sequence on ultra low carbon (ULC) steel cleanliness after RH treatment [J]. Ironmaking & Steelmaking, 2012, 39(7): 519–529.
6. The 19th Committee on Steelmaking, The Japan Society for the Promotion of Science. Steelmaking Data Sourcebook [M]. New York: Gordon and Breach Science Publishers, 1988.
7. Mitsutaka Hino, Kimihisha Ito Edited. Thermodynamic data for steelmaking [M]. Sendai, Japan: Tohoku University Press, 2010.
8. Morioka Y, Morita K, Tsukihashi F, et al. Equilibria between molten steels and inclusions during deoxidation by titanium-manganese alloy [J]. Tetsu-to-Hagané, 1995, 81(1): 40–45. (in Japanese).
9. Ohta M, Morita K. Interaction between silicon and titanium in molten steel [J]. ISIJ Int., 2003, 43(2): 256–258.
10. Ghosh A, Murthy G V R. An Assessment of thermodynamic parameters for deoxidation of molten iron by Cr, V, Al, Zr and Ti [J]. Trans ISIJ, 1986, 26(7): 629–637.
11. Cha W, Nagasaka T, Miki T, et al. Equilibrium between titanium and oxygen in liquid Fe–Ti alloy coexisted with titanium oxides at 1873 K [J]. ISIJ Int., 2006, 46(7): 996–1005.
12. Sigworth G K, Elliott J. The thermodynamics of liquid dilute iron alloys [J]. Metal Sci., 1974, 8 (1): 298–310.
13. Woo-Yeol Cha, Takahiro Miki, Yasushi Sasaki et al. Temperature dependence of Ti deoxidation equilibria of liquid iron in coexistence with 'Ti$_3$O$_5$' and Ti$_2$O$_3$ [J]. ISIJ Int., 2008, 48(6): 729–738.
14. Ohta M, Morita K. Interaction between silicon and titanium in molten steel [J]. ISIJ Int., 2003, 43(2): 256–258.
15. Pak J, Yoo J, Jeong Y, et al. Thermodynamics of titanium and nitrogen in Fe-Si melt [J]. ISIJ Int. 2005, 45(1): 23–29.

16. Morita K, Ohta M, Yamada A, et al. Interaction between Ti and Si, and Ti and Al in molten steel at 1873 K [C]. Conference Proceedings of the 3rd International Congress on the Science and Technology of Steelmaking, 2005:15–22.
17. Kimura H. Advances in high-purity steel (IF steel) Manufacturing technology [J]. Shinnittetsu Giho, 1994, (351): 59–63. (in Japanese).
18. Kawashima Y, Nagata Y, Shinme K, et al. Influence of Ti concentration on nozzle clogging on Al-Ti deoxidation. Behavior of inclusion on Al-Ti deoxidation-2 [J]. CAMP-ISIJ, 1991, 4(4): 1237. (in Japanese).
19. Basu S, Choudhary S K, Girase N U. Nozzle clogging behaviour of Ti-bearing Al-killed ultra low carbon steel [J]. ISIJ Int., 2004, 44 (10): 1653–1660.
20. Ruby-Meyer F, Lehmann J, Gaye H. Thermodynamic analysis of inclusions in Ti-deoxidised steels [J]. Scand. J. Metall., 2000, 29 (5): 206–212.
21. Jung I, Decterov S A, Pelton A D. Computer applications of thermodynamic databases to inclusion engineering [J]. ISIJ Int., 2004, 44 (3): 527–536.
22. Ito H, Hino M, Ban-ya S. Assessment of Al deoxidation equilibrium in liquid iron [J] Tetsu-to-Haganè, 1997, 83 (12): 773–778. (in Japanese).
23. Turkdogan E T. Physical Chemistry of High Temperature Technology [M]. New York: Academic Press, 1980.
24. Guo Yuanchang, Wang Changzhen, Yu Hualong. Interaction coefficients in the iron-carbon-titanium and titanium-silver systems [J]. Metall. Trans. B, 1990, 21B(3): 537–541.
25. Lupis C H P. Chemical Thermodynamics of Materials [M]. New York: North-Holland, 1983: 255.
26. Ohta M, Morita K. Thermodynamics of the Al_2O_3-SiO_2-TiO_x System at 1873 K [J]. ISIJ Int., 2002, 42 (5): 474–481.
27. Pajunen M, Kivilahti J. Thermodynamic analysis of the titanium-oxygen system [J]. Z. Metallkd., 1992, 83(1): 17–20.
28. Hiroyuki Matsuura, Wang Cong, Wen Guanghua, et al. The transient stages of inclusion evolution during Al and/or Ti additions to molten iron [J]. ISIJ Int., 2007, 47(9): 1265–1274.
29. Jung In-Ho, Gunnar Eriksson, Wu Ping, et al. Thermodynamic modeling of the Al_2O_3-Ti_2O_3-TiO_2 system and its applications to the Fe-Al-Ti-O inclusion diagram [J]. ISIJ Int., 2009, 49 (9): 1290–1297.
30. Basu S, Choudhary S K, Girase N U. Nozzle clogging behaviour of Ti-bearing Al-killed ultra low carbon steel [J]. ISIJ Int., 2004, 44 (10): 1653–1660.
31. Park D-C, Jung I-H, Rhee P C H, et al. Reoxidation of Al-Ti containing steels by CaO-Al_2O_3-MgO-SiO_2 slag [J]. ISIJ Int., 2004, 44 (10): 1669–1678.
32. Marie-Aline Van Ende, Guo Muxing, Rob Dekkers, et al. Formation and evolution of Al-Ti oxide inclusions during secondary steel refining [J]. ISIJ Int., 2009, 49(8): 1133–1140.
33. Wang Min, Bao Yanping, Yang Quan. Effect of ferro-titanium alloying process on steel cleanness [J]. Journal of University of Science and Technology Beijing, 2013, 35(6): 725–732. (in Chinese).
34. Zhang Feng, Li Guangqiang. Control of ultra low titanium in ultra low carbon Al-Si killed steel [J]. Journal of Iron and Steel Research, International. 2013, 20(4): 20–25.
35. Fabienne Ruby-Meyer, Jean Lehmann, Henri Gaye. Thermodynamic analysis of inclusions in Ti-deoxidized steels [J]. Scandinavian Journal of Metallurgy, 2000, 29(5): 206–212.
36. Gloor K. Non-metallic inclusions in weld metal [R]. IIW DOCII-A-106–63, 1963.
37. Katoh K. Investigation of nonmetallic inclusions in mild steel weld metals [R]. IIW DOC II-A-158–65, 1965.
38. Harrison P L, Farrar R A. Influence of oxygen-rich inclusions on the $\gamma \rightarrow \alpha$ phase transformation in high-strength low-alloy (HSLA) steel weld metals [J], Journal of Materials Science, 1981, 16(8): 2218–2226.
39. Takamura J I, Mizoguchi S. Role of oxides in steel performance [C]// Proceeding of the sixth international iron and steel congress. Nagoya, ISIJ. 1990: 591–597.
40. Song Yu, Li Guangqiang, Yang Fei. Impacts of Al-Ti-Mg complex deoxidation on inclusions and the microstructure of steel [J]. Journal of University of Science and Technology Beijing, 2011, 33(10): 1214–1219. (in Chinese).

41. Wang C, Noel T N, Seetharaman S. Transient behavior of inclusion chemistry, shape, and structure in Fe-Al-Ti-O melts: effect of titanium/aluminum ratio [J]. Metallurgical and materials transactions B, 2009, 40B(6): 1022–1034.
42. Ohta H, Suito H. Characteristics of particle size distribution of deoxidation products with Mg, Zr, Al, Ca, Si/Mn and Mg/Al in Fe–10mass%Ni alloy [J]. ISIJ Int., 2006, 46(1): 14–22.
43. Ohta H, Suito H. Dispersion behavior of MgO, ZrO_2, Al_2O_3, $CaO–Al_2O_3$ and $MnO–SiO_2$ deoxidation particles during solidification of Fe–10mass%Ni alloy [J]. ISIJ Int., 2006, 46 (1): 22–28.
44. Akselsen O M, Grong, Ryum N, et al. Modelling of grain growth in metals and alloys [J]. Acta Metallurgy. 1986, 34(9):1807–1811.
45. Miller O O. Influence of austenitizing time and temperature on austenite grain size of steel [J]. Tran ASM. 1951, 43:261–287.
46. Zener C. The effect of deformation on grain growth in Zener pinned systems [J]. Acta metal. 2001, 49(8): 1453–1461.
47. Li Peng. Influences of Al-Ti complex deoxidation on inclusions and the microstructure of non-quenched and tempered steel [D]. Wuhan: Wuhan University of Science and Technology, 2013. (in Chinese).
48. Zheng Wan, Wu Zhenhua, Li Guangqiang. Effect of Al content on the characteristics of inclusions in Al–Ti complex deoxidized steel with calcium treatment [J], ISIJ Int., 2014, 54 (8): 1755–1764.
49. Wu Zhenhua, Zheng Wan, Li Guangqiang, et al. Effect of inclusions' behavior on the microstructure in Al-Ti deoxidized and magnesium-treated steel with different aluminum contents [J], Metallurgical and Materials Transactions B, 2015, 46B(3): 1226–1241.

Chapter 3
Physical Metallurgical Principles of Titanium Microalloyed Steel—Dissolution and Precipitation of Titanium-Bearing Secondary Phases

Qilong Yong, Xinjun Sun, Zhaodong Li, Zhenqiang Wang and Ke Zhang

The precipitation of microalloying elements is one of the most important issues in microalloyed steels. It is well recognized that controlling the precipitation process of microallying elements in steels is an effective means to significantly improve the strength of steel material due to precipitation strengthening and grain refinement by controlling the austenite grains coarsening during reheating process and recrystallization process. Moreover, controlling the precipitation behavior of secondary phases in steel, leading to an accurately control of volume fraction, shape, size and distribution of precipitates, could effectively improve the microstructure and mechanical properties, which is a significant issue for microalloyed steel in the field of theory research and production practice.

Compared with the steels containing Nb or V, it is more difficult to control the precipitation process in Ti microalloyed steel which attributes to the more types of secondary phases in the steel and wider temperature range of precipitation. For instance, during the melting process, the micron-sized Ti_2O_3 and TiN particles will precipitate in the liquid steel that can improve the as-cast microstructure [1, 2]. During the slab cooling process, the TiN and $Ti_4S_2C_2$ with the size of tens to hundreds of

---

Q. Yong (✉) · X. Sun · Z. Li
Central Iron & Steel Research Institute, Beijing, China
e-mail: yongql@126.com

X. Sun
e-mail: fallbreeze@126.com

Z. Li
e-mail: 3172087@qq.com

Z. Wang
Harbin Engineering University, Harbin, China
e-mail: wangzhe19840203@163.com

K. Zhang
Anhui University of Technology, Ma'anshan, China
e-mail: huzhude@yeah.net

© Metallurgical Industry Press, Beijing and Springer Nature Singapore Pte Ltd. 2019
X. Mao (ed.), *Titanium Microalloyed Steel: Fundamentals, Technology, and Products*,
https://doi.org/10.1007/978-981-13-3332-3_3

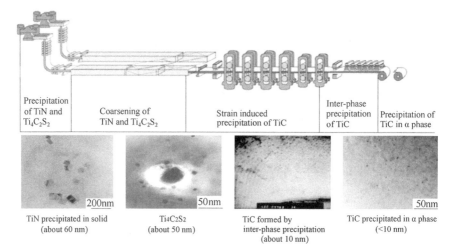

Fig. 3.1 The titanium-bearing precipitation in different stages of TSCR process

nanometer precipitated in the solid solution could play an important role in controlling the grain growth of austenite during soaking and recrystallization processes. During the hot rolling process, the deformation induced TiC precipitation with the size of several to tens nanometer could play an important role in inhibiting the recovery and recrystallization of austenite. Furthermore, during the rapid cooling and coiling process, the interphase precipitation and supersaturated precipitation of TiC with the size below 10 nm could result in the significantly precipitation hardening [1]. Figure 3.1 shows the titanium bearing secondary phase precipitated in different stages of thin slab casting and direct rolling (TSCR) process.

In order to correctly design the chemical composition and manufacturing process of titanium microalloying steels, the precipitation behavior law and its influence factors should be investigated and understood. In this chapter, the existence forms of titanium in the steel and solubility product were introduced firstly. Then the kinetic theory of secondary phase containing titanium was expounded. The effect of alloying elements, such as Mn and Mo, etc. on the deformation induced TiC precipitation were discussed. Finally, the Ostwald ripening of precipitation containing titanium was introduced.

3.1 Existence Form of Titanium in Steel and Its Solubility Products

The existence forms of titanium element in steel are mainly of two types, including the trace titanium in the form of solid solution in the iron matrix and the residual titanium in the form of various secondary phases, such as TiO_x, TiS, $Ti_4C_2S_4$ and Ti(C, N), etc. As the existence form of titanium element in steel is different while its effect and mechanism were different, the role of titanium could be widely divergent in some cases. Thus, the existence from and its content should be accurately understood firstly. Then the deep analysis and discussion were carried out according to the different existence forms of titanium element in steel.

According to the Fe–Ti equilibrium phase diagram, the titanium is a ferrite former element which can close the γ phase region and form a ring-shaped γ phase region. With the addition of titanium element, the A_4 point of iron decreases and the A_3 point of iron increases. Consequently, the γ phase region closed at about 1100 °C. And the maximum solid solubility of titanium in austenite is 0.69% at 1157 °C. Moreover, as the formation of Fe_2Ti which limits the solid solubility of titanium in ferrite, the titanium and iron can only form the limited solid solution. And the eutectic transformation of liquid phase transformed into ferrite and Fe_2Ti will occur at 1289 °C. The maximum solid solubility of titanium in ferrite is 8.7% at this temperature [3].

According to the phase diagram and Thermo-Calc database, the solubility products of Fe_2Ti in ferrite can be deduced as follows [4]:

$$\log[Ti]_\alpha = 2.458 - 2392/T \, (Fe_2Ti \text{ in paramagnetic } \alpha\text{-iron, } 850-1560.2 \, K) \quad (3.1)$$

$$\log[Ti]_\alpha = 1.074 - 1284/T \, (Fe_2Ti \text{ in ferromagnetic } \alpha\text{-iron, } 300-800 \, K) \quad (3.2)$$

Among them, the accuracy and credibility of Eq. (3.1) is higher (the limited solid solubility was expressed with dotted line in Fig. 2.1).

In addition, the variation of formation free energy of Fe_2Ti with temperature is obtained according to the related thermodynamic data [5].

$$\Delta G = -103{,}548 + 30.403T \, (800-1600 \, K) \quad (3.3)$$

Thus, the solubility product of pure titanium in ferrite can be deduced as:

$$\log[Ti]_\alpha = 0.8700 + 3017/T \quad (3.4)$$

It can be seen from Fe–Ti phase diagram and the solubility products as mentioned above, in pure Fe–Ti alloy, the solid solubility of titanium in iron matrix is larger and the titanium mainly exists in solid solution.

However, there are certain amount of carbon and nitrogen elements in titanium microalloyed steels in the industrial production practice. The titanium element has a very strong chemical affinity with carbon or nitrogen, leading to the formation of TiC, TiN or Ti(C, N) easily. And the solid solubility of titanium in steel will significantly change when the carbonitride forms. Therefore, more attention should be paid to the solid solubility of TiC and TiN in iron matrix.

The solubility products of TiC and TiN in iron matrix were obtained up to now as follows [6–20]:

$$\log\{[\text{Ti}][\text{C}]\}_\gamma = 5.33 - 10{,}475/T \text{ [6]} \quad (3.5)$$

$$\log\{[\text{Ti}][\text{C}]\}_\gamma = 5.54 - 11{,}300/T \text{ [6]} \quad (3.6)$$

$$\log\{[\text{Ti}][\text{C}]\}_\gamma = 2.75 - 7000/T \text{ [7]} \quad (3.7)$$

$$\log\{[\text{Ti}][\text{C}]\}_\gamma = 4.37 - 10{,}580/T \text{ [8]} \quad (3.8)$$

$$\log\{[\text{Ti}][\text{C}]\}_\gamma = 2.97 - 6780/T \text{ [9]} \quad (3.9)$$

$$\log\{[\text{Ti}][\text{C}]\}_\gamma = 3.23 - 7430/T + [\text{C}](-0.03 + 1300/T) \text{ [10]} \quad (3.10)$$

$$\log\{[\text{Ti}][\text{C}]\}_\gamma = 3.21 - 7480/T \text{ [11]} \quad (3.11)$$

$$\log\{[\text{Ti}][\text{N}]\}_\gamma = 0.322 - 8000/T \text{ [12]} \quad (3.12)$$

$$\log\{[\text{Ti}][\text{N}]\}_\gamma = 6.75 - 19{,}740/T \text{ [13]} \quad (3.13)$$

$$\log\{[\text{Ti}][\text{N}]\}_\gamma = 2.00 - 20{,}790/T \text{ [14]} \quad (3.14)$$

$$\log\{[\text{Ti}][\text{N}]\}_\gamma = 3.82 - 15{,}020/T \text{ [7]} \quad (3.15)$$

$$\log\{[\text{Ti}][\text{N}]\}_\gamma = 5.19 - 15{,}490/T \text{ [15]} \quad (3.16)$$

$$\log\{[\text{Ti}][\text{N}]\}_\gamma = 4.22 - 14{,}200/T \text{ [11]} \quad (3.17)$$

$$\log\{[\text{Ti}][\text{N}]\}_\gamma = 4.94 - 14{,}400/T \text{ [16]} \quad (3.18)$$

$$\log\{[\text{Ti}][\text{N}]\}_\gamma = 5.40 - 15{,}790/T \text{ [17]} \quad (3.19)$$

$$\log\{[\text{Ti}][\text{N}]\}_\gamma = 4.35 - 14{,}890/T \text{ [18]} \quad (3.20)$$

$$\log\{[\text{Ti}][\text{N}]\}_\gamma = 3.94 - 15{,}190/T \text{ [6]} \quad (3.21)$$

$$\log\{[\text{Ti}][\text{C}]\}_\alpha = 4.40 - 9575/T \text{ [19]} \quad (3.22)$$

$$\log\{[\text{Ti}][\text{C}]\}_\alpha = 5.02 - 10{,}800/T \text{ [20]} \quad (3.23)$$

$$\log\{[\text{Ti}][\text{N}]\}_\alpha = 4.65 - 16{,}310/T \text{ [18]} \quad (3.24)$$

$$\log\{[\text{Ti}][\text{N}]\}_\alpha = 5.89 - 16{,}750/T \text{ [20]} \quad (3.25)$$

According to the relevant thermodynamic data, the equilibrium solubility formula of Ti, C and N elements in ferrite were obtained firstly, while the relationships between the formation free energy and temperature of TiC and TiN were considered. The solubility products of TiC and TiN in ferrite were deduced, respectively.

$$\log\{[\text{Ti}] \cdot [\text{C}]\}_\alpha = 5.286 - 12{,}154/T \quad (3.26)$$

$$\log\{[\text{Ti}] \cdot [\text{N}]\}_\alpha = 4.179 - 15{,}776/T \quad (3.27)$$

In addition, the solubility product of TiN in liquid iron is small. The TiN particles could easily precipitate in liquid iron that resulted in a significant adverse effect on the properties of steel. Then, the solubility product of TiN in liquid iron should be considered. The correlation formula is as follows [21].

$$\log\{[\text{Ti}][\text{N}]\}_L = 4.46 - 13{,}500/T \text{ [18]} \quad (3.28)$$

$$\log\{[\text{Ti}][\text{N}]\}_L = 5.922 - 16{,}066/T \text{ [21]} \quad (3.29)$$

$$\log\{[\text{Ti}][\text{N}]\}_L = 5.90 - 16{,}586/T \; [6] \tag{3.30}$$

Figure 3.2 shows the comparison of different solubility products of TiC in austenite. The solubility products of TiC in austenite were mainly obtain from the experimental results. The experimental results of solubility products at 1200 °C were mainly in the range of $10^{-1.9}$ to $10^{-2.0}$. It can be seen that solubility product calculated by Eq. (3.8) was obvious smaller. It probably relate to the nitrogen content in steel. As the solubility product of TiN in austenite was obvious smaller than that of TiC, if the influence of nitrogen content on the solubility product didn't deducted, the calculated result of solubility product will be significantly smaller. Thus, the Eq. (3.7) will be used in this book for calculation.

Figure 3.3 shows the comparison of different solubility products of TiN in austenite. As the solubility product of TiN in austenite is small, it was mainly deduced by thermodynamics. Obviously, the result calculated by Eqs. (3.12) and (3.14) with low credibility was significant deviated from the normal result. In contrast, the Eqs. (3.16), (3.17) and (3.19) have higher credibility. And the Eq. (3.20) also has high credibility due to the deduced result based on new theory from extensive preliminary works. Moreover, compared Fig. 3.2 with Fig. 3.3, it can be found that the solubility product of TiN in austenite is 3.5 orders of magnitude lower than that of TiC at 1200 °C. And when the temperature deceased to 727 °C, the solubility product of TiN in austenite is 6 orders of magnitude lower than that of TiC. As the TiC and TiN are usually completely mutually soluble to form $\text{Ti}(C_xN_{1-x})$ and the significant difference between the solubility product of TiC and TiN, the change of coefficient x in the

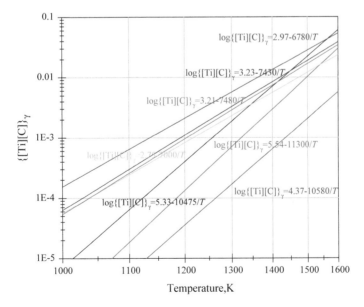

Fig. 3.2 The solubility products of TiC in austenite

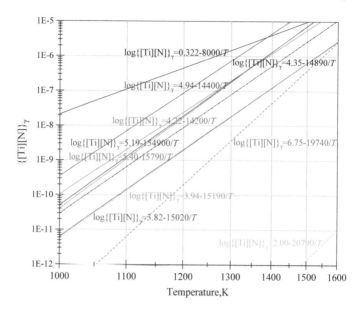

Fig. 3.3 The solubility products of TiN in austenite

Ti(C_xN_{1-x}) will be obvious in the certain temperature range, consequently leading to a C-shaped precipitation-temperature-time (PTT) curve to the ε-shaped PTT curve or even completely separated into PTT curve of TiN at high temperature and PPT curve of carbon-rich Ti(C, N) at low temperature.

Figure 3.4 shows the comparison of different solubility products of TiC in ferrite. As the solubility product of TiC in ferrite is very small and difficult to be determined by experiment, all of the solubility products of TiC in ferrite were deduced by thermodynamics. The Eq. (3.23) will be used in this book for calculation.

Figure 3.5 shows the comparison of different solubility products of TiN in ferrite. As the solubility product of TiN in ferrite even smaller than that of TiC, the solubility products of TiN in ferrite were also deduced by thermodynamics and the results calculated by different solubility product of TiN in ferrite were very close as shown in Fig. 3.5. Compared Fig. 3.4 with Fig. 3.5, it can be found that the solubility product of TiN in ferrite is about 5 orders of magnitude lower than that of TiC at 727 °C. And when the temperature deceased to 400 °C, the solubility product of TiN in ferrite is about 7 orders of magnitude lower than that of TiC. The difference between solubility product of TiN and TiC in ferrite is very large.

Figure 3.6 shows the comparison of different solubility products of TiN in liquid iron. The solubility product of TiN liquid iron can be measured directly. Thus, its reliability is higher. The Eq. (3.28) will be used in this book for calculation.

Figure 3.7 shows the results calculated by the relatively reliable solubility product as mentioned above. The solubility products of TiC and TiN in different iron matrix

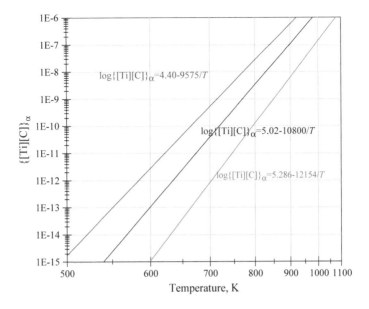

Fig. 3.4 The solubility products of TiC in ferrite

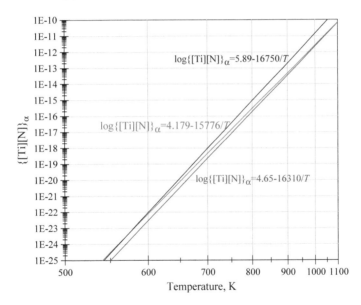

Fig. 3.5 The solubility products of TiN in ferrite

can be compared and analyzed. It can be seen that the solubility product of TiC or TiN in the high temperature phase was evident higher than that in the low temperature phase. As a result, the solubility product will rapidly decrease during the solidification

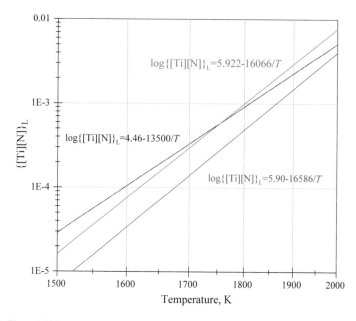

Fig. 3.6 The solubility products of TiN in liquid iron

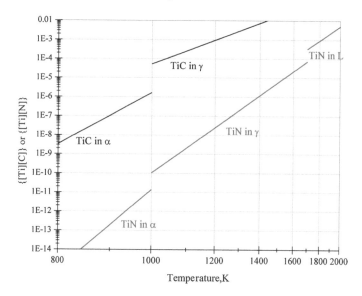

Fig. 3.7 The solubility products of TiC and TiN in different iron matrix

phase transformation and austenite to ferrite phase transformation processes. The TiC or TiN would precipitate accompanied with phase transformation of matrix.

3 Physical Metallurgical Principles of Titanium Microalloyed …

Furthermore, it can be seen from the solubility products as mentioned above, because of titanium has a strong chemical affinity with carbon and nitrogen, the TiC or TiN forms easily. As the equilibrium solubility product of TiC or TiN is very small, the mainly existence form of titanium in steels is formation of various compounds, such as TiC or TiN, etc. And the content of titanium existed in solid solution is very small.

In addition, the titanium can also form oxide, sulfide or sulfo-carbide. And the precipitation of these compounds which were harmful to the properties of steels should be avoided. Thus, its solubility products in austenite at high temperature should also be considered. The correlative solubility products were listed as follows [18, 22–31].

$$\log\{[Ti][O]\}_\gamma = 2.03 - 14{,}440/T \ [18] \tag{3.31}$$

$$\log\{[Ti][S]\}_\gamma = 8.20 - 17{,}640/T \ [22] \tag{3.32}$$

$$\log\{[Ti][S]\}_\gamma = 6.24 - 14{,}559/T \ [23] \tag{3.33}$$

$$\log\{[Ti][S]\}_\gamma = 6.75 - 16{,}550/T \ [24] \tag{3.34}$$

$$\log\{[Ti][S]\}_\gamma = -2.01 - 3252/T \ [25] \tag{3.35}$$

$$\log\{[Ti][S]\}_\gamma = 5.43 - 13{,}975/T \ [26] \tag{3.36}$$

$$\log\{[Ti][S]\}_\gamma = 6.92 - 16{,}550/T \ [27] \tag{3.37}$$

$$\log\{[Ti][S]\}_\gamma = 4.28 - 12{,}587/T \ [28] \tag{3.38}$$

$$\log\{[Ti][S]\}_\gamma = 7.74 - 17{,}820/T \ [30] \tag{3.39}$$

$$\log\{[Ti][C]^{0.5}[S]^{0.5}\}_\gamma = 6.50 - 15{,}600/T \ [22] \tag{3.40}$$

$$\log\{[Ti][C]^{0.5}[S]^{0.5}\}_\gamma = 6.03 - 15{,}310/T \ [23] \tag{3.41}$$

$$\log\{[Ti][C]^{0.5}[S]^{0.5}\}_\gamma = -0.78 - 5208/T \ [25] \tag{3.42}$$

$$\log\{[Ti][C]^{0.5}[S]^{0.5}\}_\gamma = 7.90 - 17{,}045/T \ [26] \tag{3.43}$$

$$\log\{[Ti][C]^{0.5}[S]^{0.5}\}_\gamma = 6.32 - 15{,}350/T \ [27] \tag{3.44}$$

$$\log\{[Ti][C]^{0.5}[S]^{0.5}\}_\gamma = 4.093 - 12{,}590/T \ [28] \tag{3.45}$$

$$\log\{[Ti][C]^{0.5}[S]^{0.5}\}_\gamma = 5.51 - 14{,}646/T \ [29] \tag{3.46}$$

$$\log\{[Ti][C]^{0.5}[S]^{0.5}\}_\gamma = 7.313 - 15{,}125/T \ [30, 31] \tag{3.47}$$

$$\log\{[Ti][C]^{0.5}[S]^{0.5}\}_\gamma = 0.392 - 7004/T - (4.783 - 7401/T)[Mn] \ [31] \tag{3.48}$$

Most of the formulas were obtained from experimental results, and some of them were deduced by thermodynamics. Literature [31] deeply discussed the effect of Mn on the solubility product of TiS. It was found that the effect of Mn was very significant due to the competition between MnS and TiS. As the content of Mn is only 0.006–0.01% in steel, the Eqs. (3.35) and (3.42) derived from literature [25] have significant difference with the others, and the effect of temperature on them is very small. In addition, the experiment in literature [29] was carried out in stainless

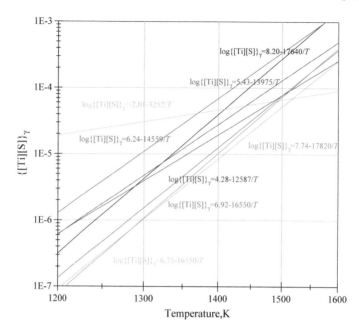

Fig. 3.8 The solubility products of TiS in austenite

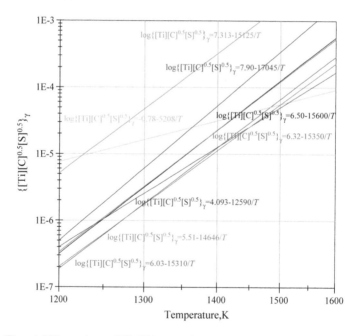

Fig. 3.9 The solubility products of Ti_2CS in austenite

3 Physical Metallurgical Principles of Titanium Microalloyed …

steels, while the experiments in the others literatures were mainly carried out in ultra low carbon interstitial free (IF) steels.

Figure 3.8 shows the comparison of solubility products of TiS in austenite. Comparatively speaking, the Eq. (3.36) has a higher accuracy which derived from the experimental results after fully considering the previous works. And it is preferred in low carbon and ultra low carbon steels as its results are very close to the Eq. (3.38) which deduced from the latest thermodynamic theory.

Figure 3.9 shows the comparison of solubility products of $Ti_4C_2S_2$ in austenite (the chemical formula was normalized as $TiC_{0.5}S_{0.5}$ for comparison). The Eqs. (3.44) and (3.41) are preferred in low carbon and ultra low carbon steels.

Figure 3.10 shows the solubility products of TiO, TiN, TiS, $Ti_4C_2S_2$ and TiC in austenite. The order of precipitation precipitated in steel can be roughly analyzed. It can be seen from Fig. 3.10, the solubility products of TiO in austenite is very small. If there is trace oxygen in steel, the TiO will precipitate firstly. Then the TiN will precipitate (In fact, because of the solubility products of TiO and TiN in austenite were very small, if there were TiO and TiN precipitated in solid solution, they must occur during the solidification process). When the temperature is higher than 1250–1300 °C, the TiS will precipitate earlier than $Ti_4C_2S_2$. When the temperature is lower than that range, the precipitation of $Ti_4C_2S_2$ will be dominant. And only when more titanium existed in steel can TiC precipitate at lower temperature. In fact, as the content of S and C were very similar in the ultra low carbon IF steel, the

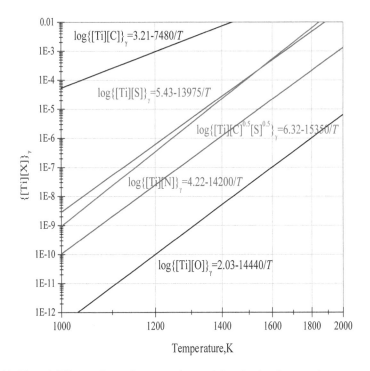

Fig. 3.10 The solubility products of compounds containing titanium in austenite

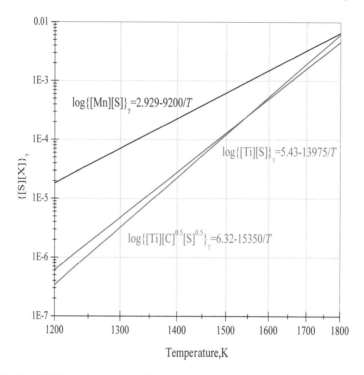

Fig. 3.11 The solubility products of sulfur containing compounds in austenite

sequence of precipitation precipitated in steel can be well explained in Fig. 3.10. And it is TiN → TiS → Ti_2CS from the experimental observation. Moreover, the effect of other elements on the sequence of precipitation precipitated in steel as shown in Fig. 3.10 should be considered in the high strength low carbon steels. In these steels, the content of C is significantly larger than the content of S, the precipitation of TiS will be suppressed. And the order of precipitation precipitated in steel is usually TiN → Ti_2CS → TiC from the experimental observation.

In order to deeply analyze the competition between TiS, MnS and $Ti_4C_2S_2$ precipitated in high strength low alloy steels. The correlative solubility products were plotted in Fig. 3.11. Similarly, the effect of other elements content on the solubility products should be considered. It can be seen from Fig. 3.11, the MnS doesn't seem to precipitate easily, but the situation will be different when the content of other elements were considered in steel. The solubility products of TiS, MnS and $Ti_4C_2S_2$ in austenite at the solidification temperature of 1763 K (1490 °C) in low carbon steel were 0.003185, 0.005136 and 0.004104, respectively. If the content of Mn and C in steel is 1 and 0.1%, respectively, the TiS could precipitate in titanium microalloying treatment steel (the Ti content is 0.02%) when the content of S is greater than 0.16%, the MnS could precipitate when the content of S is greater than 0.0051%, and the Ti_2CS could precipitate when the content of S is greater than 0.42%. The sequence

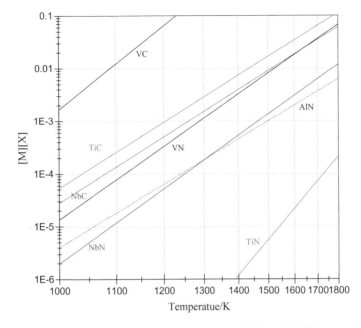

Fig. 3.12 The solubility products of various microalloying carbide and nitride in austenite

of precipitation precipitated in steel is MnS → TiS → $Ti_4C_2S_2$. That means the MnS will precipitate preferentially and the precipitation of TiS and $Ti_4C_2S_2$ is inhibited. Moreover, in the titanium microalloyed steels with the titanium content of 0.1%, the TiS could precipitate when the content of S is greater than 0.032%, the MnS could precipitate when the content of S is greater than 0.0051%, and the $Ti_4C_2S_2$ could precipitate when the content of S is greater than 0.017%. The sequence of precipitation precipitated in steel is MnS → $Ti_4C_2S_2$ → TiS. The precipitation of MnS is also preferential, but its superiority has been decreased significantly.

Generally, the precipitation of $Ti_4C_2S_2$ with large size is harmful to the properties of steels and significantly influences the precipitation behavior of TiC as it will consume the Ti and C in steel. It usually occurs in the titanium microalloyed steels with the higher titanium content. Appropriate increase of the Mn content in steel is benefit to the precipitation of MnS. And appropriate decrease of the C content in steel can inhibit the precipitation of $Ti_4C_2S_2$. When the chemical composition adjustment can't effectively inhibit the precipitation of $Ti_4C_2S_2$, the control of precipitation kinetics should be carried out.

Finally, compared the solubility products of various microalloying carbonitrides, the effect of these carbonitrides on the strength and toughness of steel can be analyzed and interpreted. Figure 3.12 shows the solubility products of various microalloying carbonitrides in austenite. It can be seen that the precipitation of TiN has the smallest solubility product but the effect of TiN on inhibiting austenite grain growth at high temperature is the strongest. The solubility product of TiC is slightly greater than that

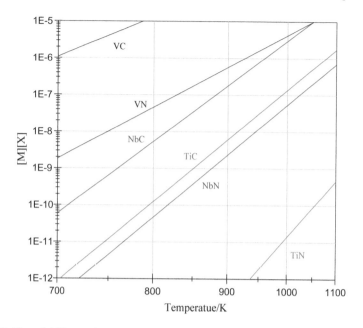

Fig. 3.13 The solubility products of various microalloying carbide and nitride in ferrite

of NbC and VN, leading to the result that the effect of strain-induced precipitation of TiC on inhibiting deformed austenite recrystallization is less than that of NbC and VN, but greater than that of VC. It means that the precipitation of TiC has a certain function to inhibit the deformed austenite recrystallization. On the other hand, there is the largest difference of solubility product between TiN and TiC (about 3–4 order of magnitudes, the difference will be further increased if the differences of C content and N content in steel were considered.). Thus, the precipitation of titanium can be divided into two parts: the pure TiN at high temperature and TiC at low temperature (when the nitrogen in steel was consumed completely) or Ti(C, N) (when the nitrogen in steel was not consumed completely). In contrast, the PTT curve of Nb(C, N) in niobium microalloyed steel usually exhibits a C-shaped due to the result that the difference between the solubility products of NbN and NbC is very small. And the variation of chemical formula coefficient of Nb(C, N) with the precipitation temperature is small. However, the difference between the solubility products of VN and VC is almost 2 order of magnitudes, consequently leading to the PTT curve of vanadium precipitation changing from C-shaped to ε-shaped. Thus, the nitrogen-rich V(C, N) will precipitate at high temperature and the carbon-rich V(C, N) will precipitate at a lower temperature (it is difficult to observe the carbon-rich V(C, N) at practice as the precipitation temperature is lower than Ar_3 temperature). It means that the chemical formula coefficient of V(C, N) will significantly change with the temperature.

Figure 3.13 shows the comparison of the solubility products of various microalloying carbide and nitride in ferrite. It can be found that the solubility products of various microalloying carbide and nitride in ferrite is very small and the microalloying elements will be almost entirely precipitated. In titanium microalloyed steels, the precipitation which precipitated in ferrite region is mainly TiC particles due to the fact that nitrogen element has been precipitated as the form of TiN in austenite region. In niobium microalloyed steels, the precipitation which precipitated in ferrite region is mainly carbon-rich Nb(C, N) particles. In vanadium microalloyed steels, the precipitation which precipitated in ferrite region is mainly carbon-rich V(C, N) particles. And in the vanadium-nitrogen microalloyed steel, the precipitation which precipitated in ferrite region is mainly nitrogen-rich V(C, N) particles. In addition, as the solubility product of NbC in ferrite is significant larger than that of TiC, the Nb is usually added into bake hardening (BH) steels because a part of NbC precipitates will dissolute during annealing process consequently leading to the baking hardening effect.

The equilibrium solid solubility of microalloying elements in iron matrix and the mass fraction of the precipitates can be calculated by solubility products. When there is only MX phase precipitated in steel, the equilibrium solid solubility of MX phase can be calculated by simultaneous equations of solubility product and the ideal chemical matching formulas of secondary phase.

$$\log\{[M] \cdot [X]\} = A - B/T \tag{3.49}$$

$$\frac{w_M - [M]}{w_X - [X]} = \frac{A_M}{A_X} \tag{3.50}$$

where A and B represent the constants in the solubility product of MX phase in iron matrix, w_M and w_X represent the content of M and X in steel, respectively, A_M and A_X represent the relative atomic mass of M and X. And the mass fraction of undissolved or precipitated MX phase in equilibrium condition can be calculated by following equation.

$$w_{MX} = (w_M + w_X) - ([M] + [X]) \tag{3.51}$$

And the complete dissolution temperature T_{AS} of MX phase can be calculated by the following equation.

$$T_{AS} = \frac{B}{A - \log(w_M \cdot w_X)} \tag{3.52}$$

The mole chemical free energy ΔG_M of precipitation MX at the temperature of T can be calculated by the following equation.

$$\Delta G_M = -19.1446B + 19.1446T\{A - \log([M]_H \cdot [X]_H)\} \tag{3.53}$$

where $[M]_H$ and $[X]_H$ represent the equilibrium solid solubility of M and X at elevated soaking temperature T_H. If the soaking temperature T_H is higher than the complete dissolution temperature T_{AS}, w_M and w_X can replace $[M]_H$ and $[X]_H$, respectively.

For the $Ti_4C_2S_2$ phase which contains three elements, the equilibrium solid solubility can be calculated by simultaneous equations of solubility product and the two ideal chemical matching formulas as follows.

$$\log\{[Ti] \cdot [C]^{0.5} \cdot [S]^{0.5}\} = A - B/T \tag{3.54}$$

$$\frac{w_{Ti} - [Ti]}{w_C - [C]} = \frac{2A_{Ti}}{A_C} \tag{3.55}$$

$$\frac{w_{Ti} - [Ti]}{w_S - [S]} = \frac{2A_{Ti}}{A_S} \tag{3.56}$$

where w_{Ti}, w_C and w_S represent the content of Ti, C and S in steel, respectively, A_{Ti}, A_C and A_S represent the relative atomic mass of Ti, C and S. The mass fraction of undissolved or precipitated $Ti_4C_2S_2$ phase in equilibrium condition can be calculated by following equation.

$$w_{Ti_2CS} = (w_{Ti} + w_C + w_S) - ([M] + [C] + [S]) \tag{3.57}$$

The complete dissolution temperature T_{AS} of $Ti_4C_2S_2$ phase can be calculated by the following equation.

$$T_{AS} = \frac{B}{A - \log(w_{Ti} \cdot w_C^{0.5} \cdot w_S^{0.5})} \tag{3.58}$$

When the temperature is T, the mole chemical free energy ΔG_M of $Ti_4C_2S_2$ phase can be calculated by the following equation.

$$\Delta G_M = -19.1446B + 19.1446T\{A - \log([Ti]_H \cdot [C]_H^{0.5} \cdot [S]_H^{0.5})\} \tag{3.59}$$

where $[Ti]_H$, $[C]_H$ and $[S]_H$ represent the equilibrium solid solubility of Ti, C and S at elevated soaking temperature T_H. If the soaking temperature T_H is higher than the complete dissolution temperature T_{AS}, w_{Ti}, w_C and w_S can replace $[Ti]_H$, $[C]_H$ and $[S]_H$, respectively.

As the TiC and TiN phases have the same crystal structure and the difference of lattice constant between them is very small, both of them can be miscible to form carbonitride of $Ti(C_x, N_{1-x})$ when the carbon and nitrogen both present in the steel. And the titanium present in $Ti(C_x, N_{1-x})$ phase should remain the ideal chemical matching with the carbon and nitrogen. Thus, the equilibrium amounts of titanium [Ti], carbon [C], nitrogen [N] in solid solution and the chemical coefficient x can be calculated by following simultaneous equations.

$$\log\{\frac{[Ti] \cdot [C]}{x}\} = A_1 - B_1/T \tag{3.60}$$

$$\log\{\frac{[Ti] \cdot [N]}{1-x}\} = A_2 - B_2/T \tag{3.61}$$

$$\frac{w_{Ti} - [Ti]}{w_C - [C]} = \frac{A_{Ti}}{xA_C} \tag{3.62}$$

$$\frac{w_{Ti} - [Ti]}{w_N - [N]} = \frac{A_{Ti}}{(1-x)A_N} \tag{3.63}$$

where A_1, B_1, A_2 and B_2 represent the constants in the solubility product of TiC and TiN phases in iron matrix, w_{Ti}, w_C and w_N represent the content of Ti, C and N in steel (mass fraction), respectively, A_{Ti}, A_C and A_N represent the relative atomic mass of Ti, C and N.

Thus, the mass fraction of undissolved or precipitated Ti(C_x, N_{1-x}) phase in equilibrium condition can be calculated by following equation.

$$w_{TiCN} = (w_{Ti} + w_C + w_N) - ([Ti] + [C] + [N]) \tag{3.64}$$

The calculation of the complete dissolution temperature T_{AS} of Ti(C_x, N_{1-x}) phase is more complicated. And it can be calculated by the following equation.

$$w_{Ti} \cdot w_C \cdot 10^{-A_1 + B_1/T_{AS}} + w_{Ti} \cdot w_N \cdot 10^{-A_2 + B_2/T_{AS}} = 1 \tag{3.65}$$

When the temperature is T, the mole chemical free energy ΔG_M of Ti(C_x, N_{1-x}) phase can be calculated by the following equation.

$$\Delta G_M = -19.1446\{xB_1 + (1-x)B_2\}$$
$$+ 19.1446T\{xA_1 + (1-x)A_2 - \log([Ti]_H \cdot [C]_H^x \cdot [N]_H^{1-x})\} \tag{3.66}$$

where $[Ti]_H$, $[C]_H$ and $[N]_H$ represent the equilibrium solid solubility of Ti, C and N at elevated soaking temperature T_H. If the soaking temperature T_H is higher than the complete dissolution temperature T_{AS}, w_{Ti}, w_C and w_N can replace $[Ti]_H$, $[C]_H$ and $[N]_H$, respectively. It is noted that the x in the Eq. (3.66) is the chemical coefficient of Ti(C_x, N_{1-x}) phase at the temperature of T. And it varies with temperature.

3.2 Basic Data of Titanium and Titanium-Bearing Phases

In order to deeply investigate the role of titanium in steel and its controlling technology, the processes of solution and precipitation kinetics of various Ti-bearing phases should be deeply analyzed besides considering the existence form of titanium in steel from the aspect of thermodynamic equilibrium. The precipitation of useful Ti-bearing phase should be promoted in the appropriate temperature range, and the precipitation of harmful Ti-bearing phase should be inhibited. The coarsening process of Ti-bearing phase at this temperature range should also be inhibited. To achieve these as mentioned above, the basic data of titanium and various titanium-bearing phase should be understand firstly.

The diffusion coefficient of Ti in austenite can be obtained by diffusion couple experiment and Matano-Boltzmann analysis method [32], cm^2/s.

$$D = 0.15\exp(-\frac{251{,}000}{RT}) \text{ (atomic fraction of titanium is 0–0.7\%, 1075–1225\,°C)} \tag{3.67}$$

The diffusion coefficient of Ti in ferrite can be obtained by using the same method [32], cm^2/s.

$$D = 3.15\exp(-\frac{248{,}000}{RT}) \text{ (atomic fraction of titanium is 0.7–3.0\%, 1075–1225\,°C)} \tag{3.68}$$

Moreover, the diffusion coefficient of Ti in austenite can also be obtained by the method of thin-layer residual radioactivity measurement [33], cm^2/s.

$$D = 2.8\exp(-\frac{242{,}000}{RT}) \text{ (atomic fraction of titanium is 2\%, 900–1200\,°C)} \tag{3.69}$$

The self-diffusion coefficient of α-Ti can be measured by the radioactive-tracer method [34], cm^2/s.

$$D = 8.6 \times 10^{-6} \exp(-\frac{150{,}000}{RT}) \text{ (690–880\,°C)} \tag{3.70}$$

And the self-diffusion coefficient of β-Ti was measured by the radioactive-tracer method as follow [35], cm^2/s.

$$D = 1.9 \times 10^{-3} \exp(-\frac{153{,}000 \pm 2100}{RT}) \text{ (900–1580\,°C)} \tag{3.71}$$

Titanium is a strong carbonitride-forming element. The titanium present in steels is mainly formed as carbonitride which plays an important role in steel. Simultaneously, the titanium is also easy to form the harmful phase in steels, such as TiS and $Ti_4C_2S_2$, etc.

Both of the atom radius ratios of C/Ti and N/Ti were smaller than 0.59 which were 0.53 and 0.50, respectively. Thus, the carbides and nitrides of titanium are interstitial phases with simple lattice structure. And the TiC and TiN in steels are also interstitial phases with the NaCl (B1) type face-centered cubic (fcc) structure as shown in Fig. 3.14. The absence of interstitial atom will exist to a certain extent, but it is a very small amount. Thus, the crystal structures of TiC and TiN can be considered as complete.

At ambient temperature, the lattice constant of TiC is 0.43176 nm, the mole volume is 1.212×10^{-5} m^3/mol, the theoretical density is 4.944 g/cm^3, the coefficient of linear expansion is 7.86×10^{-6}/K (12–270 °C), the melting point is 3017 °C, the elastic

Fig. 3.14 The crystal structure of TiC and TiN

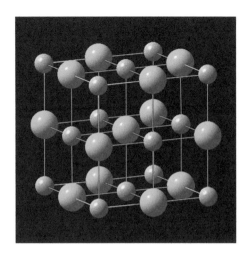

modulus E is 4.51×10^5 MPa, the micro-hardness is HV3200. And the constant-pressure specific heat can be expressed as $C_p = 49.957 + 0.962 \times 10^{-3}T - 14.770 \times 10^5 T^{-2} + 1.883 \times 10^{-6} T^2$ J/(K·mol) (298–3290 K), the formation enthalpy (ΔH) at 298 K is -184.096 kJ/mol [5].

In addition, at ambient temperature, the lattice constant of TiN is 0.4239 nm, the mole volume is 1.147×10^{-5} m³/mol, the theoretical density is 5.398 g/cm³, the coefficient of linear expansion is 79.35×10^{-6}/K (25–1100 °C), the melting point is 2950 °C, the elastic modulus E is 3.17×10^5 MPa, the micro-hardness is HV2450. And the constant-pressure specific heat can be expressed as $C_p = 49.831 + 3.933 \times 10^{-3}T - 12.385 \times 10^5 T^{-2}$ J/(K·mol) (298–3223 K), the formation enthalpy (ΔH) at 298 K is -337.858 kJ/mol [5].

As the TiC and TiN phases have the same crystal structure and the difference of lattice constant between them is very small, both of them can be miscible to form the carbonitride of TiC_xN_{1-x} ($0 \leq x \leq 1$). The carbon atom and nitrogen atom in TiC_xN_{1-x} can exchange with arbitrary ratio. The lattice constant and related physical properties vary with the chemical coefficient x. Generally, the lattice constant of TiC_xN_{1-x} can be estimated by linear interpolation method.

TiS has a hexagonal crystal structure with NiAs type (As shown in Fig. 3.15, Ti: 2a (0; 0; 0), S: 2c (1/3; 2/3; 1/4)). The lattice constant at ambient temperature are $a = 0.341$ nm, $c = 0.570$ nm. The mole volume is 1.728×10^{-5} m³/mol. The theoretical density is 4.625 g/cm³. The melting point is 1927 °C, And the constant-pressure specific heat can be expressed as $C_p = 45.898 + 7.364 \times 10^{-3}T$ J/(K·mol) (298–3223 K), the formation enthalpy (ΔH) at 298 K is -271.960 kJ/mol [5].

$Ti_4C_2S_2$ has a hexagonal crystal structure with $Ti_4C_2S_2$ type (As shown in Fig. 3.16, Ti: 4e(0; 0; z), z =0:1, C: 2a(0; 0; 0), S: 2d(1/3; 2/3; 3/4)) [36]. The lattice constants at ambient temperature are $a = 0.3209$ nm, $c = 1.1210$ nm. The mole volume is 3.0102×10^{-5} m³/mol. The theoretical density is 4.644 g/cm³. The

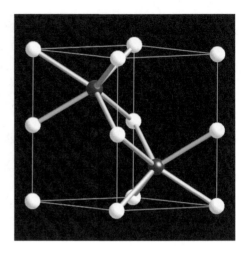

Fig. 3.15 The crystal structure of TiS. The white atom represents Ti atom, the purple atom represents S atom

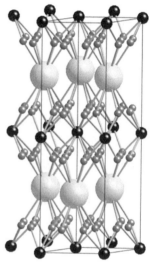

Fig. 3.16 The crystal structure of Ti$_2$CS. The black atom represents C atom, the yellow atom represents S atom, the blue atom represents Ti atom

coefficients of linear expansion at a and c directions are 8.55×10^{-6}/K (25–949 °C) and 8.82×10^{-6}/K (25–949 °C), respectively.

When the carbide, nitride and carbonitride precipitated in steel, they will have a certain orientation relationship with iron matrix, such as parallel orientation relationship with austenite [37], the Baker-Nutting relationship with ferrite [38]. And the TiC, TiN and TiCN also maintain the same orientation relationship with iron matrix as follows.

$$(100)_{MCN}//(100)_\gamma, \quad [010]_{MCN}//[010]_\gamma$$

$$(100)_{MCN}//(100)_\alpha, \quad [011]_{MCN}//[010]_\alpha$$

Therefore, according to the misfit dislocation theory, the specific interfacial energy of semicoherent interface between TiC (or TiN) and austenite can be calculated by using the interfacial misfit dislocation energy [39]. Because of the TiC and TiN have the parallel orientation relationship with austenite, and the TiC, TiN and austenite are fcc structure. Thus, the misfit degree between TiC (or TiN) and austenite in every direction is the same, leading to the specific interfacial energy in every direction is the same. The energy of dislocation is in direct proportion to the elastic modulus, and the elastic modulus of austenite has a strict linear relationship with temperature which decreases with the temperature increases. Thus, the interfacial energy between carbonitride and austenite expresses to a linear decrease trend with the temperature increasing. Therefore, the specific interfacial energy between TiC (or TiN) and austenite can be obtained by following formulas.

$$\sigma_{TiC-\gamma}(J/m^2) = 1.2360 - 0.5570 \times 10^{-3} T(K) \quad (3.72)$$

$$\sigma_{TiN-\gamma}(J/m^2) = 1.1803 - 0.5318 \times 10^{-3} T(K) \quad (3.73)$$

As the specific interfacial energy between TiC (or TiN) and austenite is the same in every direction, the TiC and TiN will exhibit a spherical or cubic shape when they precipitated in austenite. The TiN which precipitated at high temperature usually exhibits a spherical shape and the Ti(C, N) or TiC which precipitated at lower temperature usually exhibits a spherical shape. Simultaneously, the lattice constant of carbide is slightly larger than that of nitride, leading to a larger misfit degree between austenite. Thus, the interfacial energy between TiC and austenite will slightly larger than the interfacial energy between TiN and austenite.

Table 3.1 shows the calculation results of specific interfacial energy between various carbides (or nitrides) and austenite. It can be seen that the value of specific interfacial energy decreases with the temperature increasing due to the elastic modulus of austenite significant decreases with the temperature increasing, and the variation is quite large. Thus, in many literatures, it is obviously inappropriate to regard the specific interfacial energy as a constant value which is independent of temperature. If the variation of interfacial energy with the temperature was not considered, the calculation results of precipitation kinetic and Ostwald ripening process will be unreasonable. In addition, the specific interfacial energy between various carbide (or nitride) and austenite is in the range of 0.3–0.7 J/m². It is in agreement with the estimated value of specific interfacial energy of semicoherent interface and near to the upper limit. This attributes to the larger misfit degree of interplanar spacing between carbide (or nitride) and austenite. In addition, it can be also found that the specific interfacial energy of carbide is slightly larger than that of nitrogen for the same microalloying element that attribute to larger misfit degree between carbide and austenite as the lattice constant of carbide is slightly larger than that of nitride. From the comparison of various microalloying elements, it can be found that the vanadium carbide (or nitride) has the smallest interfacial energy with the austenite, followed by titanium carbide (or nitride), while the niobium carbide (or nitride) is the largest. It is related to the value of their lattice constants. Finally, the energy of

Table 3.1 The calculation results of specific interfacial energy of semicoherent interface between carbonitrides and austenite (J/m²)

Temperature (°C)	VC	TiC	NbC	VN	TiN	NbN
850	0.5578	0.6105	0.6636	0.5374	0.5831	0.6420
900	0.5324	0.5826	0.6334	0.5129	0.5565	0.6128
950	0.5069	0.5548	0.6031	0.4884	0.5299	0.5835
1000	0.4815	0.5269	0.5728	0.4639	0.5033	0.5542
1050	0.4561	0.4991	0.5426	0.4394	0.4767	0.5249
1100	0.4306	0.4712	0.5123	0.4149	0.4501	0.4956
1150	0.4052	0.4434	0.4820	0.3903	0.4235	0.4663
1200	0.3797	0.4155	0.4517	0.3658	0.3970	0.4370
1250	0.3543	0.3877	0.4215	0.3413	0.3704	0.4077
1300	0.3289	0.3598	0.3912	0.3168	0.3438	0.3784

high angle grain boundary in austenite at 1100 °C was measured as 0.756 J/m². And the specific interfacial energy between various carbides (or nitrides) and austenite is about 55–68% of the measured value (0.756 J/m²) which is consistent with the estimation range.

Moreover, the semi-coherent interfacial energy of TiC and TiN with ferrite can also be calculated by using a similar theoretical method [40]. Because of the misfit degrees in different directions are different, then the specific interfacial energy in different directions are different, leading to the TiC and TiN precipitated in ferrite exhibiting a disk shape. The bottom surface is $(100)_{MCN}//(100)_\alpha$, and the diameter to thickness ratio and the value of interfacial energy of side surface to the interfacial energy of bottom surface ratio can be regarded as constants. It is 1.816 for TiC and 2.049 for TiN. This shape characteristic can be clearly observed by means of transmission electron microscope (TEM) through rotating the electron beam leading to the side surface of precipitation paralleled with observation face. Then the diameter to thickness ratio can be measured. And actual observation results are in good agreement with the theoretical predictions. By using the difference of the shape characteristics, the carbonitride precipitated in austenite or ferrite can be effectively distinguished.

In addition, the variation of elastic modulus of ferrite with temperature is complicated which will significantly deviate from linearity in the temperature range of ferromagnetic transition. Thus, the specific interfacial energy between TiC (or TiN) and ferrite can be deduced in the temperature range where the elastic modulus varies linearly.

$$\sigma_{1TiC-\alpha} = 0.7487 - 0.2488 \times 10^{-3}T \quad (293-856\,\text{K}) \quad (3.74)$$

$$\sigma_{2TiC-\alpha} = 1.3593 - 0.4517 \times 10^{-3}T \quad (293-856\,\text{K}) \quad (3.75)$$

$$\sigma_{1TiN-\alpha} = 0.6160 - 0.2047 \times 10^{-3}T \quad (293-856\,\text{K}) \quad (3.76)$$

$$\sigma_{2TiN-\alpha} = 1.2623 - 0.4195 \times 10^{-3}T \quad (293-856\,\text{K}) \quad (3.77)$$

where σ_1 is the specific interfacial energy of bottom surface, σ_2 is the specific interfacial energy of side surface, respectively.

As the σ_2 is larger than σ_2, in order to reduce the energy, the TiC and TiN precipitation will exhibit a disk shape that the diameter is larger than thickness. Moreover, the specific interfacial energy in a higher temperature range can be calculated by using the corresponding elastic modulus of ferrite

Table 3.2 shows the calculated results of the specific interfacial energy between various carbides (or nitride) and austenite. In the Table 3.2, the results were calculated by the formulas as mentioned above when the temperature is lower than 550 °C. And when temperature is equal to 600 °C or larger than 600 °C, the results were calculated by using the actual linear interpolation of elastic modulus. It can be found from the calculated results, that the specific interfacial energy of carbide is slightly larger than that of nitrogen for the same microalloying element that be attributed to larger misfit degree between carbide and ferrite as the lattice constant of carbide is slightly larger than that of nitride. From the comparison of various microalloying elements, it can be found that the vanadium carbide (or nitride) has the smallest interfacial energy with the austenite, followed by titanium carbide (or nitride), while the niobium carbide (or nitride) is the largest. It is related to the value of their lattice constants. In addition, the specific interfacial energy varies significantly with the temperature in the range of 550–800 °C. And the temperature range is just the effective temperature range for the carbonitride precipitation. Thus, the variation of the precipitation behavior of carbonitrides is significant in this temperature range due to the specific interfacial energy varying with the temperature. Finally, the calculation results in Table 3.2 is consistent with the estimation range of specific interfacial energy of semi-coherent interface.

According to the chemical coefficient x in $Ti(C_x, N_{1-x})$, the specific interfacial energy between $Ti(C_x, N_{1-x})$ and austenite of ferrite can be calculated by linear interpolation method of lattice constant.

TiS and $Ti_4C_2S_2$ can only precipitate in austenite. Both of them were hexagonal structure. The orientation relationship between them and austenite is difficult to be accurately identified and there is no relevant report up to now. Thus, the specific interfacial energy of semi-coherent interface between TiS (or $Ti_4C_2S_2$) and austenite is difficult to be estimated. In this book, we regarded the specific interfacial energy as 0.5 J/m^2 to calculate PTT curve, while assuming that the specific interfacial energy was in direct proportion to the elastic modulus of austenite and considering the effect of temperature on specific interfacial energy.

Table 3.2 The calculation results of specific interfacial energy of semicoherent interface between carbonitrides and ferrite (J/m^2)

Temperature (°C)		VC	TiC	NbC	VN	TiN	NbN
20	σ_1	0.4453	0.6758	0.8443	0.3330	0.5560	0.7682
	σ_2	1.0637	1.2269	1.3637	0.9910	1.1394	1.2997
100	σ_1	0.4322	0.6558	0.8195	0.3232	0.5396	0.7455
	σ_2	1.0323	1.1908	1.3235	0.9618	1.1058	1.2614
200	σ_1	0.4158	0.6310	0.7884	0.3109	0.5192	0.7173
	σ_2	0.9932	1.1456	1.2733	0.9253	1.0638	1.2135
300	σ_1	0.3994	0.6061	0.7573	0.2986	0.4987	0.6890
	σ_2	0.9540	1.1005	1.2231	0.8888	1.0219	1.1657
400	σ_1	0.3830	0.5812	0.7262	0.2864	0.4782	0.6607
	σ_2	0.9149	1.0553	1.1729	0.8523	0.9800	1.1178
500	σ_1	0.3666	0.5563	0.6951	0.2741	0.4578	0.6324
	σ_2	0.8757	1.0101	1.1227	0.8158	0.9380	1.0700
550	σ_1	0.3584	0.5439	0.6796	0.2680	0.4475	0.6183
	σ_2	0.8561	0.9875	1.0976	0.7976	0.9170	1.0461
600	σ_1	0.3448	0.5233	0.6539	0.2579	0.4306	0.5949
	σ_2	0.8237	0.9502	1.0561	0.7674	0.8824	1.0065
650	σ_1	0.3177	0.4821	0.6023	0.2376	0.3967	0.5480
	σ_2	0.7588	0.8753	0.9728	0.7069	0.8128	0.9272
700	σ_1	0.2905	0.4408	0.5508	0.2172	0.3627	0.5011
	σ_2	0.6939	0.8004	0.8896	0.6464	0.7432	0.8478
750	σ_1	0.2633	0.3996	0.4992	0.1969	0.3288	0.4542
	σ_2	0.6289	0.7255	0.8063	0.5859	0.6737	0.7685
800	σ_1	0.2384	0.3619	0.4521	0.1783	0.2977	0.4114
	σ_2	0.5696	0.6570	0.7302	0.5307	0.6101	0.6960
850	σ_1	0.2276	0.3454	0.4316	0.1702	0.2842	0.3926
	σ_2	0.5437	0.6272	0.6970	0.5065	0.5824	0.6643
900	σ_1	0.2168	0.4110	0.3382	0.1621	0.2707	0.3739
	σ_2	0.5178	0.6638	0.5463	0.4824	0.5546	0.6327

3.3 Kinetics Analysis of Titanium-Bearing Precipitates

3.3.1 Kinetic Theory on Precipitation Phase Transformation

According to the classical nucleation theory, the critical size of spherical nucleus by homogeneous nucleating, d^*, is expressed as:

$$d^* = -\frac{4\sigma}{\Delta G_V + \Delta G_{EV}} \quad (3.78)$$

The critical nucleation energy ΔG^* is described as follows:

$$\Delta G^* = \frac{16\pi\sigma^3}{3(\Delta G_V + \Delta G_{EV})^2} \quad (3.79)$$

In Eqs. (3.78) and (3.79), ΔG_V is the unit volume free energy, ΔG_{EV} is the unit volume elastic strain energy produced by new phase, σ is the interfacial free energy between the new phase and the parent phase.

The homogeneous nucleation rate per unit volume can be expressed as (I):

$$I = n_V a^* pv \exp(-\frac{\Delta G^*}{kT}) \exp(-\frac{Q}{kT}) \quad (3.80)$$

In Eq. (3.80) n_V is the number of nucleation sites per unit volume, a^* denotes the number of atoms on the surface of a critical nucleus, V refers to the jumping frequency of atom in matrix, Q is the activation energy of the controlled atom diffusion, p represents the probability product of atom jumping, K is Boltzmann constant, and T is the transformation temperature.

All items independent of temperature are summarized in a constant K, $K = n_V \alpha pv$, the nucleation rate can be thus expressed as:

$$I = K \cdot d^{*2} \cdot \exp(-\frac{\Delta G^* + Q}{kT}) \quad (3.81)$$

The precipitation kinetics can be described by empirical Avrami kinetics equation:

$$X = 1 - \exp(-Bt^n) \quad (3.82)$$

In Eq. (3.82), t is the time, B is a factor related to the nucleation rate and the core growth rate, and n is a time index of nucleation rate.

When the nucleation rate and the nucleus growth rate are constants and independent of time, $n = 4$, Johnson Mehl equation is obtained. While the nucleation rate rapidly decreases to zero, the time exponent will be reduced to 1. For the diffusion-controlled transformation, because the unidirectional dimension is proportional to the square root of time, while the growth rate is inversely proportional to square root of time and the diffusion time exponent of each dimension is 0.5. When the nucleation rate is a constant, the time exponent of homogeneous nucleation process is 2.5 for the three dimensions growth, and the time exponent is reduced by 0.5 when the growth dimension is reduced to one dimension. However, while the nucleation rate rapidly reduced to 0, the time index of the homogeneous nucleation is 1.5 for 3D growth and the time index is reduced by 0.5 for 1D growth.

The critical nucleation energy is significantly reduced when the new phase is heterogeneous nucleation on the grain boundaries. In the early stage, The microalloyed carbonitrides are usually nucleated on the grain boundaries. At this time,

$$\Delta G_g^* = A_1 \Delta G^* = \frac{1}{2}(2 - 3\cos\theta + \cos^3\theta)\Delta G^* \qquad (3.83)$$

$$\cos\theta = \frac{1}{2} \cdot \frac{\sigma_B}{\sigma} \qquad (3.84)$$

where, the subscript g is nucleated along grain boundary, θ is the contact angle between the new phase nucleus of the biconvex lens and the grain boundary of patent phase, σ_B is the grain boundary energy of parent phase, σ is specific interfacial energy between the new phase and the parent phase.

If δ is the thickness of grain boundary and L is the average grain size, then, the volume fraction of grain boundaries in the parent phase is roughly proportional to (δ/L). Hence, the nucleation rate of grain boundary (I_g) which can be expressed in the following equation form.

$$I_g = n_V a^* p v \frac{\delta}{L} \exp(-\frac{Q_g}{kT}) \exp(-\frac{A_1\Delta G^*}{kT}) = K d^{*2} \frac{\delta}{L} \exp(-\frac{A_1\Delta G^*}{kT}) \cdot \exp(-\frac{Q_g}{kT}) \qquad (3.85)$$

where Q_g is the activation energy of diffusion along grain boundaries of the controlled atoms, and it is usually assumed that the diffusion activation energy along grain boundaries is equal to half of the bulk diffusion activation energy inside grains. The latter was determined from the following equation:

$$I_g/K = \frac{\delta}{L} d^{*2} \exp(-\frac{Q}{2kT}) \exp(-\frac{A_1\Delta G^*}{kT}) \qquad (3.86)$$

Microalloying carbonitrides precipitated from ferrite matrix are mainly nucleated on dislocation lines. Based on our correction of the nucleation theory on the dislocation line proposed by Yong [41].

$$d_d^* = -\frac{2\sigma}{\Delta G_V}[1 + (1+\beta)^{1/2}] \qquad (3.87)$$

$$\Delta G_d^* = (1+\beta)^{3/2}\Delta G^* \qquad (3.88)$$

$$\beta = \frac{A\Delta G_V}{2\pi\sigma^2} = \frac{1}{4}\alpha \qquad (3.89)$$

where the subscript d represents precipitates nucleated on dislocation line; β is a nucleated parameter introduced for a dislocation line, which is substituted for α parameter in the Cahn's theory, $\beta = -1$–0; A is a coefficient, $A = Gb^2/[4\pi(1-$

$\nu)]$ (edge dislocation), or $A = Gb^2/(4\pi)$ (screw dislocation); G is the shear elastic modulus of matrix; ν is the poisson's ratio; b is the Burgers vector.

If $\beta = 0$, the precipitates are homogeneous nucleation; If $\beta \leq -1$, $\beta = -1$ is then taken, and the critical size of the nucleus is half of that of homogeneous nucleation. The critical nucleation energy is thus zero, and the nucleation is spontaneous.

If the dislocation density is ρ and the diameter of dislocation nucleus is $2b$, the volume fraction of dislocations in the patent phase is about $\pi \rho b^2$, and the dislocation nucleation rate I_d can be written in the following equation form.

$$\begin{aligned} I_d &= n_V a_d^* p \nu \cdot \pi \rho b^2 \exp(-\frac{Q_d}{kT}) \exp(-\frac{(1+\beta)^{3/2} \Delta G^*}{kT}) \\ &= K \pi \rho b^2 d_d^{*2} \exp(-\frac{Q_d}{kT}) \exp(-\frac{(1+\beta)^{3/2} \Delta G^*}{kT}) \end{aligned} \quad (3.90)$$

where, Q_d is the diffusion activation energy of controlling atoms along dislocation lines. It is assumed that the diffusion activation energy along dislocation lines is equal to 2/3 of the bulk diffusion activation energy inside grains. The following expression is obtained:

$$I_d/K = \pi \rho b^2 d_d^{*2} \exp(-\frac{2Q}{3kT}) \exp\{-\frac{(1+\beta)^{3/2} \Delta G^*}{kT}\} \quad (3.91)$$

Equation (3.82) is taken in the form of iterated logarithm, and the amount of transformation 5% is defined as the start time of transformation $t_{0.05}$:

$$\log t_{0.05} = \frac{1}{n}(-1.28994 - \log B) \quad (3.92)$$

When the new phase is homogeneous nucleation and the nucleation rate is constant, then the dimension of the precipitation nucleus is 3D, and $n = 2.5$, so B is now expressed in the following equation.

$$B = \frac{8\pi}{15} I \lambda^3 D^{3/2} = \frac{8\pi}{15} K D_0^{3/2} [\frac{2(C_0 - C_M)}{(C_N - C_M)}]^{3/2} d^{*2} \exp(-\frac{\Delta G^* + 2.5Q}{kT}) \quad (3.93)$$

$$\log t_{0.05} = \frac{2}{5}(-1.28994 - \log C - 2 \log d^* + \frac{1}{\ln 10} \frac{\Delta G^* + 2.5Q}{kT}) \quad (3.94)$$

where, λ is a scale coefficient, $\lambda = (-k)^{1/2} = [\frac{2(C_0 - C_M)}{C_N - C_M}]^{1/2}$; C_0 is the mean atomic concentration of solute atoms; C_M and C_N are atom concentrations of solute atoms along phase interface in the parent phase and the precipitate phase, respectively; D is the diffusion coefficient of solute atoms in matrix; D_0 is a constant independent of temperature in diffusion coefficients; C is a constant independent of temperature (It is assumed that the function of the degree of supersaturation of solutes is basically independent of temperature).

$$C = \frac{8\pi}{15} K D_0^{3/2} [\frac{2(C_0 - C_M)}{(C_N - C_M)}]^{3/2}$$

When $\log t_0 = 2/5 \log C$, so:

$$\log \frac{t_{0.05}}{t_0} = \frac{2}{5}(-1.28994 - 2\log d^* + \frac{1}{\ln 10} \frac{\Delta G^* + 2.5Q}{kT}) \quad (3.95)$$

When the new phase is homogeneously nucleated in the matrix and the nucleation rate dramatically decreased to 0, the time exponent decreases to 1 and $n = 1.5$. The following expression is obtained:

$$B_a = \frac{4\pi}{3} I \tau_1 \lambda^3 D^{3/2} = \frac{4\pi}{3} K D_0^{3/2} [\frac{2(C_0 - C_M)}{(C_N - C_M)}]^{3/2} \tau_1 d^{*2} \exp(-\frac{\Delta G^* + 2.5Q}{kT}) \quad (3.96)$$

where, the subscript a represents the condition that the nucleation rate rapidly reduce to 0; τ is effective nucleation time which still can not be calculated accurately.

When $\log t_{0a} = \frac{2}{3} \log C + \frac{2}{3} \log \frac{5\tau_1}{2}$, then, the following expression is given as:

$$\log \frac{t_{0.05a}}{t_{0a}} = \frac{2}{3}(-1.28994 - 2\log d^* + \frac{1}{\ln 10} \frac{\Delta G^* + 2.5Q}{kT}) \quad (3.97)$$

When the new phase is nucleated at grain boundary and the nucleation rate is constant, solute atoms diffuse from the matrix to grain boundary as the nucleus grows up. Then, the dimension of the nucleus growth is 1D and $n = 1.5$. The B_g is thus given by:

$$B_g = \frac{4}{3} I_g A \lambda_1 D^{1/2}$$
$$= \frac{4}{3} K A D_0^{1/2} [\frac{2(C_0 - C_M)}{\pi^{1/2}(C_N - C_M)}] \frac{\delta}{L} d^{*2} \exp(-\frac{A_1 \Delta G^* + Q_g + 0.5Q}{kT}) \quad (3.98)$$

where λ_1 is scale coefficient; $\lambda_1 = -k/\pi^{1/2} = \frac{2(C_0-C_M)}{\pi^{1/2}(C_N-C_M)}$; A is grain boundary area.

Assuming the diffusion activation energy of the solute along grain boundary is 1/2 of that in the matrix, Thus, if $\log t_{0g} = \frac{2}{3} \log C + \frac{2}{3} \log \frac{5A\delta}{2\pi^{3/2} L D_0 \cdot [\frac{2(C_0-C_M)}{C_N-C_M}]^{1/2}}$, then, the following equation is expressed as:

$$\log \frac{t_{0.05g}}{t_{0g}} = \frac{2}{3}(-1.28994 - 2\log d^* + \frac{1}{\ln 10} \frac{A_1 \Delta G^* + Q}{kT}) \quad (3.99)$$

The time exponent decreases to 1 when the new phase nucleated on the grain boundaries, and the nucleation rate rapidly decreases to 0 and $n = 0.5$. So, the following expression is suggested as:

$$B_{\text{ga}} = 2I_g\tau_1 A\lambda D^{1/2} = 2K\tau_{1g}AD_0^{1/2}[\frac{2(C_0 - C_M)}{\pi^{1/2}(C_N - C_M)}]\frac{\delta}{L}d^{*2}\exp(-\frac{A_1\Delta G^* + Q}{kT}) \quad (3.100)$$

When $\log t_{0\text{ga}} = 2\log C + 2\log \frac{15A\delta\tau_{1g}}{4\pi^{3/2}LD_0\cdot[\frac{2(C_0-C_M)}{C_N-C_M}]^{1/2}}$, it is given as:

$$\log\frac{t_{0.05\text{ga}}}{t_{0\text{ga}}} = 2(-1.28994 - 2\log d^* + \frac{1}{\ln 10}\frac{A_1\Delta G^* + Q}{kT}) \quad (3.101)$$

When a new phase nucleated on dislocation lines and its nucleation rate is a constant, solute atoms diffuse from the matrix to dislocation lines as the nucleus grows up. Then, the dimension of the nucleus growth is 2D, and $n = 2$. The B_d can be thus written as:

$$B_d = \frac{\pi}{2}I_d l\lambda_2^2 D = \frac{\pi^2}{2}K\rho b^2 D_0 l\lambda_2^2 d_d^{*2}\exp(-\frac{(1+\beta)^{3/2}\Delta G* + Q_d + Q}{kT}) \quad (3.102)$$

where, l is the length of dislocation line; λ_2 is a rate coefficient of the nucleus growth in 2D (which is associated with the supersaturation and has some difference from the expression of λ and λ_1).

Supposing that the diffusion activation energy on the dislocation lines is 2/3 of the bulk diffusion activation energy, and let $\log t_{0d} = \frac{1}{2}\log C + \frac{1}{2}\log \frac{15\pi\rho b^2 l\lambda_2^2}{16D_0^{1/2}\lambda^3}$. Then, it is obtained as:

$$\log\frac{t_{0.05d}}{t_{0d}} = \frac{1}{2}(-1.28994 - 2\log d_d^* + \frac{1}{\ln 10}\frac{(1+\beta)^{3/2}\Delta G^* + \frac{5}{3}Q}{kT}) \quad (3.103)$$

When the secondary phase nucleated on the dislocation lines and its nucleation rate dramatically decreases to 0, the dimension of the nucleus growth is 2D and $n = 1$:

$$B_{\text{da}} = \pi I_d\tau_1 l\lambda_2^2 D = \pi^2 K\rho b^2 D_0 l\lambda_2^2 d_d^{*2}\exp(-\frac{(1+\beta)^{3/2}\Delta G^* + Q_d + Q}{kT}) \quad (3.104)$$

where, the subscript a represents the condition that the nucleation rate rapidly reduces to 0; τ_1 is the effective nucleation time and hard to be accurately calculated, which is remarkably different from those of homogeneous nucleation and grain boundary nucleation.

Similarly, $-\log t_{0\text{da}} = \log C - \log \frac{15\pi\tau_1\rho b^2 l\lambda_2^2}{8D_0^{1/2}\lambda^3}$. It is obtained as:

$$\log\frac{t_{0.05\text{da}}}{t_{0\text{da}}} = -1.28994 - 2\log d_d^* + \frac{1}{\ln 10}\frac{(1+\beta)^{3/2}\Delta G^* + \frac{5}{3}Q}{kT} \quad (3.105)$$

The precipitation process of various Ti containing precipitates of steels is that precipitates firstly nucleated on grain boundaries and their nucleation rate rapidly decreases to 0. However, grain boundary nucleation can be only partially finished because of the limited amount of solute atoms on or near the grain boundaries. After that, the nucleus is mainly nucleated on dislocation lines and its nucleation rate rapidly decreases to 0. Therefore, the homogeneous nucleation rarely happens. Thus, using Eq. (3.91) to calculate the relative nucleation rate as a function of temperature of different Ti containing precipitates, and then the nucleation-temperature (NrT) curve can be obtained by connecting the various relative nucleation point of different temperatures. The start time of transformation has been calculated at different temperatures using Eq. (3.105), and PTT curve can be plotted by the values of various start times of transformation. If necessary, NrT and relative nucleation rate of precipitates nucleated on grain boundaries can be calculated using Eqs. (3.86), and (3.101) can also be used to determine PTT curve and the start time of transformation.

As the precipitates nucleated on the dislocation and their nucleation rates rapidly decreases to 0, the average size of precipitates is inversely proportional to the $1/(2n)$ power of nucleation rate when the phase transformation is complete. So the nucleation rate can be written as:

$$\bar{d}_f \propto I_d^{-\frac{1}{2n}} \qquad (3.106)$$

Therefore, the finest precipitates could be obtained if the precipitation process occurs at the temperature of maximum nucleation rate (that is the nose temperature in NrT curve).

The effective precipitation temperature of TiN or the N-rich Ti(C, N) is very high and the atoms diffuse significantly fast, so the precipitation transformation is easily to finished, and the precipitates continue to grow as the high holding time increases. Therefore, the final size of the precipitates depends on the actual precipitation temperature. The higher the temperature, the bigger size the precipitates. At this moment, increasing the cooling rate will decrease the precipitation temperature of TiN and Ti(C, N) particles, and it is beneficial and important to achieve fine precipitates.

3.3.2 Analysis and Calculation About Kinetics of Titanium-Bearing Carbides

For the calculation of 0.10%C–0.02%Ti–0.004%N steel and 0.10%C–0.02%Ti–0.008%N steel, TiS and $Ti_4S_2C_2$ are hardly to precipitate due to the low content of Ti, so the main consideration is the precipitation behaviors of Ti(C, N). Using the solubility products of Eqs. (3.7) and (3.17), the total solid solution temperatures of two steels with different compositions were 1438 and 1501 °C, respectively. Variation of the chemical formula of $Ti(C_x, N_{1-x})$ as a function of temperature has been calculated via the formulas of the solubility products and the ideal chemical proportioning, as indicated in Fig. 3.17.

Fig. 3.17 Variation of x value of TiC_xN_{1-x} as a function of temperature in a trace Ti-treated steel

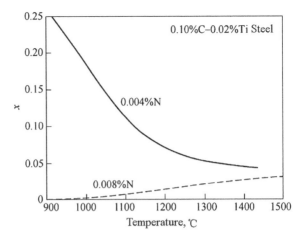

Fig. 3.18 PTT curves of Ti(C, N) in a trace Ti-treated steel

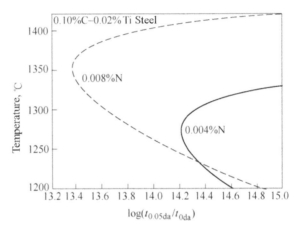

When the ratio of Ti/N is higher than that of the ideal chemical ratio, the chemical coefficients of $Ti(C_x, N_{1-x})$ decreases with the increase of temperature. As the temperature is near the solid solution temperature, the x value is very small and the precipitates are almost TiN. While the ratio of Ti/N is lower than that of the ideal chemical ratio, the chemical coefficient x is very small and it increases with the increase of temperature, and the precipitates are very close to TiN at all temperatures. Figure 3.18 shows the PTT curve of Ti(C, N), and it is seen that the nose temperatures of 0.004%N steel and 0.008%N steel are 1272 and 1353 °C, respectively, which is 166 and 148 °C lower than their solid solution temperature. Moreover, the precipitation start time in 0.008%N steel is 0.835 order of magnitude faster than that of 0.004%N steel.

The precipitation behaviors of Ti microalloyed steel with higher content of Ti are very complex [42]. Taking the calculation results of 0.10%C–0.08%Ti–0.004%N steel and 0.10%C–0.08%Ti–0.008%N as examples, the solid solution temperatures

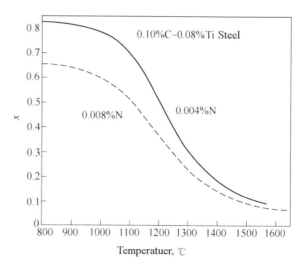

Fig. 3.19 Variation of coefficients of mole fraction of Ti(C, N) versus temperature in Ti microalloyed steels

of the two steels were 1577 and 1650 °C, respectively. Meanwhile, the change of coefficient of chemical formula Ti(C, N) as a function of temperature has been calculated by the solubility products formula and the ideal stoichiometry on the composition design, as shown in Fig. 3.19. Since the ratio of Ti/N is greater than 2, the coefficient of chemical formula x decreases monotonically as the temperature increases. The x value is less than 0.06 when the temperature reaches the total solid solution temperature. In other words, the precipitates are N-rich particles which are close to TiN.

Figure 3.20 shows the PTT diagram of Ti(C, N). The nose temperatures with 0.004%N steel and 0.008%N steel are 1443 and 1544 °C, which is 134 and 106 °C lower than their total solid solution temperatures, respectively. In addition, the fastest precipitation start time of the steel with 0.008%N is 0.898 order of magnitude faster than that of the steel with 0.004%N. Furthermore, the fastest precipitation start time of Ti microalloyed steel is 1.6 order of magnitude faster than that of the previously mentioned trace Ti-treated steel. This suggests that Ti(C, N) will be quickly precipitated. Obviously, since the size of Ti(C, N) precipitated at high temperature is mainly dependent on the precipitation temperature, the size of Ti(C, N) precipitated at high temperature of Ti microalloyed is larger than that of the trace Ti-treated steel.

Ti-microalloyed steels are usually rolled after heating at 1250 °C, and the Ti(C, N) precipitates and matrix can roughly be balanced during the soaking process. The solid solubility of [Ti], [C] and [N] and the amount of Ti(C, N) precipitated can be calculated by the Eqs. (3.60)–(3.63). For the 0.10%C–0.08%Ti–0.004%N and 0.10%C–0.08%Ti–0.008%N steels, the soluble contents of [Ti], [C] and [N] at 1250 °C are 0.05780703, 0.0977908, 0.000082915 and 0.04222654, 0.0972680, 0.000132754%, respectively, and they can be used as the initial concentrations of carbonitride precipitated in the subsequent rolling process.

Fig. 3.20 Precipitation-time-temperature curves of Ti(C, N) precipitated at high temperature of Ti microalloyed steels

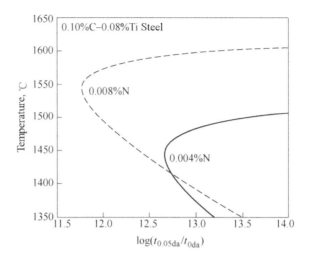

The amounts of Ti(C, N) precipitated (undissolved Ti(C, N)) in 0.004%N steel and 0.008%N steel are 0.02472 and 0.04117%, respectively. Despite of the relatively larger size of Ti(C, N) at high temperature, in comparison with that of the trace Ti-treated steel, it is essential to restrict the austenite grain growth because of the larger amount of Ti(C, N) in Ti microalloyed steel.

Taking the equilibrium solid solubility of [Ti], [C] and [N] at 1250 °C as the initial concentrations, it is clear that the coefficient of chemical formula Ti(C, N) with the temperature can be calculated in subsequent cooling and rolling process. As shown in Fig. 3.21, it is seen that the x value of Ti(C, N) precipitates is less than 0.98 as the temperature below 1000 °C. Therefore, it is usually believed that the precipitates are pure TiC during the subsequent rolling process.

Figure 3.22 shows the PTT curve of Ti(C, N) during the cooling and rolling process for the 0.10%C–0.08%Ti–0.004%N and 0.10%C–0.08%Ti–0.008%N steels. The content of Ti in Ti-containing precipitates is more than that in solid solution when the temperature is above 1250 °C, and the precipitation kinetics of Ti(C, N) is somewhat slow in subsequent cooling process. Moreover, the fastest precipitation start time of 0.008%N steel is 1.052 order of magnitude slower than that of 0.004%N steel, and the nose temperature of 0.008%N steel is relatively low. The nose temperature of 0.008%N steel and 0.004%N steel are 725 and 692 °C. Obviously, the effective temperature range of Ti(C, N) nucleated on the dislocation lines is lower than that of austenite matrix. This characteristic of Ti microalloyed steel is significantly different from that of Nb microalloyed steel, but similar to that of V microalloyed steel.

However, previous studies [42] showed that some Ti(C, N) particles existed at the temperature above 900 °C in Ti microalloyed steel with the Ti content less than 0.08%. It is presumed that this was related to Ti(C, N) nucleated on grain boundary. To explain it, the PTT curve of Ti(C, N) precipitates in austenite grain boundary also has been calculated, as shown in Fig. 3.23. Meanwhile, the carbonitrides of the steels with 0.004%N steel occurred earlier than that of the 0.008%N steel because of the

Fig. 3.21 Variation of coefficients of chemical formula Ti(C, N) versus temperature in Ti microalloyed steels during the cooling process after rolling

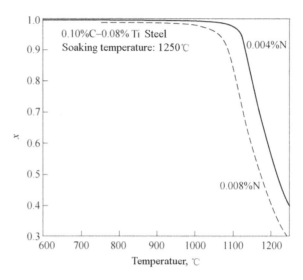

Fig. 3.22 PTT curves of Ti(C, N) precipitated on the dislocation lines during the cooling process after soaking at 1250 °C

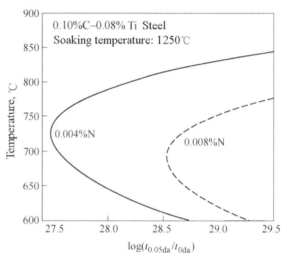

amount of precipitates at high temperature. The nose temperatures of the 0.004%N steel and 0.008%N steel are 943 and 910 °C, respectively. In other words, a certain amount of Ti(C, N) precipitates would precipitate at the temperature above 900 °C. Furthermore, the deformation stored energy introduced during the rolling process not only promotes the process of precipitation, but also improves the nose temperature of PTT curve of precipitates nucleated on the dislocation line or grain boundary. This is the reason why a amount of Ti(C, N) precipitates could be seen at the temperature above 900 °C.

The experimental results show that a large amount of Ti(C, N) precipitated from the austenite matrix and distributed uniformly inside the grains in the Ti micoralloyed

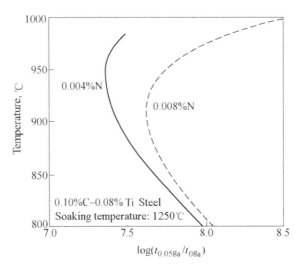

Fig. 3.23 PTT curves of Ti(C, N) precipitated at grain boundaries during the cooling process after soaking at 1250 °C

steel with high content of Ti [42]. It suggests that a considerable amount of Ti(C, N) are precipitated on the dislocation lines. For this purpose, the calculation of PTT curves of Ti(C, N) precipitated on the dislocation lines from austenite for the steels with 0.004%N and different Ti contents has been carried on. The mean solid solubility of N of the steels with different Ti contents were 0.000082195, 0.000037452, 0.000019902, 0.000011950 and 0.000007950%, which were lower than 0.0001%. The equation for solubility products of TiC in austenite was actually determined via the steel with the 0.0001%N, which had already considered the influence of trace Ti. As a result, the subsequent correlation calculation can be entirely based on the precipitation process of pure TiC. Combining the equation of solubility products with the equation of the ideal stoichiometric ratio, it is established that the solid solubilities of [Ti] and [C] (%) with steels with different Ti content were 0.08, 0.10; 0.12, 0.10; 0.147219, 0.0967929; 0.158888, 0.0896841; 0.171878, 0.0829065. Obviously, the total solid solution temperatures of steels with 0.08%Ti and 0.12%Ti were below 1250 °C, and the solid solubilities of [Ti] and [C] at 1250 °C are the same as their actual contents. The result of the PTT diagram of TiC nucleated on the dislocation lines during the cooling process after soaking at 1250 °C is shown in Fig. 3.24. The result of PTT curve of TiC nucleated at grain boundaries is given in Fig. 3.25. When the total solid solution temperature of TiC is higher than 1250 °C, the PTT curves of steels with different 0.16, 0.20, 0.24% Ti contents of were exactly the same, and the nose temperature of TiC nucleated on the dislocation lines is 855 °C, whereas the nose temperature in the grain boundaries was 1072 °C. The molar free energy of 0.08%Ti and 0.12%Ti steels was significantly reduced, so the PTT curve was obviously moving to the longer time and lower temperature.

The nose temperatures of precipitates on the dislocation lines of both steels were decreasing to 827 and 765 °C, and the fastest precipitation times were 0.76 and 2.55 orders of magnitudes slower than those of partially soluble steels. However, the

Fig. 3.24 Precipitation-time-temperature curves of TiC precipitated on the dislocation lines during the cooling process after soaking at 1250 °C

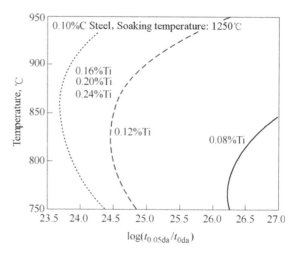

Fig. 3.25 PTT curves of TiC precipitated at grain boundaries during the cooling process after soaking at 1250 °C

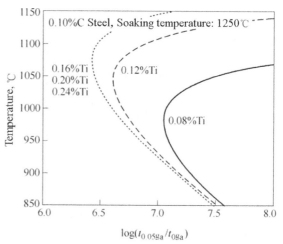

nose temperatures of precipitation on grain boundaries of two steels were reduced to 1044 and 983 °C, and the fastest precipitation times were 0.185 and 0.622 orders of magnitudes slower than those of the partially soluble steels. There is a centain amount of deformation stored energy in the austenite during the rolling process, and the deformation stored energy in Ti microalloyed steel is slightly less than that in Nb microslloyed steel. Here, the deformation stored energy of 3000 J/mol is adopted for calculation, and the PTT curves of TiC nucleated on the dislocation lines or on grain boundaries can be calculated during cooling process after soaking at 1250 °C and were shown in Figs. 3.26 and 3.27. Similarly, the steels with 0.16%Ti, 0.20%Ti and 0.24%Ti have the same PTT curves which the total solid solution temperature is above 1250 °C. The nose temperatures on dislocations and at grain boundaries of these steels were 902 and 1115 °C, respectively.

Fig. 3.26 PTT curves of TiC precipitated on dislocation lines during the cooling process after soaking at 1250 °C (with deformation stored energy)

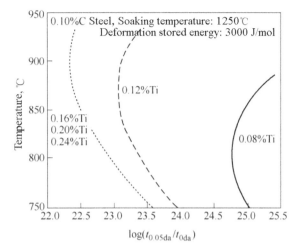

Fig. 3.27 PTT curves of TiC precipitated on grain boundaries during the cooling process after soaking at 1250 °C (with deformation stored energy)

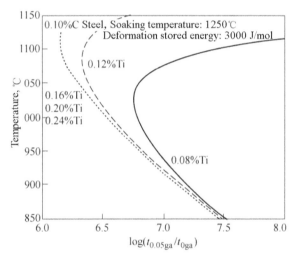

The molar free energies of 0.12%Ti and 0.08%Ti steels were significantly reduced, so the PTT curves were obviously moved to the longer time and lower temperature. The nose temperatures on the dislocation lines of both steels were decreased to 872 and 807 °C, and the fastest precipitation times were 0.725 and 2.43 orders of magnitude slower than those of partially soluble steels. However, the nose temperatures at grain boundaries of the two steels were reduced to 1086 and 1022 °C, and the fastest precipitation times were 0.180 and 0.603 orders of magnitudes slower than those of the partially soluble steels.

From the calculation results, the deformation stored energy has a obvious impact on the behaviors of TiC, especially for the steel with the Ti content higher than 0.16%, and the nose temperatures of TiC nucleated on the dislocation lines and on grain boundaries increases by 47 and 43 °C, respectively. The fastest precipitation start time is 1.35 and 0.18 time orders earlier respectively. For the 0.12%Ti and 0.08%Ti steels, the nose temperatures of PTT curves of precipitates nucleated on the dislocations and on grain boundaries increase by 45, 42 and 42, 39 °C, respectively, and the fastest precipitation start time is earlier by 1.39, 0.29 and 1.47 and 0.30 time orders of magnitude, respectively.

By analyzing the experimental and actual production results that extensively reported in literatures and the above-mentioned theoretical calculation results deeply, some laws and the technical points of controlling the precipitation of Ti(C,N) in austenite can be obtained as follows.

(1) Ti mainly exists in the TiN or N-rich Ti (C, N) forms in the temperature range of 1250–1400 °C in a trace Ti-treated steel. In order to effectively prevent the growth of austenite grains at soaking temperature, it is necessary to increase the cooling rate and decrease the precipitation temperature of carbonitrides.

(2) In the process from solidification to the soaking process, most of Ti atoms precipitated in the form of TiN or N-rich Ti(C, N) and inhibits the growth of the austenite grain. Because the supersaturation of Ti microalloyed steel is larger than that of a trace Ti-treated steel, the equilibrium precipitation temperature is higher and the precipitation start time is shorter, so a faster cooling rate is necessary to ensure the proper refinement of precipitates.

(3) The precipitation of Ti element is mostly in the form of TiC of Ti microalloyed steel during the process of cooling and rolling after soaking.

(4) The total solid solution temperature of TiC is higher than its soaking temperature due to the higher contents of Ti and C. The precipitation behavior of TiC is almost identical during the cooling and rolling process, and a further increase of Ti content has no effect on the precipitation behavior of TiC.

(5) The total solid solution temperature of TiC is higher than that of the soaking solid solution temperature. TiC can effectively precipitate in austenite matrix, and the fastest precipitation temperature on the dislocation lines is above 900 °C under the rolling deformation condition, which can prevent the recrystallization of the deformed austenite or inhibit the growth of recrystallization grains.

(6) Because the contents of Ti and C are high, the total solid solution temperature of TiC is higher than the soaking temperature of TiC, and the effective precipitation temperature of TiC is lower than that of Nb(C, N) in Nb microalloyed steels, but higher than that of V(C, N) in the V–N microalloyed steel. Thus, the Zener effect of Ti prevents deformed austenite recrystallization is less than that of Nb, but greater than that of V.

(7) The complete solid solution temperature of TiC is lower than the soaking solid solution temperature of TiC due to the higher contents of Ti and C. The precipitation behavior of TiC is obviously influenced by the contents of Ti and C. With the smaller contents of Ti and C in steel, the temperature range of TiC effectively precipitated is obiviously lower and even below the temperature range of austenite phase, and the fastest precipitation start time is remarkably

prolonged. When the contents of Ti and C is low, TiC can be hardly obtained inside austenite grains except for a small amount of TiC on grain boundaries.

(8) The deformation will obviously promote the precipitation process of TiC in austenite, so that the nose temperature of PTT curve becomes obviously higher and the fastest precipitate starting time becomes remarkably shorter. The effect of the deformation on TiC nucleated on the dislocation lines is significantly greater than that of the TiC nucleated on the grain boundaries.

Because of the shorter time of rolling process, the precipitation process of TiC in austenite matrix can hardly reach the equilibrium condition, and no TiC precipitated from austenite matrix in the lower Ti content steel, so a considerable portion of Ti will be precipitated in ferrite during subsequent coiling process. Nano-sized TiC particles precipitated from ferrite can produce a large precipitation hardening increment. Therefore, it is necessary to understand and master the precipitation behavior of TiC in ferrite.

The pearlite transformation usually occurs preferentially in the cooling process of the Ti microalloyed steel, so the C amount of solid solution in the ferrite is basically fixed, which can be considered to be 0.0218%. The theoretical calculation of PTT curves of the steels with different residual Ti contents (0.02, 0.04, 0.06, 0.08%) is based on the solid solubility formula of Eq. (3.23), as shown in Fig. 3.28. With the increase of the solid solubility of Ti (with the increase of solute supersaturation), the nose temperatures of PTT curves were 633, 668, 690, 705 °C. The general rule was that the fastest precipitation temperature of TiC increases by 35 °C as the solute supersaturation was doubled. In addition, the precipitation start time becomes shorter as the supersaturation of Ti increases. The rule is that the fastest precipitation start time was 0.85 order of magnitude earlier with the solute supersaturation doubled.

The PTT curves of TiC precipitated from ferrite matrix were calculated in Ti microalloyed steels with a certain extent of deformation. The deformation stored

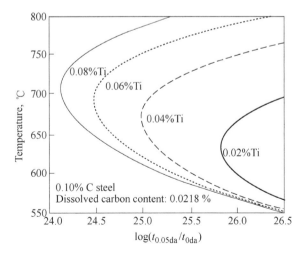

Fig. 3.28 PTT curves of TiC nucleated on the dislocation lines from ferrite matrix of different residual Ti content steels (without deformation stored energy)

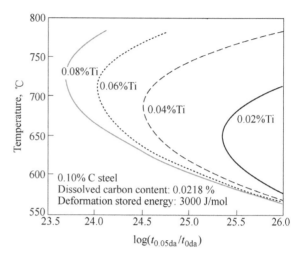

Fig. 3.29 The PTT curves of TiC nucleated on the dislocation lines in different content of Ti microalloyed steels

energy was still considered to be 3000 J/mol (the resistance of deformation increases but accumulated deformation decreases at low temperature), and the PPT curves of different residual Ti content (0.02, 0.04, 0.06, 0.08%) steels are shown in Fig. 3.29. The nose temperatures of these different steels were 651, 686, 708, 724 °C with the increase of the solid solubility of Ti (with the increase of solute supersaturation). The general rule was the fastest precipitation temperature of TiC increases by 35 °C as the solute supersaturation was doubled. Meanwhile, the precipitation start time was moving earlier as the supersaturation of Ti increases. The rule is that the fastest precipitation start time was 0.83 order of magnitude earlier with the solute supersaturation doubled.

Compared with the steel without deformation stored energy, the nose temperature of PTT curve of TiC was increasing by 18 °C in Ti microalloyed steel with a deformation stored energy of 3000 J/mol. Meanwhile, the fastest precipitation start time was 0.44-0.48 order of magnitude rapider than that without deformation. It is clear that deformation stored energy has an important promoting role on the precipitation behavior of TiC. However, the effect of the identical deformation stored energy was obviously reduced compared with the precipitation behavior in austenite.

By analyzing the experimental and actual production results that extensively reported in literatures and the above-mentioned theoretical calculation results deeply, some laws and the technical points of controlling the precipitation of Ti(C,N) in ferrite can be obtained as follows.

(1) Ti mainly precipitated in the form of TiC at the temperature range of 650–750 °C, and the precipitation mode of TiC is mainly on the dislocation lines.
(2) The size of TiC precipitated from ferrite is usually several nanometers and they are uniformly distributed in ferrite, which can produce a large precipitation hardening increment.

(3) The volume fraction of TiC precipitated from ferrite is mainly determined by the remaining contents of Ti and C. Therefore, the precipitation of TiC is significantly related with the thermal history and deformation history of the Ti microalloyed steel.
(4) The PTT curve of TiC precipitated from ferrite was significantly influenced by the residual content of Ti and C. The general rule was that the fastest precipitation temperature of TiC increases by 35 °C as the solute supersaturation was doubled. In addition, the fastest precipitation start time was advanced by 0.85 order of magnitude. This is the main reason why the Ti microalloyed steel has the fluctuation in mechanical properties. For the actual production, the coiling temperature must be adjusted according to the remaining Ti content and C content in the steel, and the stable precipitation hardening and mechanical properties can be obtained.
(5) The deformation on the steel will remarkably enhance the precipitation process of TiC in ferrite. Moreover, the nose temperature of PTT curve will be raised and the fastest precipitation start time will be shortened. However, with the same deformation stored energy, the effect of strain induced TiC precipitation in ferrite is not as effective as that of austenite matrix.

3.4 Effects of Manganese and Molybdenum on Strain Induced Precipitation of TiC

The strain induced precipitation of TiC exerts significant effect on the recovery and recrystallization of deformed austenite, which is beneficial to the grain refinement of resulting microstructure. Moreover, it can influence the amount of subsequent precipitation e.g. interphase precipitation and supersatured precipitation in ferrite occurring at lower temperatures. The strain induced TiC precipitate has a particle size of approximately several ten nanometers. In contrast, the precipitates via interphase precipitation or supersatured precipitation own much finer particle size, usually smaller than 10 nm [43]. According to Orowan mechanism [4], the contribution to the yield strength from precipitation hardening is roughly inversely proportional to the particle size of precipitate. Namely, refining the particle size down to a half can improve the precipitation hardening by a factor of 2. Therefore, a feasible approach to obtain more precipitates formed at lower temperatures is how to control the strain induced precipitation. In the above sections, the effects of several carbide formers including Ti, C and N on the precipitation kinetics were discussed. This section will introduce the latest achievement on the effects of Mn and Mo on the kinetics of strain induced TiC precipitation. These achievement will have important guiding significance for the control of TiC precipitation and further improvement of mechanical properties of Ti microalloyed steels.

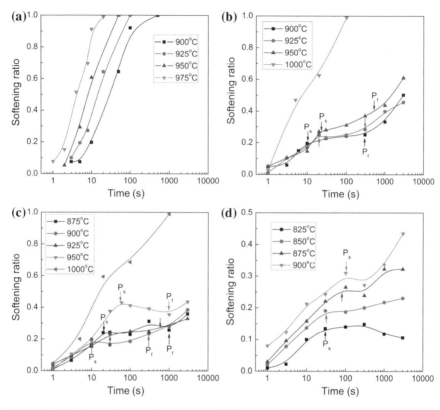

Fig. 3.30 Softening ratio versus time curves of the investigated steels: **a** reference steel, **b** 0.5Mn steel, **c** 1.5Mn steel and **d** 5.0Mn steel. Here P_s and P_f denote the start and finish time of precipitation, respectively

3.4.1 Effect of Manganese on Strain Induced Precipitation of TiC

3.4.1.1 Experimental Study

Softening Kinetics Curves and PTT Diagram

Using two-stage interrupted method, the softening kinetics curves of a plain carbon steel and three Ti microalloyed steel with different Mn concentrations are obtained. The start and finish times of TiC precipitation were determined by analyzing softening kinetics curves. The softening kinetics curves of the experimental steels are shown in Fig. 3.30. The curves of the plain carbon steel exhibit typical "S" shape in the whole temperature range. However, those for Ti microalloyed steels have softening plateaus at relatively lower temperatures, indicating the occurrence of precipitation.

By defining the start and finish times as the start and end time of precipitation, respectively, we can obtain the PTT diagrams, as shown in Fig. 3.31. As Mn concen-

Fig. 3.31 Precipitation-time-temperature diagrams of the investigated steels

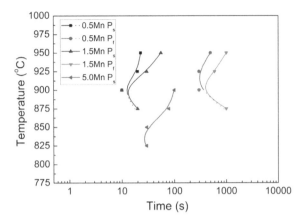

tration increased, the upper parts of the PTT diagram were shifted to longer times and lower temperatures. The nose at the minimum times of the PTT diagram was about 10 s at 900–925 °C for 0.5Mn steel, 10 s at 900 °C for 1.5Mn steel and 30 s at 825–850 °C for 5.0Mn steel. However, the retard effect of Mn on TiC precipitation weakened with decreasing testing temperature. At 950 °C, the incubation time of precipitation was prolonged by about 0.4 order of magnitude, as Mn concentration increased from 0.5 to 1.5%. But the value reduced to about 0.2 order of magnitude when the temperature decreased to 925 °C. Even at 900 °C, there was little retardation effect on precipitation from 0.5%Mn to 1.5%Mn. When Mn concentration further increased from 1.5 to 5.0%, the incubation time of precipitation at 900 °C was prolonged from 10 to 100 s, i.e. 1.0 order of magnitude. However, this value reduced to 0.6 when the temperature decreased to 875 °C. That is to say, increasing Mn concentration in Ti microalloyed steel decreased the rate of precipitation at higher temperatures, but had weaker effect on that at lower temperatures. This result was slightly different from that of the previous researchers [44, 45], who observed that the obvious delay effect of Mn on the precipitation kinetics existed at all studied temperatures. However, this kind of temperature dependence of the quaternary element effect (Mn, Si et al.) on the precipitation kinetics of microalloying carbonitride was also obtained in a recent work reported by Dong et al. [46] Their results showed that increasing Si concentration can accelerate the rate of precipitation of Nb(C, N) in austenite at higher temperatures, but have little effect on that at lower temperatures.

Precipitates

In order to understand the precipitation behavior in Ti-microalloyed steel, and to further confirm the delay effect of Mn on the rate of precipitation at higher temperatures, TEM study on precipitates was performed. Figure 3.32 is TEM image showing the evolution of strain induced precipitates in 0.5Mn steel and 1.5Mn steel with holding time at 925 °C. As shown in Fig. 3.32a–c, the particle size of TiC precipitates in 0.5Mn steel increased with holding time, from 10.9 ± 7.9 nm for 60 s to 16.4 ± 10.6 nm for 100 s, and to 39.5 ± 22.9 nm for 200 s. But the number density

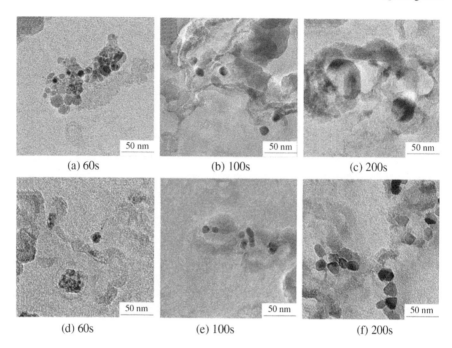

Fig. 3.32 TEM images showing TiC precipitates in 0.5Mn and 1.5Mn steel at 925 °C for various isothermal holding times: **a** 60 s in 0.5Mn, **b** 100 s in 0.5Mn, **c** 200 s in 0.5Mn, **d** 60 s in 1.5Mn, **e** 100 s in 1.5Mn and **f** 200 s in 1.5Mn

of precipitates decreased with time, possibly due to the earlier coarsening of precipitates. In the case of 1.5Mn steel, the strain induced precipitates were very difficult to be observed by TEM at the initial stage of precipitation as shown in Fig. 3.32d, which showed a small amount of tiny precipitates in a local area of carbon replica film. However, the precipitates could be observed easily by TEM until 200 s isothermal holding as shown in Fig. 3.32e, and their sizes increased with time, from 5.2 ± 1.2 for 60 s to 19.0 ± 4.3 nm for 200 s, and to 55.0 ± 28.6 nm for 600 s, but were smaller than those in 0.5Mn steel for the same isothermal holding time. These above TEM studies indicated that the strain induced TiC precipitation was delayed by more Mn addition at 925 °C, which corresponded well with the results obtained by two-stage interrupted compression test.

3.4.1.2 Effect of Manganese on Precipitation Thermodynamics and Kinetics of TiC

Precipitation Thermodynamics

The retardation of precipitation kinetics by more Mn addition at higher temperatures can be attributed to the effect of Mn on the activities of the precipitating species,

Fig. 3.33 Calculated solubility curves for TiC in austenite. The dashed line represents the stoichiometry of TiC. The square, circle and triangle denote the locations of the initial compositions ([Ti]$_0$, [C]$_0$) before TiC precipitation of 0.5Mn, 1.5Mn and 5.0Mn steels, respectively

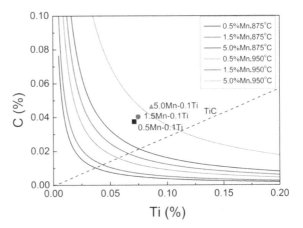

as a result of which TiC solubility is affected. According to Yong's theory [4] and Irvine's report [47], the solubility product of TiC in austenite with Mn concentration can be described by the following equation:

$$\log([\text{Ti}][\text{C}]_\gamma) = 2.75 - \frac{7000}{T} - \frac{56}{100 \times 55 \times \ln 10}(e_{\text{Ti}}^{\text{Mn}} + e_{\text{C}}^{\text{Mn}})(\%\text{Mn} - 1.0) \quad (3.107)$$

where $e_{\text{Ti}}^{\text{Mn}}$, e_{C}^{Mn} represent Mn–Ti and Mn–C interaction coefficients, respectively, and %Mn represents Mn concentration in solution (wt%).

According to Akben et al. [45], $e_{\text{Ti}}^{\text{Mn}} = -31{,}164/T$ and $e_{\text{C}}^{\text{Mn}} = -5070/T$, hence Eq. (3.107) is changed as

$$\log([\text{Ti}][\text{C}]_\gamma) = 2.75 - \frac{7000}{T} - \frac{160}{T}(\%\text{Mn} - 1.0) \quad (3.108)$$

Equation (3.107) indicates that increasing Mn concentration in steel can raise the solubility product of TiC in austenite. This relation was plotted in Fig. 3.33.

By deducting the amounts of Ti and C in the undissolved TiN and Ti$_4$S$_2$C$_2$ particles, the initial soluble Ti and C before precipitation, marked in Fig. 3.33, can be approximately estimated by the following equation:

$$[\text{Ti}]_0 = \text{total}\%\text{Ti} - \frac{48 \times 4}{32 \times 2}\%\text{S} - \frac{48}{14}\%\text{N} \quad (3.109)$$

$$[\text{C}]_0 = \text{total}\%\text{C} - \frac{12 \times 2}{32 \times 2}\%\text{S} \quad (3.110)$$

where total%Ti and total%C are the initial compositions of Ti and C in steel, respectively, %S and %N are the concentrations of S and N elements in steel, respectively. The increases in Mn content from 0.5 to 1.5%, and to 5.0% lead to a significant decrease of the solution temperature of TiC calculated by Eqs. (3.108)–(3.110), from 1062 to 1043 °C, and to 960 °C. Therefore, the chemical supersaturation degree of the solutes at a given temperature below solution temperature was reduced by increasing Mn concentration, hence causing the decrease in the driving force for precipitation.

Precipitation Kinetics

The results shown in Fig. 3.31 indicated that increasing Mn content in steel can result in the decrease of the rate of TiC precipitation at higher temperatures, but have little effect on that at lower temperatures. The PTT diagram for precipitation is mainly controlled by two independent factors. The first one is the driving force, playing the dominant role at higher temperatures. The second one is the diffusivity of the precipitate species, playing the major role at lower temperatures. At a certain intermediate temperature, the combined effect reaches the maximum, so that the PTT diagram exhibits a "C" shape curve with a nose. For the steel with a higher solution temperature of TiC, for example 0.5Mn steel, the driving force for TiC precipitation at higher temperatures (certainly below solution temperature) is larger, thus resulting in the faster precipitation kinetics. However, with temperature decreasing, the effect of the driving force on the precipitation kinetics becomes less marked. Instead, the diffusivity of precipitation controlling element, here Ti atoms, plays the dominant role. A greater diffusivity of atom can lead to a faster precipitation kinetics. Whereas, the little effect of Mn on the precipitation kinetics at lower temperatures as shown in Fig. 3.31 strongly indicated that Mn content within the studied range ($\leq 5.0\%$) had little influence on the diffusivity of Ti atom in austenite. In other words, Mn influences TiC precipitation kinetics mainly by changing the solubility of TiC in austenite.

The effect of Mn on TiC precipitation kinetics can be further clarified by model calculation. According to the analytical model built in Sect. 3.3 and Eq. (3.105), the relative PTT diagram for TiC precipitation can be calculated, as shown in Fig. 3.34. It can be seen that the upper part locations of the relative PTT diagrams were shifted to longer times and lower temperatures as Mn concentration was increased, but the lower part were relatively independent of chemical composition. This result is in a good agreement with the experimental result. The nose temperature of the relative PTT diagram reduced from 910 to 890 °C, and to 810 °C, when Mn concentration increased from 0.5 to 1.5%, and to 5.0%. This result is in a very good agreement with experimental result considering the scatter involved in the experiment.

Combined Effect of Manganese and Titanium

During hot rolling under industrial conditions, the TiC precipitates formed at higher temperatures, for example during the first several passes, usually exhibit larger particle sizes, thus having less advantageous contribution to the final mechanical properties of steel plate. According to our results, an appropriate increase in Mn content in

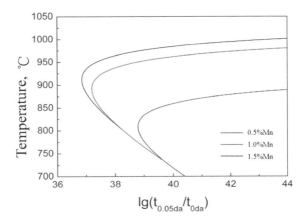

Fig. 3.34 Calculated PTT diagrams of the investigated steels

Ti microalloyed steel can delay TiC precipitation at higher temperatures, thus leading to the availability of Ti and C for the formation of finer TiC particles at lower temperatures or in the ferrite, and therefore to an additional component of yield strength at ambient temperatures. The tiny TiC precipitates formed at lower temperatures, for example during the final several passes, together with Ti in solution can inhibit the recrystallization of deformed austenite, thus favoring the refinement of final ferrite grain size. In addition, an appropriate increase in Mn content can lower the critical phase transformation temperature [48], hence allowing the finish rolling carried out at lower temperatures, which is beneficial to the refinement of final microstructure. Meanwhile, increasing Mn concentration can raise the carbon concentration in supersatured ferrite, thus promoting the supersaturated precipitation of TiC in ferrite. Wang et al. [49] studied the mechanical properties of Ti micralloyed steels produced by compact strip production (CSP), and found that the amount of TiC precipitates smaller than 10 nm increased with Mn concentration increasing from 0.4 to 1.4%, hence leading to the increases by 80 MPa in the precipitation strengthening, and by 100 MPa in the yield strength.

3.4.2 Effect of Molybdenum on Strain Induced Precipitation of TiC

3.4.2.1 Experimental Study

Softening Kinetics and Stress Relaxation Curves

Figure 3.35 shows the softening kinetics curves of unary Ti microalloyed steel (abbreviated as Ti steel) and binary Ti–Mo microalloyed steel (abbreviated as Ti–Mo steel). The softening kinetics curves of two steels below 1000 °C both exhibit "S" shape, indicating that no detected precipitation occurs after deformation. The finish times

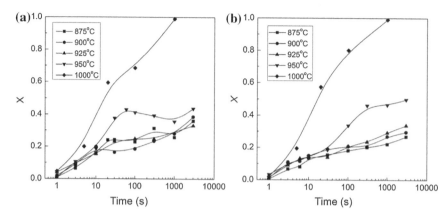

Fig. 3.35 Softening ratio versus time curves of the investigated steels: **a** Ti steel and **b** Ti–Mo steel

for full recrystallization are both approximately 1000 s for two steels. However, as the temperature is 950 °C or below, the plateaus appear in the softening kinetics curves, implying the occurrence of the precipitation. The start and end time of the plateaus are determined, as shown in Fig. 3.35.

By comparing Fig. 3.35a with b, it is found that at 925 °C or below, the softening ratios of Ti–Mo steel are almost lower than those of Ti steel. This result indicates that Mo addition plays more obvious role in inhibiting the softening of deformed austenite at relatively lower temperatures, but had little effect on that at 950 °C or above. In addition, the softening ratios of Ti–Mo steel increase more slowly at 925 °C or below for a long holding time. It indicates that Mo addition shows a persistence effect on delaying austenite recrystallization of Ti microalloyed steel, which is possibly associated with the coarsening kinetics of strain induced precipitates and discussed in details later.

The stress relaxation curves of the two steels at different temperatures are shown in Fig. 3.36. the stress relaxation curves at 950 °C or below exhibit an initial linear stress drop, which corresponds to a static recrystallization process. After the linear stress drop process, a relaxation stress plateau appears in these curves, indicating the onset of strain induced precipitation, which halts the progress of recrystallization. The stress relaxation curve at 975 °C does not show a relaxation plateau. Instead, a fast drop in stress value appears after a linear-drop stage, which indicates the occurrence of a fast static recrystallization as marked by SRX in Fig. 3.36.

According to Liu and Jonas [50], the stress versus log(time) relationship under this condition can be described as follows:

$$S = S_0 - \alpha \log(1 + \beta t) + \Delta S \quad (3.111)$$

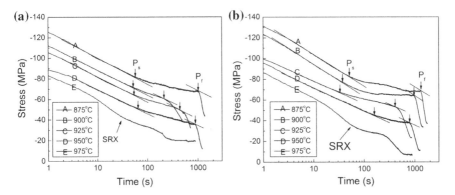

Fig. 3.36 Stress relaxation curves of the investigated steels: **a** Ti steel, and **b** Ti–Mo steel

where S is the relaxation stress, S_0 is the initial stress, t is the holding time after deformation in second, α and β are experimental constants and ΔS is the stress increment due to precipitation. For sufficiently long times, $\beta t \gg 1$, and Eq. (3.111) becomes:

$$S = A - \alpha \log t + \Delta S \qquad (3.112)$$

Here $A = S_0 - \alpha \log \beta$. According to Eq. (3.112), ΔS equals 0 during the initial linear stress drop stage. As relaxation proceeds, ΔS increases gradually and finally reaches the maximum value at a certain time which is marked by arrow (P_f) in Fig. 3.36. The maximum value of ΔS is also a function of testing temperature and it decreased with increasing temperature as shown in Fig. 3.37. It shows that the maximum of ΔS is higher for Ti–Mo steel than Ti steel in the whole testing temperature range. As the temperature increases, the difference in the maximum of ΔS becomes smaller. The maximum of ΔS is 17 MPa at 875 °C, and decreases down to 4 MPa at 925 °C. At 950 °C, it is only 0.5 MPa. This result indicates that the precipitates inhibit the softening of deformed austenite more significantly in Ti–Mo steel than Ti steel. But, with decreasing temperature, the difference becomes smaller. In other words, Mo addition markedly delays the recrystallization of deformed austenite at relatively low temperature, which is consistent with the results of two-stage interrupted tests.

PTT Diagram

According to Figs. 3.35 and 3.36, the PTT diagrams are obtained, as shown in Fig. 3.38. In Fig. 3.38, the PTT diagrams measured by the two methods both exhibit typical "C" shape, with the nose temperatures locating at 900–925 °C. The PTT diagrams obtained by two-stage interrupted method shows sharper "C" curves than those by stress relaxation method. This is possibly related to the high sensitivity of the former method in the detection of precipitation, or to the difference in the strain rate, which is 0.1 s^{-1} for the former and 1.0 s^{-1} for the later. Higher strain rate results in faster precipitation rate, in turn shorter incubation time for Ti–Mo steel.

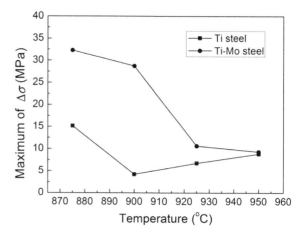

Fig. 3.37 Maximum of $\Delta\sigma$ as a function of testing temperature

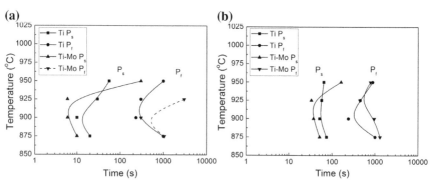

Fig. 3.38 PTT diagram: **a** two-stage deformation and **b** stress relaxation

More importantly, the PTT diagrams obtained by the two methods display similar trend. At 925 °C or below, the incubation of precipitation is shortened due to Mo addition. However, it is delayed at 950 °C after Mo addition. The effect of Mo on TiC precipitation is in good agreement with those obtained by Akben on the effect of Mo on Nb(C, N) precipitation kinetics at higher temperatures, but corresponds well with Watanabe's report at lower temperatures [45, 51]. Compared with Ti steel, the finish time of precipitation for Ti–Mo steel is delayed in a wide temperature range, especially at low temperatures. According to the discussion in Sect. 3.2, P_s represents the onset of precipitation, and P_f denotes the start time at which the pinning force of precipitation becomes weaker due to the coarsening of precipitates such that the precipitation cannot inhibit the austenite recrystallization effectively. The results in Fig. 3.38 indicate that the precipitates have larger pinning force in Ti-Mo steel than Ti steel, which leads to no occurrence of recrystallization of austenite for a very long time in Ti–Mo steel.

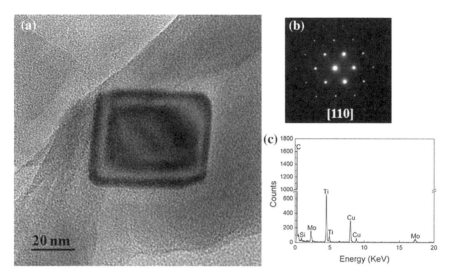

Fig. 3.39 a TEM image showing a (Ti, Mo)C precipitate, **b** selected area electron diffraction (SAD) and **c** EDS

Characterization of Precipitates in Ti–Mo Microalloyed Steel

Figure 3.39 is a TEM image showing a carbide particle formed in the specimen deformed at 925 °C followed by a hold for 1800 s and it was identified as Ti-rich MC carbide with the NaCl-type crystal structure containing a certain amount of Mo by selected area electron diffraction (SAED) and energy dispersive spectrometer (EDS) analysis shown in Fig. 3.39b, c. The atomic ratio of Ti/Mo quantified by EDS, in the carbide is about 8.0, which is larger than that of the (Ti, Mo)C particle in ferrite reported by Jang et al. [52] and Funakawa et al. [53].

The nucleation site of the precipitates formed in the Ti–Mo microalloyed steel during stress relaxation can be confirmed indirectly by analyzing the distribution of particles on the carbon replica film. A high-angle annular dark-field scanning transmission electron microscopy (HAADF-STEM) image shown in Fig. 3.40 clearly exhibited a cell-like distribution of precipitates on the carbon replica, implying that the precipitates nucleated on dislocations or on dislocation sub-structures. The sizes of these "sub-grains", which were produced by deformation and subsequent recovery process, were measured to be in the range of 0.2–1.0 μm, which is much smaller than the prior austenite grain size of −100 μm. In addition, the shape of the most precipitates as seen in Fig. 3.40 is polyhedron rather than a cube that seems to be identified from TEM image in Fig. 3.39. A precipitate which looks like a cube observed from TEM image was selected as an example to clarify the difference.

Figure 3.41 shows the TEM image, STEM image of the precipitate and the corresponding SAED with a magnetic rotation angle of 90° relative to the TEM and STEM images. It can be seen that precipitate seems to have an almost perfect cube-like shape as shown in Fig. 3.41a. However, the strong three-dimensional contrast

Fig. 3.40 HAADF-STEM image showing the cell-like distribution of (Ti, Mo)C carbide on carbon replica film, formed in the specimen deformed at 900 °C and held for 1800 s

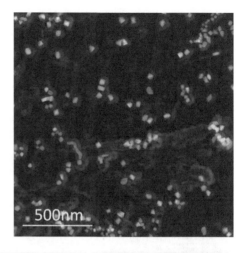

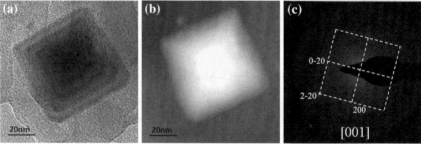

Fig. 3.41 A carbide particle formed in the specimen deformed at 925 °C and held for 1800 °C: **a** TEM image, **b** HAADF-STEM image and **c** the corresponding SAED. It should be noted that the shape of this precipitate is rectangular pyramid or an octahedron as observed in (**b**) rather than a cube as seen in (**a**). In addition, the two diagonals of this pyramid-like or octahedron-shaped precipitate are on plane {200}, which is identified by SAD in (**c**)

in Fig. 3.41b together with the annular equal thickness fringes in Fig. 3.41a indicate that the precipitate has a pyramid-like shape or an octahedron shape if the precipitate has a symmetry with respect to the observation plane, rather than a cube-like shape. Moreover, the two diagonals of this pyramid-like or octahedron-shaped precipitate were determined to be parallel with plane {200} of the precipitate.

A high-resolution transmission electron microscopy (HRTEM) investigation of the carbides formed in the Ti-Mo microalloyed steel was performed in this study, and an example of the lattice image of a carbide is presented in Fig. 3.42. It is identified as a (Ti, Mo)C particle with a NaCl-type crystal structure using a two-dimensional fast Fourier transformation (FFT) analysis, which is consistent with SAED result (see Fig. 3.39). The lattice image obtained by an inverse fast Fourier transformation (IFFT) shown in Fig. 3.42b has been identified to have a lattice parameter of 0.43 nm, which is very close to that (0.423–0.430 nm) of the inter-phase precipitated (Ti, Mo)C particles reported by other researchers [53, 54].

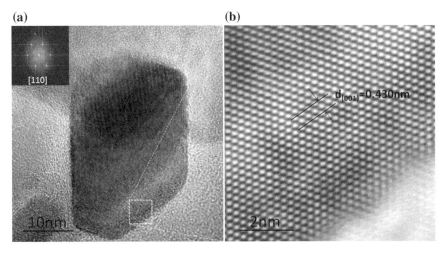

Fig. 3.42 **a** HRTEM image showing a carbide particle in the specimen deformed at 925 °C and held for 3000 s and **b** IFFT lattice image of the area marked by a white-line frame on the carbide particle in (**a**)

Growth, Coarsening and Compositional Evolution of Precipitates

The particle size evolutions of (Ti, Mo)C particle in the Ti–Mo microalloyed steel and TiC in a reference Ti steel with isothermal holding time were investigated, and the results at the testing temperature of 925 °C are presented in Fig. 3.43. It can be clearly seen that the particle sizes of carbide in both steel were very close with each other for 200 s holding time (see Fig. 3.43a, e). However, with the increase of holding time to 600 s, to 1800 s and to 3000 s, (Ti, Mo)C particle in Ti-Mo steel exhibited smaller size and higher density than TiC in Ti steel (see Fig. 3.43b–d, f–h).

In order to better understand the growth and coarsening behavior of carbide in the two steels, the average particle sizes for various holding times at 925 °C were measured and plotted in Fig. 3.44. At the initial stage of precipitation, the carbide in both steel grows very quickly, and the growth rate of particle in Ti–Mo steel is slightly faster than that in Ti steel. This is consistent with the acceleration of the precipitation kinetics due to Mo shown in Fig. 3.38. The growth of carbide at this stage approximately followed a parabolic law. With the increase of holding time, the growth rate of carbide slowed down for both steel, especially for (Ti, Mo)C particle in Ti–Mo steel being almost unchanged from 200 to 1800 s. During this stage, precipitate lies in the coarsening stage i.e. Ostwald ripening stage. The transition time from growth to coarsening was −200 s for (Ti, Mo)C particle and −600 s for TiC. Therefore, (Ti, Mo)C particle in Ti–Mo steel exhibit a superior coarsening resistance to TiC particle in Ti steel.

The compositional evolution of (Ti, Mo)C particle in Ti–Mo steel was studied at the testing temperature of 925 °C. As shown in Fig. 3.45, the atomic ratio of Ti/Mo in (Ti, Mo)C particle increased with isothermal holding time or particle growing. This

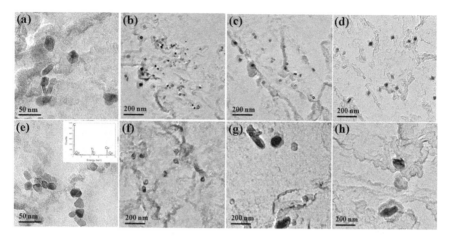

Fig. 3.43 **a–d** TEM images showing carbide particles formed at 925 °C for various holding time in Ti-Mo micro-alloyed steel: **a** 200 s, **b** 600 s, **c** 1800 s and **d** 3000 s. **e–h** TEM images showing carbide particles formed at 925 °C for various holding time in Ti microalloyed steel: **e** 200 s, **f** 600 s, **g** 1800 s and **h** 3000 s

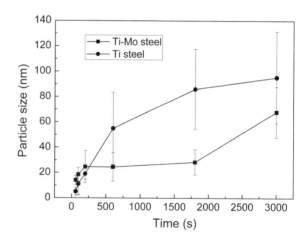

Fig. 3.44 Average particle sizes of carbides at 925 °C as functions of isothermal holding time

phenomenon seems to indicate that on the one hand, the incorporation of Mo into TiC favors the nucleation and growth during the early stages of precipitation, on the other hand, the high level replacement of Ti by Mo in TiC lattice is not stable enough or can be called as "metastable" with respect to equilibrium precipitation of (Ti, Mo)C phase. This phenomenon is in a good agreement with the compositional evolution of precipitates in Ti–Mo steel reported by Jang et al. [52] and Seto et al. [55].

The relation between average particle size and holding time at the initial stage of precipitation, before 200 s for Ti–Mo steel and before 600 s for Ti steel, is roughly determined as $d = kt^{\frac{1}{2}}$, where d is the average size of the particle, k is a constant and t is isothermal holding time.

Fig. 3.45 Atomic ratio of Ti/Mo in carbide as a function of isothermal holding time. The figures above Ti/Mo-log(time) curve represent the average particle sizes of the measured precipitates

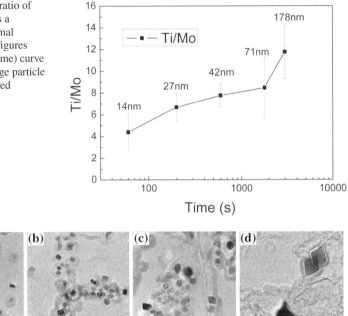

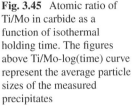

Fig. 3.46 Morphology of particles in the specimens deformed at different temperature and held for 30 min: **a** 875 °C, **b** 900 °C, **c** 925 °C and **d** 950 °C

Figure 3.46 show the variations of morphology and size of particles in the specimen held for 30 min at different temperatures after deformation. From the precipitation kinetics curve presented in Fig. 3.38, the growth time for particle is almost the same at different temperatures because of the long holding time of 30 min (1800 s). It can be seen in Fig. 3.46 that with the increase of holding temperature, the particle sizes of precipitates increase and their number density reduces, especially for 950 °C at which the particle size is extremely larger.

Figure 3.47 shows the change of Ti/Mo atomic ratio with the temperature after deformation followed by holding for 30 min. It is seen that the temperature or of particle size raises, Ti/Mo atomic ratio exhibits increase trend.

Physical and Chemical Phase Analysis (PCPA)

Figure 3.48 show the X-ray Diffraction (XRD) patterns of the precipitate powder extracted from the specimen by PCPA. It is seen that the precipitates in Ti-added steel consist of TiC as well as a trace amount of TiN and $Ti_4S_2C_2$ (undissolved particles during soaking). In Ti–Mo-added steel, only (Ti, Mo)C precipitates are detected by XRD (TiN and $Ti_4S_2C_2$ can only be observed by TEM in this case), which has the same NaCl-type lattice structure as TiC, consistent with the TEM results. In addition, the crystal parameters of the precipitates were determined accurately by using large-

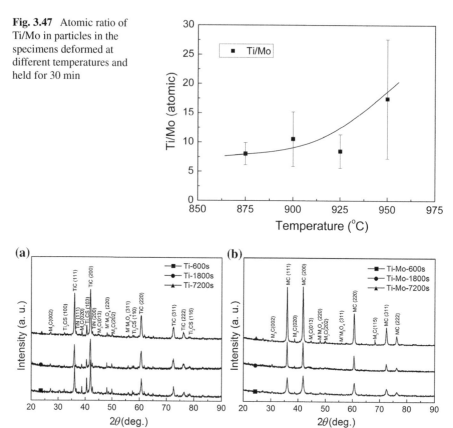

Fig. 3.47 Atomic ratio of Ti/Mo in particles in the specimens deformed at different temperatures and held for 30 min

Fig. 3.48 XRD patterns of the precipitates: **a** Ti steel, and **b** Ti–Mo steel

angle peak in XRD patterns [56]. Results show that the crystal parameter is 0.431 nm for TiC in Ti steel, which is similar to that (0.432 nm) for (Ti, Mo)C in Ti–Mo steel. These agree with TEM results. The similar crystal parameters mean that the two types of precipitates have a similar misfit with austenite matrix, thus resulting in the close interfacial structural energy.

Table 3.3 and Fig. 3.49 show M (M = Ti or Mo) mass fractions in the precipitates measured by PCPA method in the electrolyzed specimens, which are subjected to hot rolling at 920–940 °C and then held at 925 °C for various holding times. As shown in Fig. 3.49, regardless of steel, the mass fraction of Ti and Mo of precipitates increase with the prolongation of holding time. After deducting the amount of Ti in TiN and $Ti_2S_2C_4$, the mass fractions of Ti and Mo in MC precipitate can be obtained. No matter in the sum of Ti and Mo or Ti, Ti–Mo steel has relatively larger mass fraction compared with Ti steel, indicating that the precipitation is accelerated due to Mo addition. This is in a good agreement with two-stage interrupted method and stress relaxation test (see Fig. 3.38).

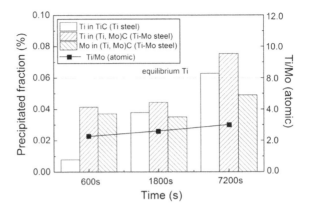

Fig. 3.49 Quantitative results of PCPA

Moreover, Ti/Mo atomic ratio increases during holding (i.e., with the particle growth), which is in accordance with that obtained by TEM/EDS. Notably, Ti/Mo atomic ratio determined by PCPA is significantly lower than that measured by TEM/EDS. This could be attributed to the difference in the particle size range measured by the two methods. For the case of PCPA test, a wide size range of precipitates are included in the final result output, which is caused by the interconnection of very fine particles (<10 nm) and relatively large ones. However, the TEM/EDS analysis is only conducted on the precipitates in relatively large size (several tens of nanometers, see Fig. 3.43), due to the limited ability of replica to extract ultrafine precipitates. Since larger particles contain lower fraction of Mo (also indicated by Fig. 3.45), Ti/Mo atomic ratio determined by PCPA is smaller, compared with that measured by TEM/EDS. Even so, the two methods show the similar variation trend in Ti/Mo atomic ratio. Figure 3.49 also shows the equilibrium precipitation amount of Ti calculated by solubility product of TiC in austenite (Eq. 3.5). It can be seen that the mass fraction of Ti in Ti steel specimen, which was subjected to hot rolling followed by holding for 7200 s, is almost equivalent to the equilibrium value. However, that in Ti–Mo steel is significantly higher the equilibrium value. This experimental fact further indicates that addition of Mo enhances the precipitation of MC in austenite.

Table 3.3 PCPA results of Ti steel and Ti–Mo steel deformed at 920–940 °C and held at 925 °C for different times (%)

Steel-time (s)	Ti in precipitate	Ti in MC	Mo in MC	Ti/Mo in MC	Soluble Ti	Soluble Mo
Ti-600	0.034	0.008	–	–	0.063	–
Ti-1800	0.064	0.038	–	–	0.033	–
Ti-7200	0.089	0.063	–	–	0.008	–
Ti–Mo-600	0.056	0.041	0.037	2.22	0.044	0.173
Ti–Mo-1800	0.059	0.044	0.035	2.51	0.041	0.175
Ti–Mo-7200	0.090	0.075	0.049	3.06	0.01	0.161

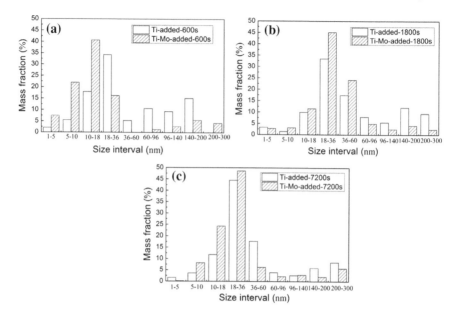

Fig. 3.50 Particle size distribution measured by small angle X-ray diffraction: **a** t = 600 s; **b** t = 1800 s; **c** t = 7200 s

Figure 3.50 shows the particle mass fraction determined by SAXS in the specimen after undergoing hot rolling and isothermal holding for different holding periods (600, 1800, and 7200 s) at 925 °C. The mass fraction of finer precipitates (e.g., <60 nm) in Ti–Mo steel is obviously higher than that in Ti variant, which has a good agreement with the TEM observation as shown in Fig. 3.43. These results indicate that (Ti, Mo)C particles in Ti–Mo added steel exhibit a superior coarsening resistance compared to TiC particles in Ti-added steel.

3.4.2.2 Thermodynamic Analysis for (Ti, Mo)C Phase

As shown in Figs. 3.45 and 3.49, the atomic ratio of Ti/Mo in (Ti, Mo)C particle increased with isothermal holding time or particle growing. This phenomenon seems to indicate that the high level replacement of Ti by Mo in TiC lattice is not stable enough or can be called as "metastable" with respect to equilibrium precipitation of (Ti, Mo)C phase.

Figure 3.51 shows the variations of the equilibrium atomic fractions of Ti, Mo and C in MC phase with temperature calculated by the Thermo-calc (TCFE6 database). FCC and BCC phases were selected as Thermo-calc calculation was carried out. This means that MC phase is in equilibrium with austenite at higher temperature and with ferrite at lower temperature. In addition, the calculated MC phase contains a small

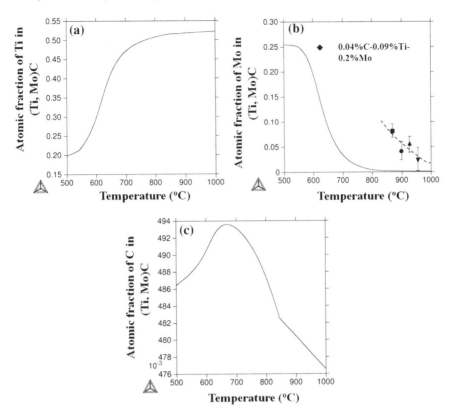

Fig. 3.51 Variations of atomic fractions of Ti (**a**), Mo (**b**) and C (**c**) in MC carbide particle with temperature calculated by Thermo-calc with database TCFE6. In (**b**), the measured atomic fraction of Mo in the MC particle were given by assuming that no vacancies are present at C atom positions in the NaCl-type crystal structure, namely perfect NaCl-type structure. The data marked by symbols in the temperature range of 875–950 °C are the present measured data under the condition of 1800 s isothermal holding. The data marked by a quadrangle at 620 °C is from the work of Funakawa et al., whose steel has a similar chemical composition with that of the present studied steel, and was isothermal held at 620 °C for 3600 s after finish rolling and furnace cooled to room temperature

amount of Fe and Mn elements, thus leading to the sum of atom factions of Ti and Mo being not 0.5 as shown in Fig. 3.51a, b, especially at lower temperatures.

For comparison, the solubility product equations of TiC and MoC in austenite and ferrite are utilized to calculate the change of the equilibrium sublattice fraction of Mo atoms in (Ti, Mo)C particle in the model of ideal solution condition. As shown in Figs. 3.51 and 3.52, the equilibrium concentration of Mo in MC phase in austenite phase region (temperature higher than 850 °C) are significantly low using both Thermo-calc and solubility product equations. The equilibrium chemical composition of MC phase with the fcc structure approaches the pure TiC. As the temperature decreases, the equilibrium atomic fraction of Mo in MC phase increases. When the temperature decreases to below 700 °C, the increase of the atomic fraction

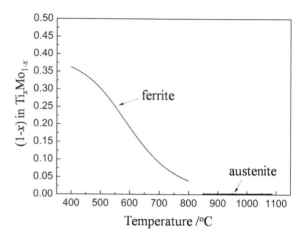

Fig. 3.52 Sub-lattice fraction of Mo in (Ti, Mo)C calculated based on the ideal solution model as a function of temperature

Table 3.4 Parameters for the solubility products[a] of TiC and MoC in austenite and ferrite [58, 59]

Phase	Carbide	A	B
Austenite	TiC	2.75	7000
	MoC	1.29	523
Ferrite	TiC	4.40	9575
	MoC	3.19	4649

[a] General expression of solubility product: $\log([M][C]) = A - \frac{B}{T}$

of Mo with temperature becomes faster. A relatively large amount of Mo exists in MC phase in the temperature range of 500–700 °C. The measured atomic fractions of Mo by EDS in MC phase at different temperatures for 1800 s isothermal holding were also shown in Fig. 3.51b. Here, it is assumed that no vacancies are present at C atom positions in the MC lattice, namely a perfect NaCl-type MC phase. In this case, the atomic fraction of Mo in MC reaches the minimum. In addition, the experimental result from Funakawa et al. [53], of a Ti–Mo microalloyed steel with a similar chemical composition to that of the present steel was given in Fig. 3.51b for a comparison (see Fig. 3.52).

As the temperature decreases, the measured atomic fraction of Mo in MC phase increases. However, they are much higher than those calculated by Thermo-calc. As discussed above, the site fraction of Mo in the (Ti, Mo)C lattice decreases with isothermal holding time or with particle growing. Such a finding together with the comparison between the experimental results and theoretical calculation indicates that the greater replacement of Ti by Mo in the TiC lattice is energetically unfavorable with respect to the equilibrium precipitation, when the other energy conditions including interfacial energy and elastic strain energy between precipitate and matrix are not considered.

However, relevant calculation results show that the partial replacement of Ti by Mo in the TiC lattice possibly decreases the chemical interfacial energy between (Ti,

Mo)C carbide and austenite [57]. Therefore, this effect, in fact, favors the decrease of the nucleation energy barrier, thus enhancing the nucleation of (Ti, Mo)C phase during the early stages of precipitation. Accordingly, smaller (Ti, Mo)C particle contains more Mo as shown in Fig. 3.45. However, with particle growing, the surface area to volume ratio of particle becomes not considerable, i.e. the interface effect becomes not so significant. At the moment, the incorporation of Ti into (Ti, Mo)C phase is more beneficial to the decrease of the total free energy of system, thus resulting in the decrease of the fraction of Mo in the (Ti, Mo)C phase, as shown in Fig. 3.45. However, the incorporation rate of Ti atom into the (Ti, Mo)C phase depends on the diffusivity of Ti in austenite, which is a function of isothermal temperature. The diffusion rate of Ti atom is slower at lower temperature, thus leading to the larger difference in terms of the faction of Mo in the (Ti, Mo)C phase, between the experimental measurements and the equilibrium results calculated by Thermo-calc at lower temperatures.

3.4.2.3 Analysis of Effect of Molybdenum on Precipitation Kinetics of Carbide

Experimental results show that Mo incorporates into TiC lattice during precipitation, but Mn does not. Hence, the effect mechanism of Mo addition on the strain induced precipitation in Ti microalloyed steel is different from that of Mn addition. This disagrees with the previous understanding [59–61]. The retardation of precipitation kinetics by more Mn addition can be attributed to the effect of Mn on the activities of the precipitating species, as a result of which TiC solubility is affected. However, the effect mechanism of Mo is more complex. According to classical nucleation and growth theory [4], three factors are considered as follows:

(1) Variation of driving force of TiC precipitation i.e. change of Gibbs free energy due to the incorporation of Mo into TiC lattice.
(2) Variation of nucleation barrier of TiC precipitate i.e. interfacial energy of TiC/austenite due to the incorporation of Mo into TiC lattice.
(3) Effect of Mo addition on the recovery and recrystallization of austenite, thus affecting the nucleation and growth of TiC.

Regarding the effect of Mo on the Gibbs free energy of TiC precipitation, simple theoretical model calculations are carried out to clarify it. Figure 3.53 shows the effect of the incorporation of a small fraction of Mo on the free energy of precipitation. In this calculation, ideal solution models are employed for the mixing of TiC and MoC. The equation for calculation can be found in Ref. [4]. It can be seen that the incorporation of a small amount of Mo decreases the free energy of TiC precipitation i.e. driving force. As Mo fraction increases, the free energy gradually decreases. While, with the variation of temperature, the extent of decrease varies. Below 950 °C, the Gibbs free energy of $Ti_{0.9}Mo_{0.1}C$ decreases to approximately 85% of that of pure TiC. As the temperature raises from 950 °C to higher temperature, reduction of Gibbs free energy becomes faster and faster. In the case of $Ti_{0.8}Mo_{0.2}C$, due to the

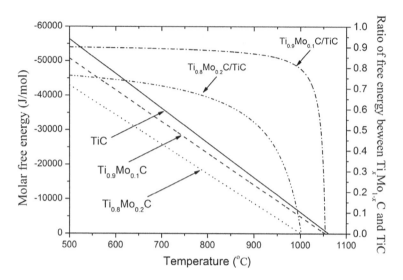

Fig. 3.53 The effect of a small amount of Mo on the Gibbs free energy of precipitation of TiC

incorporation of Mo, the Gibbs free energy decreases to 60–70% of that of pure TiC below 850 °C. Also, with the temperature increasing from 850 °C to higher temperature, the decrease of Gibbs free energy becomes faster. From the above analysis, the corporation of a small amount (sub-lattice fraction $-10-20\%$) of Mo into TiC slightly decreases the absolute value of Gibbs free energy of precipitation at relative low temperatures. However, as the temperature increases to a high value, the decrease of Gibbs free energy due to Mo incorporation become very significant. Hence, the substitute of Ti by Mo in TiC lattice is difficult at elevated temperatures, which is in consistent with the experimental results (see Fig. 3.47).

The incorporation of Mo into TiC was reported to decrease the interfacial energy of precipitate/austenite matrix. This is beneficial to the early nucleation of precipitates. For convenience, the critical nucleus radius of homogenous nucleation is used to clarify this problem:

$$r^* = -\frac{2\sigma}{\Delta G_V} \quad (3.113)$$

where σ is the interfacial energy of nucleus/matrix, ΔG_V is the volume free energy.

As shown in Fig. 3.53, ΔG_V decreases slightly due to Mo incorporation into TiC, which results in the increase of r^*. On the contrary, σ become smaller as results of Mo incorporation. Obviously, these are two conflicting factors. On the one hand, as the temperature is relatively low, due to the corporation of a small amount of Mo into TiC, the decrease of ΔG_V is not significant, but σ reduces remarkably, which in turn increases r^*. The nucleation rate thus increases, and the precipitation is accelerated. This is in a good agreement with the experimental result of low temperature, as shown in Fig. 3.38. On the other hand, as the temperature raises up to high value, the incorporation of Mo considerably decreases ΔG_V, thus increasing the critical nucleus radius r^* (although the interfacial energy also decreases, it is not the dominant factor). The nucleation is difficult to occur, and the precipitation is decelerated. Accordingly, Mo in TiC should be less in amount at high temperatures. This agrees with the experimental results shown in Fig. 3.38.

The PCPA results in Fig. 3.49 show that both the precipitation amount of total Ti and Mo, and that of Ti are enhanced due to Mo addition. Additionally, it is noted that the precipitation amount of Ti exceeds the equilibrium value calculated by solubility product of TiC in austenite. This is related to the role of Mo in the recovery and recrsyatllization of austenite. According to Ref. [62], Mo atoms in solution can effectively retard the recovery and recrystallization of deformed austenite, in turn a large number of structural defects such as dislocations remained. The dislocation and dislocation substructures serve as the nucleation sites of precipitates. Mo addition possibly increases the effective nucleation sites of precipitates, thus accelerating the precipitation. Moreover, experimental results indicate that the recrystallization kinetics in Ti–Mo steel becomes slower as compared with Ti steel, which results in a large number of structural defects being remained. According to Ref. [4], structural defects can promote the super equilibrium precipitation. That is to say, the actual amount of precipitation is greater than the equilibrium amount of precipitation. As a result, Mo addition increases the amount of precipitation of Ti at relatively lower temperatures, as shown in Fig. 3.49.

3.5 Ostwald Ripening of Titanium-Bearing Phase

After the second phase precipitation is completed, it will immediately follow the process of accumulation and growth, which is Ostwald Ripening process. The driving force in the process is the interface energy between the second phase and the matrix. When the volume fraction of the second phase remains constant, the total interface area will decrease with the increase of size of second phase, which results in reduced system interface energy. When the temperature is high enough and holding time is long enough, the second phase may be severely coarsened, resulting that the actual second phase size is much larger than the size at which the precipitation process is completed. Thereby, it may weaken or even lose the advantages of second phase at high temperatures.

In the second phase system, the Ostwald ripening rules and the ripening rates are often very different, resulting in a significant difference in the actual size of the second phases. Many of the second phases have a very small size in precipitation. But once a certain degree of Ostwald ripening occurs, the size will grow very rapidly and thus lose the relevant function; while other second phases may be can still maintain a very small size at very high temperature and still maintains the corresponding function.

In order to study deeply and calculate theoretical-quantitatively the behavior of various titanium-containing phases in titanium-containing steels, especially the high-temperature behavior, it is necessary to master the Ostwald ripening rule of titanium-containing phase, that is, the quantitative law of particle size with temperature and time.

The Ostwald ripening rule of the second phase under different control mechanisms has been studied in depth, and the relationship between the average size and the high temperature holding time has a one-second power relationship, one third of the power relations, one quarter Party relations and one-fifth power relations [63–65]. However, the results of the related experimental studies show that the process of generating the second phase by the solute atom reaction is easy for the second phase, which is more evenly distributed in the matrix and has strong chemical stability (the formation of free energy values of the second phase with strong chemical stability are large), and the solute atoms are few at rapid the diffusion channel, which will soon be depleted. Therefore, the high-temperature agglomeration and growth behavior must depend on the diffusion process of controlled solute atoms in the matrix. The Ostwald maturation process mainly follows the law of one-third power relationship, which is:

$$r_t^3 = r_0^3 + \frac{8D\sigma V_p^2 C_0}{9V_B RT} t = r_0^3 + m^3 t \qquad (3.114)$$

In the equation, r_0 and r_t represents the average size of second phase at initial time and after t seconds, respectively; D represents the diffusion coefficient of control elements in matrix phase; σ represents the interface energy ratio; V_P represents the molar volume of the second phase; C_0 represents the equilibrium solute concentration of the control element dissolved in the matrix phase; V_B represents the molar volume of control elements; T represents the temperature, K.

TiC, TiN and Ti (C, N) are very stable second phases. They maintain small sizes at very high temperatures for a long time and the average size of the coarsening rule is available in Eq. (3.114) when they are uniformly distributed in the iron matrix.

According to the formula of solid solubility or the formula of solid solubility in the iron matrix, the equilibrium solid solution [M] of the control element M in the matrix can be calculated at the certain temperature in the steel with the certain chemical composition. But the calculated solid solution [M] is the mass percentage, and the corresponding conversion must be made to obtain the atomic concentration C_0 of the control element M in the matrix:

$$C_0 = \frac{[M]\overline{A_{Fe}}}{100 A_M} \qquad (3.115)$$

In the equation, $\overline{A_{Fe}}$ and A_M are the average atomic weight of the iron matrix and the atomic weight of the control element M, respectively. When the content of solid solution alloying element in the iron matrix is small, $\overline{A_{Fe}} \approx A_{Fe}$, in which A_{Fe} A represents the iron atomic weight, in that case:

$$C_0 = \frac{[M]A_{Fe}}{100A_M} \quad (3.116)$$

According to the corresponding calculation, when the steel composition meets the ideal chemical ratio, the various micro-alloy carbonitride in austenite coarsening rate with temperature changes is shown in Fig. 3.54. It can be seen that the coarsening rate of microalloyed carbonitrides increased with the increase of temperature due to the increase of diffusion coefficient and solute concentration with the increase of temperature; the coarsening rate of microalloyed nitrides is significantly smaller than that of microalloyed carbides due to the difference in equilibrium solid solubility between the control elements (microalloying elements) in austenite; while the roughening rate of titanium carbide or nitride is smaller than that of niobium in comparison with various microalloying elements; the high temperature dimensional stability of TiN is particularly excellent. When the temperature is 900 °C, the coarsening rate m of VC, NbC, TiC, VN, NbN, TiN are 0.547, 0.328, 0.350, 0.204, 0.139, 0.054 nm/s$^{1/3}$, respectively; when the temperature is 1200 °C, the coarsening rate m of VC, NbC, TiC, VN, NbN, TiN are 5.36, 2.86, 2.65, 1.90, 1.51, 0.66 nm/s$^{1/3}$.

Therefore, the radius of VC, NbC, TiC, VN, NbN, TiN precipitated during the rolling process at 900 °C for about 125 min (about 2 min) will grow by 2.74, 1.64, 1.75, 1.02, 0.70, 0.27 nm, which can effectively ensure that the strain-induced precipitation of microalloyed carbonitride size maintained about 10 nm (radius of 5 nm). The radius of VC, NbC, TiC, VN, NbN, TiN will grow 117, 57, 53, 38, 30, 13 nm respectively after the austenitizing homogenization temperature of 1200 °C for 8000 s (2.2 h). It is clear that the microalloyed carbides are not competent because the size of the second phase particles controlling the grain growth must be less than 100 nm, even if the volume fraction of the unfilled microalloyed carbonitride is not taken into account. The size of VN and NbN can basically meet the requirements, and TiN is a significant surplus. As can be seen from Fig. 3.54, TiN can still maintain a relatively small size at 1300 °C, while other microalloyed carbonitrides will be sub-

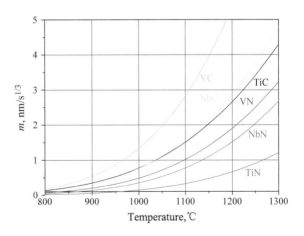

Fig. 3.54 Comparison of the coarsening rates of various carbides and nitrides in austenite (ideal stoichiometric ratio)

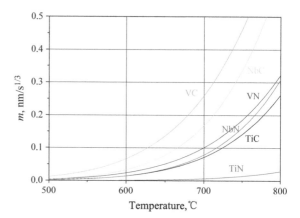

Fig. 3.55 Comparison of the coarsening rates of various carbides and nitrides in ferrite (ideal stoichiometric ratio)

stantially coarsened, which is the main reason why trace Ti treated steel can stabilize the austenite grain growth at high temperature.

Usually the content of microalloying elements in the steel composition is lower than the ideal chemical ratio, so the coarsening rate will be reduced to some extent.

Similarly, when the steel composition meets the desired chemical composition, the coarsening rate of the various microalloyed carbazides in the ferrite changed with temperature can be calculated, as shown in Fig. 3.55.

It can be seen from Fig. 3.55 that the coarsening rate of microalloyed nitrides is significantly smaller than that of the corresponding microalloyed carbides; and compared the carbides and nitrides of microalloying elements, titanium carbide or nitride roughening rate is less than niobium and far less than vanadium. When the temperature is 700 °C, the coarsening rate m of VC, NbC, TiC, VN, NbN, TiN are 0.259, 0.168, 0.071, 0.101, 0.079, 0.0064 nm/s$^{1/3}$, respectively. Therefore, the radius of precipitated VC, NbC, TiC, VN, NbN, TiN will grow up to 0.518, 0.336, 0.142, 0.201, 0.157, 0.013 nm, respectively, at the coiling temperature of 700 °C for 8000 s (2.2 h), which grows insignificantly and results in maintaining a few nanometers in size to produce a strong precipitation strengthening effect. In contrast, the effective precipitation temperature of Nb (C, N) is relatively low, so the size of the precipitated phase is smaller; the roughening rate of TiC is smaller with respect to V (C, N) and Nb (C, N) and able to acquire a smaller size.

Finally, the coarsening rate of the composite microalloyed precipitation phase in Ostwald ripening is related to the diffusion and coordination of various alloying elements, and thus the coarsening rate is obviously reduced. Therefore, in the complex microalloyed steel, both in the austenite region or ferrite region, smaller microalloy carbonitride precipitation phase can be acquired and results in a much more significant effect.

References

1. Mao X P, Sun X J, Kang Y L, Lin Z Y. Physical metallurgy for the titanium microalloyed strip produced by thin slab casting and rolling process [J]. Acta Metallurgica Sinica, 2006, 42(10), 1091–1095.
2. Wang M L, Cheng G Q, Qiu S T, Zhao P, Gan Y. Behavior of precipitation containing titanium during solidification [J]. Journal of Iron and Steel Research, 2007, 19(5), 44–53.
3. Hansen M, Anderko K. Constitution of Binary Alloys [M]. New York: McGraw-Hill, 1958.
4. Yong Q L. Secondary Phase in Steel Materials [M]. Beijing: Metallurgy Industry Press, 2006.
5. Ye D L, Hu J H, Manual of Thermodynamic Data for Inorganic Substance [M]. Beijing: Metallurgy Industry Press, 2002.
6. Narita K. Physical Chemistry of the Groups IVa(Ti,Zr), Va(V,Nb,Ta) and the Rare Earth Elements in Steel [J]. Trans ISIJ, 1975, 15: 145–152.
7. Irvine K J, Pickering F B, Gladman T. Grain Refined C-Mn Steels [J]. JISI, 1967, 205: 161–182.
8. Chino H, Wada H. Jawata Tech Rep., 1965, 251: 5817.
9. Williams R, Harries W. Met Soc., 1974: 152.
10. Hillert M, Jonsson S. An Assessment of the Al-Fe-N System [J]. Metall Trans., 1992, 23A: 3141–3149.
11. Akamatsu S, Hasebe M, Senuma T, Matsumura Y, Akisue O. Thermodynamic Calculation of solute Carbon and Nitrogen in Nb and Ti Added Extra-low Carbon Steels [J]. ISIJ Inter., 1994, 34: 9–16.
12. Matsuda S, Okumura N. Effect of Distribution of TiN Precipitate Particle on the Austenite Grain Size of Low Carbon Low Alloy Steels [J]. Trans ISIJ, 1978, 18: 198–202.
13. Gurevic J G. Gernaya Metallurgija, 1960(6): 59.
14. Adachi A, Mizukawa K, Kanda K. Tetsu-to-Hagane, 1962, 48: 1436.
15. Kunze J. Solubility product of titanium nitride in gamma-iron [J]. Met. Sci., 1982, 16: 217–218.
16. [16]Wada H, Pehlke R D. Nitrogen Solubility and Nitride Formation in Austenitic Fe-Ti Alloys [J]. Metall. Trans., 1985, 16B: 815–822.
17. Turkdogan E T. Causes and effects of nitride and carbonitride precipitation during continuous casting [J]. Iron Steelmaker, 1989, 16: 61–75.
18. Inoue K, Ohnuma I, Ohtani H, Ishida K, Nishizawa T. Solubility Product of TiN in Austenite [J]. ISIJ Inter. 1998, 38: 991–997.
19. Tailor K A. Solubility Products for Titanium-, Vanadium- and Niobium-Carbides in Ferrite [J]. Script Metall. Mater., 1995, 32: 7–12.
20. Akamatsu S, Hasebe M, Senuma T, Matsumura Y, Akisue O. Thermodynamic Calculation of solute Carbon and Nitrogen in Nb and Ti Added Extra-low Carbon Steels [J]. ISIJ Inter., 1994, 34: 9–16.
21. Chen J X, Manual of Figures and Tables for Steelmaking [M]. Beijing: Metallurgy Industry Press, 1984.
22. Liu W J, Yue S, Jonas J J. Characterization of Ti Carbosulfide Precipitation in Ti Microalloyed Steels [J]. Metall Trans., 1989, 20A: 1907–1915.
23. Liu W J, Jonas J J, Bouchard D. Gibbs Energies of Formation of TiS and $Ti_4C_2S_2$ in austenite [J]. ISIJ Inter., 1990, 30: 985–990.
24. Swisher J H. Sulphur Solubility and Internal Sulfidation of Iron-Titanium Alloys [J]. Trans. Metall. Soc. AIME, 1968, 242: 2433.
25. Yoshinaga N, Ushioda K, Akamatsu S, Akisue O. Precipitation Behavior of Sulfides in Ti-added Ultra Low-carbon Steels in austenite [J]. ISIJ Inter., 1994, 34:24–32.
26. Yang X, Vanderschueren D, Dilewijns J, Standaert C, Houbaert Y. Solubility Products of Titanium Sulphide and Carbosulfide in Ultra-low Carbon Steels [J]. ISIJ Inter., 1996, 36: 1286–1294.
27. Copreaux J, Gaye H, Henry J. Relation Précipitation-Propriétés Dans Les Aciers Sans Interisticiels Recuits en Continu [R]. ECSC Report, EUR17806 FR, 1997.
28. Mitsui H, Oikawa K, Onuma I. Phase Stability of TiS and Ti4C2S2 in Steel [J]. CAMP-ISIJ, 2004, 17: 1275.

29. Iorio L E, Garrison W M. Solubility of Titanium Carbosulfide in Austenite [J]. ISIJ Inter., 2002, 42: 545–550.
30. Yamashita T, Okuda K, Yasuhara E. Thermodynamic Analysis of Precipitation Behaviors of Ti, Mn Sulphide in Hot-rolled Steel Sheets [J]. Tetsu-to-Hagane, 2007, 93: 538–543.
31. Mizui N, Takayama T, Sekine K. Effect of Mn on Solubility of Ti-sulfide and Ti-carbosulfide in Ultra-low C Steels [J]. ISIJ Inter., 2008, 48: 845–850.
32. Moll S H, Ogilvie R E. Trans. Metall. Soc. AIME, 1959, 215: 613–618.
33. Lai D Y F, Borg J. USAEC Rept. UCRL 50314, 1967.
34. Dyment F, Libanati C M. Self-diffusion of Ti, Zr, and Hf in their HCP phases, and diffusion of in HCP Zr [J]. Mater. Sci., 1968, 3: 349–359.
35. Walsoe de Reca N E, Libanati C M. Acta Met., 1968, 16: 1297.
36. Kulkarni S R, Merlini M, Phatak N, Saxena S K, Artioli G, Amini S, Barsoum M W. Thermal expansion and stability of Ti_2SC in air and inert atmospheres [J]. Alloys Compounds, 2009, 463(1–2): 395–400.
37. Davenport A T, Brossard L C, Miner R E. Metals, 1975, 27(6): 21.
38. Baker R G, Nutting J. ISI Special Report, No. 64, London: ISI, 1959: 1.
39. Zener C. quoted by Smith C S, Grains, Phases, and Interfaces: An Interpretation of Microstructure [J]. Trans AIME, 1948, 175:47.
40. Cahn R W. Physical Metallurgy [M]. Netherlands: North-Holland, 1970.
41. Yong Q. Theory of Nucleation on Dislocations [J]. Chin J Met. Sci. Tech., 1990, 6: 239–243.
42. Liu W J, Jonas J J. Ti(C,N) Precitated in Microalloyed Austenite during Stress Relaxation [J]. Met. Trans. A., 1988, 19A: 1415–1424.
43. Yong Q L, Ma M T, Wu B R, Physical and Mechanical Metallurgy of Microalloyed Steel [M]. Beijing: China Machine Press, 1989.
44. Akben M G, Weiss I, Jonas J J. Dynamic precipitation and solute hardening in a V microalloyed steel and two Nb steels containing high levels of Mn [J]. Acta Metall., 1981, 29(4): 111–121.
45. Akben M G, Chandra T, Plassiard P, et al. Dynamic precipitation and solute hardening in a titanium microalloyed steel containing three levels of manganese [J]. Acta Metall., 1984, 32(4):591–601.
46. Dong J X, Siciliano J F, Jonas J J, et al. Effect of silicon on the kinetics of Nb(CN) precipitation during the hot working of Nb-bearing Steels [J]. ISIJ Int., 2000, 40: 613–618.
47. Irvine K J, Pickering F B, Gladman T. Grain Refined C-Mn Steels [J]. JISI, 1967, 205: 161–182.
48. Zurob H S, Zhu G, Subramanian S V, Purdy G R, Hutchinson C R, Brechet. Y. Analysis of the effect of Mn on the Recrystallization Kinetics of High Nb steel: An example of physical-based alloy design [J]. ISIJ Int., 2005, 45(5): 713–722.
49. Wang C J, Yong Q L, Sun X J, Mao X P, Li Z D, Yong X, Effect of Ti and Mn contents on the precipitate characteristics and strengthening mechanism in Ti microalloyed steels produced by CSP [J]. Acta Metall., 2011, 47(12), 1541–1549.
50. Liu W J, Jonas J J. A Stress Relaxation Method for Following Carbonitride Precipitation in Austenite at Hot Working Temperatures [J]. Metall. Trans. A, 1988, 19A: 1403–1413.
51. Watanabe H, Smith Y E, Pehlke R D. Precipitation kinetics of niobium carbonitride in austenite of high-strength low-alloy steels. The Hot deformation of austenite [M]. New York: TMS-AIME, 1977: 140–168.
52. Jang J H, Lee C H, Heo Y U, et al. Stability of (Ti, M)C (M = Nb, V, Mo and W) carbide in steels using first-principles calculations [J]. Acta Mater., 2012, 60: 208–217.
53. Funakawa Y, Shiozaki T, Tomita K, et al. Development of High Strength Hot-rolled Sheet Steel Consisting of Ferrite and Nanometer-sized Carbides [J]. ISIJ Int., 2004, 44: 1945–1951.
54. Yen H W, Huang C Y, Yang J R. Characterization of interphase-precipitated nanometer-sized carbides in a Ti-Mo-bearing steel [J]. Scripta Mater., 2009, 61: 616–619.
55. Seto K, Funakawa Y, Kaneko S. Hot Rolled High Strength Steels for Suspension and Chassis Parts "NANOHITEN" and "BHT® Steel" [J]. JFE Technical Report, 2007, 10:19–25.
56. Zhou Y, Materials Analysis Method [M]. Beijing: China Machine Press, 2011.
57. Pavlina E J, Speer J G, Van T C J. Equilibrium solubility products of molybdenum carbide and tungsten carbide in iron [J]. Scripta Mater., 2012, 66: 243–246.

58. Matsuda S, Okumura N. Effect of Distribution of TiN Precipitate Particle on the Austenite Grain Size of Low Carbon Low Alloy Steels [J]. Trans ISIJ, 1978, 18: 198–202.
59. Akben M G, Bacroix B, Jonas J J. Effect of Vanadium and Molybdenum Addition on High Temperature Recovery, Recrystallization and Precipitation Behavior of Niobium-based Microalloyed steels [J]. Acta Mater., 1983, 31: 161–174.
60. Lee W B, Hong S G, Park C G, et al. Influence of Mo on Precipitation Hardening in hot Rolled HSLA Steels containing Nb [J]. Scripta Mater., 2000, 43: 319–324.
61. Lee W B, Hong S G, Park C G, et al. Carbide Precipitation and High-Temperature Strength of Hot-rolled High-Strength, Low-Alloy Steels Containing Nb and Mo [J]. Metall Mater Trans A, 2002, 33A: 1689–1698.
62. Pereda B, Fernandez A I, Lopez B, Rodriguez.ibabe J M. Effect of Mo on Dynamic Recrystallization Behavior of Nb-Mo Microalloyed Steels [J]. ISIJ Int., 2007, 47(6): 860–868.
63. Lifshitz I M, Slyozov V V. The Kinetics of Precipitation from Supersaturated Solid Solutiions [J]. J. Phys. Chem. Solids, 1961, 19: 35–50.
64. Yong Q L. Ostwald ripening of second-phase particles in dilute solution-I. Universal differential equation [M]. Journal of Iron and Steel Research, 1991, 3(4), 51–60.
65. Yong Q L. Ostwald ripening of second-phase particles in dilute solution-I. Analytic solution [M]. Journal of Iron and Steel Research, 1992, 4(1), 59–66.

Chapter 4
Physical Metallurgy of Titanium Microalloyed Steel—Recrystallization and Phase Transformation

Xinjun Sun, Zhaodong Li, Xiangdong Huo and Zhenqiang Wang

High-strength Ti microalloyed steels are mainly precipitation hardened ferritic steels. The good combination of ferrite grain refinement and TiC precipitation strengthening plays a key role in obtaining both high strength and high toughness simultaneously for those steels. The ferrite grain refinement, on the one hand, depends on the refinement of austenite grain size and on the other hand depends on the control of transformation temperature. The refinement of austenite grain size mainly depends on the control of the austenite grain growth before hot rolling, and the morphology and size of the recrystallized austenite grains during hot rolling. The principle of controlling the actual transformation temperature of ferrite is to reduce the temperature as much as possible by the control of cooling process and the alloying elements under the condition that TiC can sufficiently precipitate. In addition, the properties of cold-rolled high-strength Ti microalloyed steels are also affected by the recrystallization behavior of cold-rolled ferrite. Therefore, this chapter firstly introduces the grain refinement control and the recrystallization behavior of austenite of hot-rolled Ti microalloyed steels. Secondly, introduce the characteristics of austenite-ferrite transformation and the influence of deformation and alloying elements on hot-rolled

---

X. Sun (✉) · Z. Li
Central Iron & Steel Research Institute, Beijing, China
e-mail: fallbreeze@126.com

Z. Li
e-mail: 3172087@qq.com

X. Huo
Jiangsu University, Zhenjiang, China
e-mail: hxdustb@163.com

Z. Wang
Harbin Engineering University, Harbin, China
e-mail: wangzhe19840203@163.com

© Metallurgical Industry Press, Beijing and Springer Nature Singapore Pte Ltd. 2019
X. Mao (ed.), *Titanium Microalloyed Steel: Fundamentals, Technology, and Products*,
https://doi.org/10.1007/978-981-13-3332-3_4

Ti microalloyed steels. Finally, the recrystallization temperature and time of cold rolled Ti microalloyed steels and the effect of titanium-containing phases on the recrystallization behavior of ferrite are introduced.

4.1 Austenite Recrystallization

There are two main technical measures to control the grain refinement of deformed austenite: one is to obtain fine equiaxed austenite grains by recrystallization controlled rolling; the other is to obtain pancaked austenite by non-recrystallization controlled rolling. Both can be used independently or simultaneously. For microtitanium treated steels (without niobium and vanadium) with Ti/N atomic ratio less than the stoichiometric ratio, a small amount of TiN or nitrogen-richedTi(CN) precipitated at high temperatures is difficult to inhibit the recrystallization of the deformed austenite because the driving force for austenite recrystallization is about twice to the pinning force of TiN or Ti(CN) [1]. Under this condition, a moderately accumulated deformation is not sufficient for the refinement of the austenite and the transformed ferrite, therefore recrystallization controlled rolling is commonly used. For simplex high-Ti microalloyed steel, austenite deformation can induce recrystallization, and also induce precipitation of nano-sized TiC to inhibit the recrystallization so it can affect the austenite grain morphology and size.

In this section, the recrystallization behavior of austenite in single Ti microalloyed steels is introduced. The recrystallization behavior of deformed austenite is affected by the deformation temperature, strain and strain rate as well as the austenite grain size before rolling.

4.1.1 Austenite Grain Refinement During Heating

In microalloyed steels with different titanium and nitrogen contents, the difference in coarsening rate of TiN will affect the austenite grain size control during heating process. Figure 4.1 shows the relationship between the calculated coarsening rate of TiN and the temperature in the 0.02, 0.04, 0.10% Ti steels at 0.0034% N level [see Chap. 3, Eq. (3.114)]. With the increase of heating temperature and titanium content, the coarsening rate (m) of TiN is increased. Therefore, the effect of the inhibition of austenite grain growth by TiN in the micro-titanium treated steels is better than that in the high Ti microalloyed steels, from the coarsening point of view.

As shown in Table 4.1, Medina et al. [2] studied the effect of titanium content on TiN particle size and austenite grain size during heating at different nitrogen levels. When the nitrogen content is low (about 45 ppm, the control level of nitrogen content in converter steelmaking), increasing titanium content from 0.021 to 0.047% leads to a significant increase in the fine TiN particle size and a small increase in the coarse TiN particle size. When the nitrogen content level is high (about 80 ppm, the

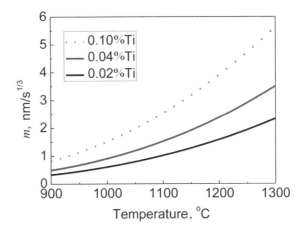

Fig. 4.1 Relationship of TiN coarsening rate (m) and temperature, titanium content in steel (0.0034% N)

Table 4.1 Influence of titanium and nitrogen contents (Ti/N mass ratio) on TiN particle size, volume fraction and austenite grain size [2]

Nitrogen content (wt%)	0.0046	0.0043	0.0080		0.0083	
Titanium content (wt%)	0.021	0.047	0.018		0.031	
Ti/N mass ratio	4.56	10.93	2.25		3.73	
Temperature (°C)	1300	1300	1300	1100	1300	1100
Mean size of fine TiN particles (nm)	23	65.2	13.8	6.5	14.1	7.6
Mean size of coarse TiN particles (μm)	2.32	2.60	1.28	0.49	2.32	1.35
Calculated TiN volume fraction ($\times 10^{-4}$)	1.75	2.36	2.42	3.34	3.94	5.22
Austenite grain size (μm)	302	213	38	23	64	32

control level of nitrogen content in electric furnace steelmaking), titanium content increasing from 0.018 to 0.031% leads to a small increase in the fine TiN particle size and a significant increase in the coarse TiN particle size; the TiN volume fraction increases in both the above two cases. According to Table 4.1 and Fig. 4.2, micro-titanium treatment or single high-Ti microalloying both can inhibit the austenite grain growth during heating but the inhibition effect of single high-Ti microalloying is not as good as micro-titanium treatment. As shown in Fig. 4.2, when the mass ratio of Ti/N is about 2–3, the effect of grain growth inhibition is better.

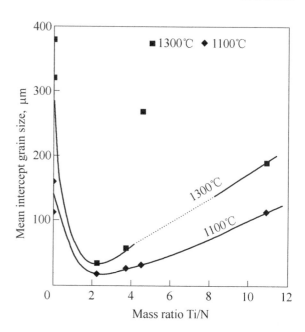

Fig. 4.2 Influence of Ti/N mass ratio on the austenite grain size during heating in the micro-titanium treatment steel [2]

4.1.2 Coarse Austenite Recrystallization

Due to the different thermal histories, the austenitic microstructure before rolling for the thin slab casting and rolling (TSCR) process is significantly different from that for the traditional process. The pre-austenite grain size for the traditional hot rolling process is generally 150–300 μm, while for the TSCR process reaches 700–1000 μm, which is 2–3 times to the traditional one. The original coarse austenite grain significantly increases the difficulty of recrystallization, combining with the inhibition effect of Ti microalloying on the austenite recrystallization, which may lead to incomplete recrystallization and thus reduce the performance of the final product. Therefore, the complete recrystallization of as-cast coarse austenite grains is necessary for thin slab cast and rolled steels microalloyed with titanium under recrystallization controlled rolling.

The authors studied the recrystallization behavior of coarse as-cast austenite grains of a single high-Ti microalloyed steel (0.055%C–1.53%Mn–0.11%Ti) by using thermal simulation experiment. Combining with the actual processing conditions, the experiments were designed, as shown in Fig. 4.3.

The steel samples were performed at 1350 °C for 30 s and then austenitized at 1150 °C for 10 min to obtain coarse austenitic microstructure with grain size of about 800 μm, comparable to that of cast thin slabs, as shown in Fig. 4.4.

The stress and strain curves at different temperatures and strain rates are shown in Fig. 4.5 by using the single pass compression experiment to simulate the heavy deformation of the F1 frame at high temperatures. As can be seen from Fig. 4.5,

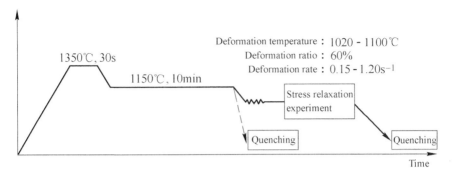

Fig. 4.3 Stress relaxation experiment of the Ti microalloyed steel

Fig. 4.4 Coarse-grained austenite obtained by thermal simulation test

the stress increases with strain under all deformation conditions, which shows a strong work hardening, indicating that the studied steel does not undergo dynamic recrystallization.

Figure 4.6 compares the stress-strain curves of the high-Ti microalloyed steel and the carbon steel with similar prior austenite grain size. The peak stress appeared in the carbon steel shows that the dynamic recrystallization was occurred during deformation. Therefore, high-Ti microalloying has a significant inhibition effect on the dynamic recrystallization of austenite.

Figure 4.7 shows the stress relaxation curves after deformation with 60% reduction of the studied steels at different temperatures and the static recrystallization kinetics calculated from the stress relaxation curves. The recrystallization kinetics curves in

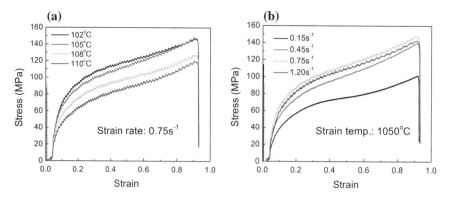

Fig. 4.5 Stress-strain curves of the high-Ti microalloyed steel deformed with 60% reduction (at true strain of 0.92) **a** at different temperatures and **b** at different strain rates

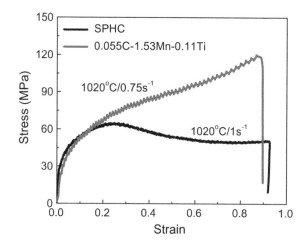

Fig. 4.6 Comparison of stress-strain curves of high-Ti microalloyed steel and carbon steel SPHC

Fig. 4.7 show typical S-type shape, and with the deformation temperature increases, the recrystallization process is accelerated. In spite of the abnormally coarse original austenite grains, the recrystallization rate is still very fast at high temperatures under heavy deformation. The recrystallization is basically finished in 8 s at 1020 °C deformation, and less than 6 and 4 s when the deformation temperature increases to 1050 and 1080 °C, respectively.

Figure 4.8 shows the microstructures of the studied steel (held for 4 and 180 s, respectively, at 1050 °C after 60% deformation, then quenched into water). After one pass of heavy deformation, the grain size is decreased rapidly from about 800 to 30 μm, and increased to about 100 μm after holding for 180 s. The

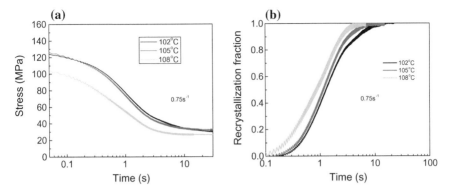

Fig. 4.7 **a** Stress relaxation curves and **b** static recrystallization kinetics curves for the studied steel

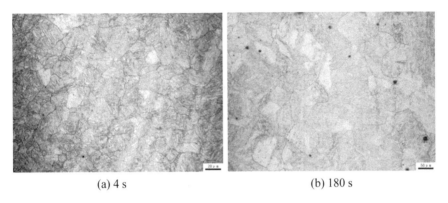

Fig. 4.8 Quenched microstructures of the studied steel at different holding time

above results were further verified by industrial experiments on a TSCR line (0.05%C–0.48%Mn–0.10%Ti, deformation temperature is 1050 °C, reduction for each pass is 57%, and the interval time is 5–7 s). The results show that complete recrystallization of as-cast coarse austenite grains for the high Ti microalloyed steel can be realized by high temperature heavy deformation during F1 pass, the grain size after recrystallization is about 108 μm, as shown in Fig. 4.9. Based on the above analysis, it can be concluded that the static recrystallization of Ti microalloyed steels can be basically completed between F1 and F2 under the TSCR process, and mixed grain structure due to incomplete recrystallization can be avoided.

Niobium microalloyed steel is different from Ti microalloyed steel because it is quite difficult for completion of static recrystallization. The initial austenite grain size of a Nb-microalloyed steel (0.06%C–1.20%Mn–0.045%Nb) is about 500 μm, as shown in Fig. 4.10a. After 50% deformation at a strain rate of 3 s^{-1} at 1050 °C followed by holding for 10 s, a certain amount of coarse original austenite is still

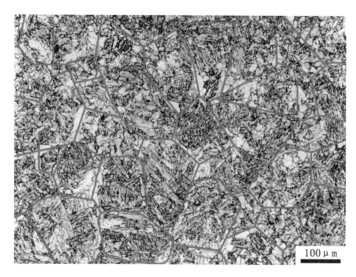

Fig. 4.9 Austenitic microstructure of the high Ti microalloyed steel after F1 rolling

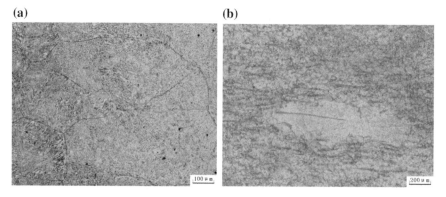

Fig. 4.10 Austenite microstructure of Nb microalloyed steel before and after deformation: **a** coarse austenite grains and **b** incompletely recrystallized grains

retained due to incomplete recrystallization, as shown in Fig. 4.10b, which is the reason for the final heterogeneous microstructure in niobium microalloyed steel produced by TSCR. Based on the above analysis, it can be seen that the combination of Ti microalloying technology and TSCR process is better than that for niobium microalloyed steel.

4.1.3 Recrystallization of Conventional Grained Austenite

As mentioned earlier, the grain size of coarse cast austenite of Ti microalloyed steel is rapidly refined after F1 rolling. The recrystallization in subsequent rolling process is the recrystallization of refined austenite by F1 rolling, that is, conventional grained austenite recrystallization. Due to the difficulty of dynamic recrystallization of Ti microalloyed steel, the static recrystallization will be mainly discussed hereinafter.

Austenite static recrystallization is a thermally activated process, so the recrystallization fraction can be described in an Avrami equation:

$$f = 1 - \exp\left(-0.693\left(\frac{t}{t_{0.5}}\right)^n\right) \quad (4.1)$$

where f is the recrystallization fraction, t is the time, $t_{0.5}$ is the time required for 50% recrystallization, and n is the constant.

The relation between $t_{0.5}$ and the original austenite grain size d_0, strain rate $\dot{\varepsilon}$, strain ε, and deformation temperature T can be expressed as:

$$t_{0.5} = Ad_0^m \varepsilon^{-p} \dot{\varepsilon}^{-q} \exp\left(\frac{Q_{rex}}{RT}\right) \quad (4.2)$$

The recrystallization of two Ti microalloyed steels (0.055%C–1.53%Mn–0.11%Ti and 0.052%C–0.44%Mn–0.08%Ti) were studied by stress relaxation test, the recrystallization kinetics equation was determined to provide a theoretical basis for the optimization of the rolling process.

The stress relaxation curves and recrystallization kinetics curves of the studied steel a (0.055%C–1.53%Mn–0.11%Ti) and steel b (0.052%C–0.44%Mn–0.08%Ti) deformed to different strain with a strain rate of 1 s^{-1} at 1020 °C are shown in Figs. 4.11 and 4.12, respectively. In general, with the increase of the strain, the recrystallization kinetic curve shifts to the left, indicating that the recrystallization rate is increased. The reason is that high density of defects in the largely deformed austenite results in an increase in the driving force for recrystallization.

According to the static recrystallization kinetics curves in Figs. 4.11 and 4.12, $t_{0.5}$ of the two studied steels are obtained, as shown in Fig. 4.13. The p values in Eq. (4.2) obtained by linear fitting are 2.12 for steel a and 2.06 for steel b, respectively. The $t_{0.5}$ value of steel a is larger than that of steel b under the same deformation condition, this is mainly due to the higher Ti content, that is, titanium has an inhibition effect on recrystallization. For steel b, when the strain is greater than 0.8, $t_{0.5}$ is almost unchanged, which is an important sign for dynamic recrystallization occurred during the deformation and sub-dynamic recrystallization occurred after the deformation.

The stress relaxation curves and recrystallization kinetics curves of steel a and steel b deformed to a strain of 0.52 with a strain rate of 1 s^{-1} at different temperatures are shown in Figs. 4.14 and 4.15, respectively. With the increase of the deformation temperature, the recrystallization kinetic curve shifts to the left, indicating that the recrystallization rate is increased.

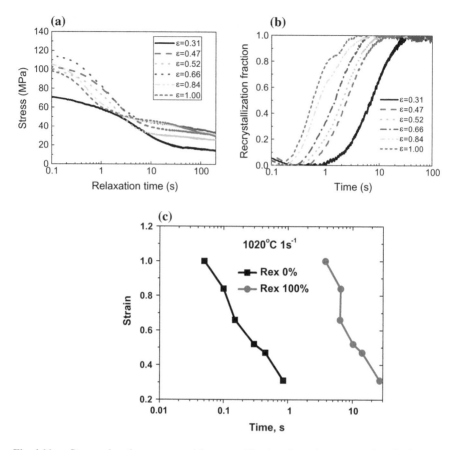

Fig. 4.11 a Stress relaxation curves and **b** recrystallization dynamics curves of studied steel a deformed to different strain and **c** influence of strain on recrystallization time

The density of defects in austenite is lower at higher temperatures, leading to a decrease in the driving force for recrystallization. However, the increased rates of atom diffusion and grain boundary migration result in an increase in the recrystallization rate.

According to the recrystallization dynamics curves in Figs. 4.14 and 4.15, the relationship between $t_{0.5}$ and $1/T$ of the two steels can be obtained, as shown in Fig. 4.16. The recrystallization activation energies of steel a and steel b obtained by linear fitting are 302189 and 235184 J/mol, respectively. The activation energy is an important parameter to characterize the ability of atom diffusion. The larger the activation energy is, the higher the energy barrier that the atomic diffusion needs to pass, the more difficult the diffusion is, and the slower the recrystallization is. Obviously, the steel a has a higher recrystallization activation energy, which is caused by the interaction between Ti and Fe atoms, the inhibition effect of titanium on recrystallization can be reflected through the recrystallization activation energy.

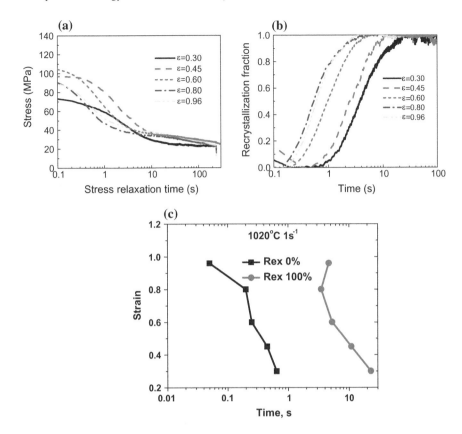

Fig. 4.12 **a** Stress relaxation curves and **b** recrystallization dynamics curves of studied steel (**b**) deformed to different strain and **c** influence of strain on recrystallization time

The curves of stress relaxation and recrystallization kinetics of steel a and steel b deformed to a strain of 0.52 at 1020 °C under different strain rates are shown in Figs. 4.17 and 4.18. It can be seen that the increase of strain rate will accelerate the recrystallization process. The recovery of the dislocation caused by the deformation of austenite cannot be occurred at higher strain rates, resulting in a higher dislocation density under the same strain, thus increasing the driving force for recrystallization, and leading to the increased recrystallization rate.

According to the recrystallization dynamics curves of Figs. 4.17 and 4.18, the relationship between the $t_{0.5}$ of the two steels and the strain rate can be obtained, as shown in Fig. 4.19. After the linear fitting, the q values, in Eq. (4.2), are 0.296 for steel a and 0.31 for steel b, respectively. The two values are similar.

Finally, according to the Eq. (4.1), the time index n can be obtained from the slope of the curve $(\log[-\ln(1-f)] \sim \log t)$ by linear fitting, and the recrystallization kinetics equation are obtained:

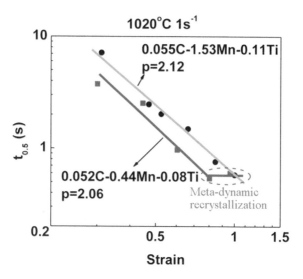

Fig. 4.13 Relationship between the $t_{0.5}$ and strain

For steel a:

$$f = 1 - \exp\left[-0.693\left(\frac{t}{t_{0.5}}\right)^{1.10}\right] \quad (4.3)$$

$$t_{0.5} = 2.8091 \times 10^{-15} d_0 \varepsilon^{-2.12} \dot{\varepsilon}^{-0.296} \exp\left(\frac{302189}{RT}\right) \quad (4.4)$$

For steel b:

$$f = 1 - \exp\left[-0.693\left(\frac{t}{t_{0.5}}\right)^{1.35}\right] \quad (4.5)$$

$$t_{0.5} = 8.385 \times 10^{-13} d_0 \varepsilon^{-2.14} \dot{\varepsilon}^{-0.31} \exp\left(\frac{235184}{RT}\right) \quad (4.6)$$

Medina and Mancilla [3] studied the static recrystallization of a Ti microalloyed steel (0.145%C–1.10%Mn–0.075%Ti–0.0102%N) by torsional deformation test, and a similar recrystallization kinetics equation was obtained:

$$f = 1 - \exp\left[-0.693\left(\frac{t}{t_{0.5}}\right)^{4.81\exp(-\frac{20000}{RT})}\right] \quad (4.7)$$

$$t_{0.5} = 3.702 \times 10^{-12} d_0 \varepsilon^{-2.15} \dot{\varepsilon}^{-0.44} \exp\left(\frac{22700}{RT}\right) \quad (4.8)$$

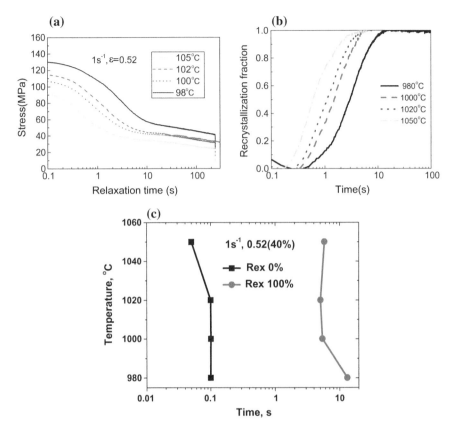

Fig. 4.14 **a** Stress relaxation curves and **b** recrystallization kinetics curves of steel a at different temperatures and **c** influence of strain on recrystallization time

Using the above equations [Eqs. (4.3) and (4.4)] of static recrystallization kinetics, and combining with the continuous hot strip rolling line in a steel mill, the interstand static recrystallization behavior of high-Ti microalloyed high strength steels during rolling is calculated and analyzed, the results are shown in Table 4.2. It can be seen that complete recrystallization is occurred between F1–F4, only partial recrystallization is occurred after F4, and non-recrystallization rolling becomes dominant after F5.

It should be noted that the recrystallization kinetic equation described above only involves the delayed effect of solid solution titanium on recrystallization. In the actual rolling process conditions, especially during the last rolling passes, strain induced TiC precipitation may occur, thereby the recrystallization will be further delayed or inhibited.

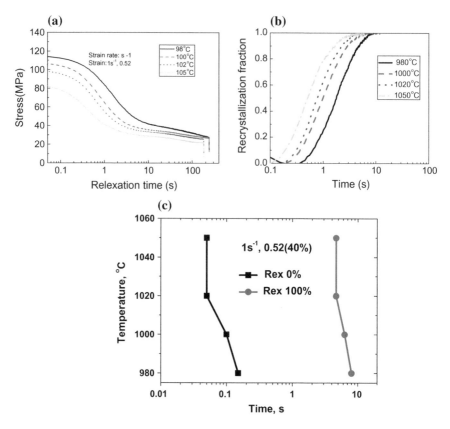

Fig. 4.15 **a** Stress relaxation curves and **b** recrystallization kinetics curves of steel b at different temperatures and **c** influence of strain on recrystallization time

4.1.4 Interaction Between Strain Induced Precipitation and Austenite Recrystallization

When the deformation temperature is reduced to a certain level, the austenite deformation will induce precipitation of nano-sized TiC, thereby inhibiting the dynamic or static recrystallization of austenite. Addition of 0.1% titanium significantly increases the temperature of the static recrystallization of deformed austenite and delays the recrystallization time (Fig. 3.30a, b), but it is beneficial for the austenite recrystallization with increasing the manganese content in the Ti microalloyed steel (Fig. 3.30c, d and Fig. 3.31). Addition of 0.20% Mo enhances the recrystallization inhibition effect of TiC precipitated at 925 °C and below (Fig. 3.35). Figure 4.20 shows the austenite morphology of Ti and Ti–Mo microalloyed steels after hot rolling at 925 °C for different time. It can be seen that the recrystallization is mainly completed

Fig. 4.16 Relationship between $t_{0.5}$ and $1/T$ of austenite recrystallization

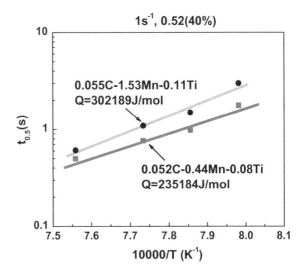

after holding for 600 s in the Ti microalloyed steel, and the deformed microstructure can be hardly observed. When the holding time reaches to 1800 s, the austenite grain growth takes place obviously. However, for the Ti–Mo microalloyed steel, the recrystallization does not occur significantly at short holding time of 600 s or even long holding time of 7200 s, and the deformed grains are still remained. The phenomenon indicates that the addition of Mo significantly inhibits the austenite recrystallization of Ti microalloyed steel after rolling.

The influence of microalloying elements on the inhibition of dynamic recrystallization of deformed austenite or the increase in non-recrystallization temperature is shown in Fig. 1.2. Under the same atom numbers, Niobium shows the strongest effect, and then does the Titanium, Vanadium is the weakest. The non-recrystallization temperature can be estimated as follows [4]:

$$T_{nr} = 887 + 464\,C + \left(6445\,Nb - 644\,Nb^{1/2}\right) + \left(732\,V - 230\,V^{1/2}\right) \\ + 890\,Ti + 363\,Al - 357\,Si \quad (4.9)$$

Equation (4.9) can be applied to the following composition systems: $0.04 \leq C \leq 0.17\%$, $0.41 \leq Mn \leq 1.90\%$, $0.15 \leq Si \leq 0.50\%$, $0.002 \leq Al \leq 0.650\%$, $Nb \leq 0.060\%$, $V \leq 0.120\%$, $Ti \leq 0.110\%$, $Cr \leq 0.67\%$, $Ni \leq 0.45\%$.

For the inhibition of static recrystallization of deformed austenite, Medina and Mancilla [3] compared the static recrystallization of deformed austenite and the kinetics of strain induced precipitation of 0.105%C–0.0112%N–0.042%N, 0.113%C–0.0144%N–0.095%V and 0.145%C–0.0102%N–0.075%Ti steels. The results showed that the critical temperature of static recrystallization in niobium and vanadium microalloyed steels is about 100 °C lower than the solid solution temperature, while in Ti microalloyed steel the temperature is about 200 °C. The fastest

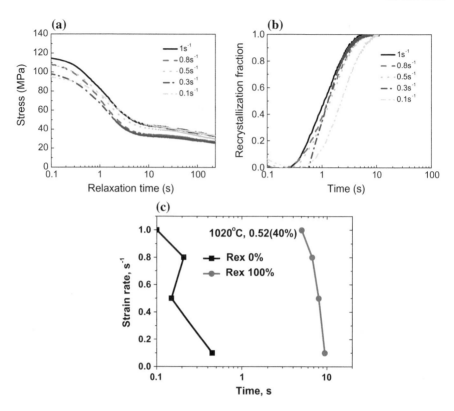

Fig. 4.17 **a** Stress relaxation curves and **b** recrystallization kinetics curves of steel a at different strain rate and **c** influence of strain on recrystallization time

precipitation temperature for niobium microalloyed steel is higher than that for titanium and vanadium microalloyed steels, while the fastest precipitation temperature for vanadium microalloyed steel is similar to that for Ti microalloyed steel, but the fastest precipitation time of vanadium microalloyed steel is the shortest. Therefore, compared with niobium and vanadium microalloyed steels, the static recrystallization of deformed austenite in Ti microalloyed steel is easily to occur. Certainly, the above cases may change when the content of Nb, V, or Ti changes.

4.2 Austenite-Ferrite Transformation

Due to the high cost performance of precipitation strengthening, Ti microalloying is widely used in hot-rolled coil with ferrite microstructure in the past decade. In addition to the precipitation of TiC in austenite, interphase precipitation during

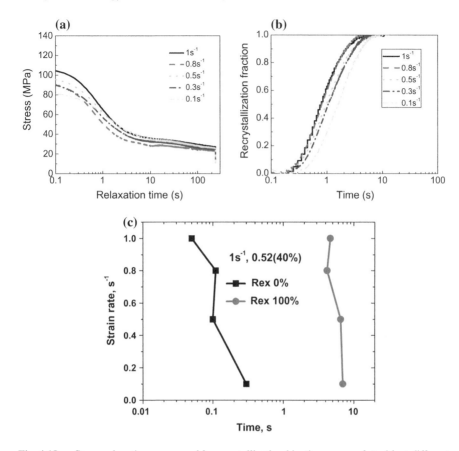

Fig. 4.18 **a** Stress relaxation curves and **b** recrystallization kinetics curves of steel b at different strain rate and **c** influence of strain on recrystallization time

austenite-ferrite transformation and precipitation in ferrite may also occur. The interphase precipitation of TiC is strongly influenced by the type of ferrite transformation and the cooling rate. The precipitation of TiC in ferrite is very sensitive to the ferrite type and the coiling temperature. Therefore, the isothermal transformation behavior of Ti microalloyed steels at different temperatures, the continuous cooling transformation behavior at different cooling rates, the type of microstructure and the hardness level are worthy of attention.

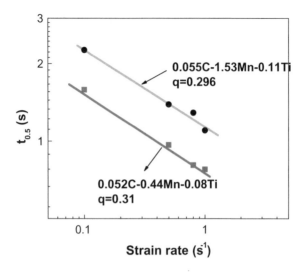

Fig. 4.19 Relationship between austenite recrystallization $t_{0.5}$ and strain rate

Table 4.2 Deformation parameters of mill housing and interstand static recrystallization fraction

No.	Deformation temperature (°C)	Entry thickness (m)	Exit thickness (m)	Entry speed(m/s)	Exit speed (m/s)	Strain (%)	Strain rate (s^{-1})	Interval time (s)	f_{rex}
F1	1010	0.060	0.030	0.30	0.60	50	2.25	9.17	1
F2	990	0.030	0.015	0.60	1.20	50	6.38	4.58	1
F3	970	0.015	0.009	1.20	2.00	40	14.86	2.75	0.98
F4	950	0.009	0.006	2.00	2.86	30	25.78	1.93	0.42
F5	930	0.006	0.004	2.86	4.08	30	51.40	1.35	0.20
F6	910	0.004	0.004	4.08	5.10	20	67.25	/	/

4.2.1 Isothermal Transformation

4.2.1.1 Microstructure Characteristics of Phase Transformation

The isothermal transformation of undeformed austenite was studied by Gleeble in a Ti microalloyed steel (0.046%C–1.5%Mn–0.1%Ti). The steel samples were heated to 1200 °C for austenitizing, then fast cooled to 550–725 °C for 30 min isothermal treatment, and finally, quickly cooled to room temperature. Figure 4.21 shows the type of the microstructure transformed from low temperature to high temperature: bainite (550–575 °C) → bainite + massive transformation ferrite (600 °C) → massive transformation ferrite (625 °C) → polygonal ferrite (650–725 °C). The polygonal ferrite grain size decreases with the decrease of isothermal temperature except at 750 °C. In Fig. 4.21 the microstructure in black is martensite (the same below), which is formed from the untransformed austenite during rapid cooling. The higher the isothermal temperature is, the greater the amount of martensite is. The reasons

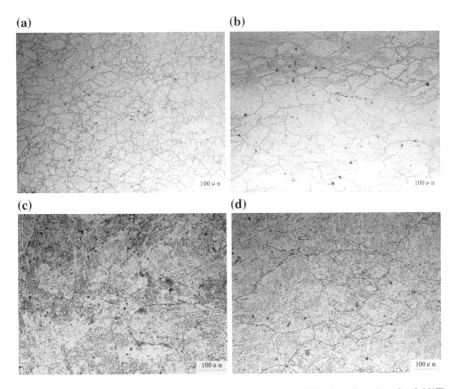

Fig. 4.20 Effect of Mo on recrystallization of deformed austenite of Ti microalloyed steel: **a** 0.1%Ti steel, 600 s, **b** 0.1%Ti steel, 1800 s, **c** 0.1%Ti–0.2%Mo steel, 600 s and **d** 0.1%Ti–0.2%Mo steel, 7200 s

for the massive transformation at lower temperatures are: (1) poor hardenability due to the low carbon content of the steel; (2) inhibition of the polygonal or proeutectoid ferrite transformation due to the large-grained austenite before transformation.

Figure 4.22 shows the transformation temperature dependence of micro-hardness of the studied steel. At higher temperatures, hardness is related to the proeutectoid ferrite or massive ferrite, while at lower temperatures, hardness is associated with the bainite (the same below). As the ferrite grain size increases, the difference of TiC precipitates in ferrite can be reflected by the hardness values with small loading. Combining with Fig. 4.21, the peak hardness corresponding to the polygonal ferrite is higher than HV260, which indicates that the TiC particles are sufficiently precipitated in the polygonal ferrite matrix and therefore leading to a significant precipitation strengthening effect. The hardness below the peak temperature decreases with the isothermal temperature decreases. This is because the TiC particles at rapid transformation rate are not sufficiently precipitated, though the dislocation density in massive transformation ferrite or bainite is high. The lower the temperature is, the less the amount of TiC precipitates. The hardness of the proeutectoid polygo-

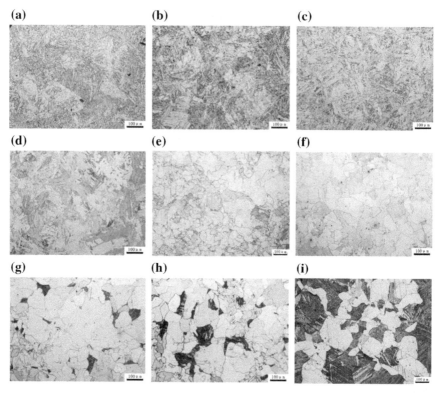

Fig. 4.21 Optical micrographs showing microstructure transformed from undeformed austenite of the studied steel at different temperatures: **a** 550 °C, **b** 575 °C, **c** 600 °C, **d** 625 °C, **e** 650 °C, **f** 675 °C, **g** 700 °C, **h** 725 °C, **i** 750 °C

nal ferrite above the peak temperature decreases more rapidly with the isothermal temperature decreases.

Wang et al. [5] investigated the influence of isothermal treatment on nano-sized carbide precipitation of a Ti microalloyed steel (<0.10%C–0.11%Ti). The steel sample was austenitized at 1200 °C for 3 min after homogenization treatment, and then quickly cooled to 650–750 °C, annealed for 30 or 60 min, microstructural characteristics and hardness are shown in Fig. 4.23. The hardness evolution of the polygonal ferrite is consistent with that at high temperatures in Fig. 4.22. As shown in Fig. 4.24, transmission electron microscopy (TEM) images show that the interphase precipitation of TiC is mainly occurred at temperatures higher than 700 °C (containing 700 °C), and the TiC particles are coarsened due to the increase in annealing temperature, thus the hardness is decreased. At temperatures lower than 675 °C (containing 675 °C), fine TiC particles are randomly distributed in the ferrite matrix, which are precipitated from the supersaturated ferrite solid solution. Thus, the high strength of the ferrite is attributed from the nano-sized TiC strengthening and solid solution strengthening.

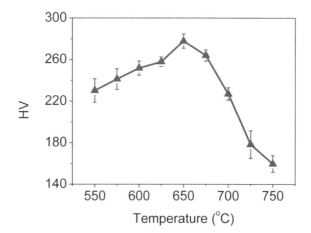

Fig. 4.22 The relationship between the isothermal temperature and the hardness transformed from undeformed austenite of the studied steel

The isothermal transformation of deformed austenite was also studied by using Gleeble for a Ti microalloyed steel (0.046%C–1.5%Mn–0.1%Ti). The steel was heated to 1200 °C for austenitization, then rapidly cooled to 975 °C at a strain rate of 1 s^{-1} for one pass compression of 30%, then rapidly cooled to 550–725 °C for 30 min isothermal treatment and finally rapidly cooled to room temperature. Compared with the undeformed condition, the relation between transformation microstructure and annealing temperature is almost unchanged when the austenite is deformed at 975 °C. But the ferrite transformation at high temperatures is accelerated, more fraction of polygonal ferrite are formed, and the ferrite grain size is refined significantly, however, the microhardness peak of ferrite is not obviously affected. According to the change of microstructure type, morphology and hardness, it can be deduced that the deformation of austenite at 975 °C results in grain refinement by recrystallization, which affects the isothermal transformation of ferrite, but hardly affects the strengthening effect of TiC.

Zhao [6] studied the isothermal transformation after double pass deformation of three Ti microalloyed steels (0.065%C–1.8%Mn–(0.08, 0.13, 0.17)%Ti) by Gleeble. The steel samples were heated to 1250 °C for austenitization, then quickly cooled to 1050 and 900 °C for two pass compression of 30%, at a strain rate of 1 and 5 s^{-1}, respectively, followed by rapid cooling to 550–725 °C for 15 min annealing and finally cooled to room temperature. As shown in Fig. 4.25, when the temperature is increased from 550 to 725 °C, the microstructural evolution of the steel with 0.08% Ti is: granular bainite (550–575 °C) → quasi-polygonal ferrite (600 °C) → polygonal ferrite (600–725 °C). In the temperature range of 600–675 °C, the grain size of polygonal ferrite decreases with the increase of isothermal temperature. However, the grain size of polygonal ferrite at the isothermal temperature of 700 and 725 °C is smaller than that at a lower temperature, which is the same as that of 0.046%C–1.5%Mn–0.1%Ti undeformed Ti microalloyed steel (Fig. 4.21). This may be related to the short time for phase transformation which leads to a large amount of untransformed austenite, and the interphase precipitates at high temperature which

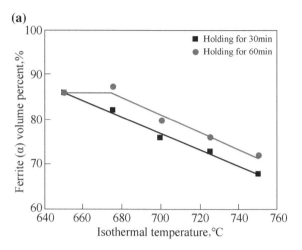
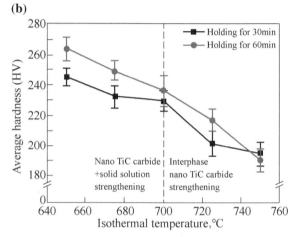

Fig. 4.23 The relationship between the annealing temperature of Ti-microalloyedsteel (<0.10%C–0.11%Ti) at different times and **a** the ferrite volume fraction and **b** the average microhardness of the ferrite [5]

inhibits grain growth. The differences between the two steels are that, the strain of 0.08% Ti steel is larger than that of 0.1% Ti steel, and massive ferrite transformation does not occur and polygonal ferrite can be obtained at lower temperatures in the 0.08% Ti steel.

Increasing the titanium content from 0.08 to 0.17% has little influence on the microstructural evolution with the isothermal temperature, but has a significant effect on the microhardness, as shown in Fig. 4.26. The hardness of 0.08% Ti steel generally decreases with the increase of the final cooling temperature, and the highest hardness value actually appears at 550 °C in fully bainitic microstructure. While when the titanium content is increased, that is, the peak hardness of 0.13 and 0.17 %Ti steels is found in the polygonal or quasi-polygonal ferrite. With the increase of titanium content, the peak hardness gradually increases. The hardness of polygonal or quasi-polygonal ferrite is higher than that of bainite, bainite/martensite and

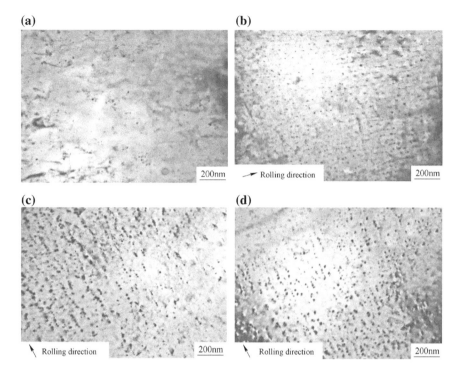

Fig. 4.24 TEM images showing the TiC preciptates at **a** 675 °C, **b** 700 °C, **c** 725 °C and **d** 750 °C in the ferrite matrix of Ti-microalloyed steel for 1 h of heat treatment [5]

ferrite/martensite, indicating that TiC particles are precipitated sufficiently in the microstructure or temperature range, resulting in strong precipitation strengthening. The TiC particles are not sufficiently precipitated during bainite transformation at lower temperatures, and coarsened during ferrite transformation at higher temperatures, resulting in weak precipitation strengthening.

4.2.1.2 Effects of Alloying Element

The isothermal transformation of undeformed austenite was also studied in 0.04%C–0.5%Mn–0.1%Ti and 0.041%C–1.0%Mn–0.1%Ti microalloyed steels by using the same Gleeble process as 0.046%C–1.5%Mn–0.1%Ti microalloyed steel. The steel samples were austenitized at 1200 °C and then cooled to 550–725 °C for either 30 min or 60 min annealing, and finally cooled to room temperature.

Figure 4.27 shows the type of the microstructure transformed from low temperature to high temperature of the 0.5% Mn steel: bainite + massive ferrite (550 °C) → massive ferrite (575–650 °C) → massive ferrite + polygonal ferrite (675 °C) → polygonal ferrite (700–750 °C).

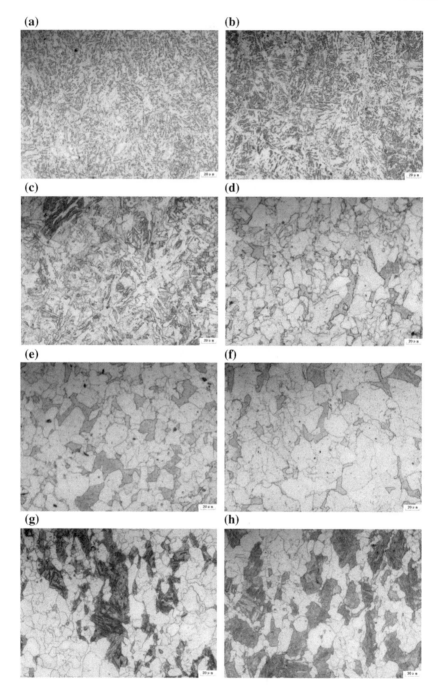

Fig. 4.25 Microstructure isothermally transformed from deformed austenite of Ti microalloyed steel (0.065%C–1.8%Mn–0.08%Ti) at different temperatures: **a** 550 °C, **b** 575 °C, **c** 600 °C, **d** 625 °C, **e** 650 °C, **f** 675 °C, **g** 700 °C, **h** 725 °C [6]

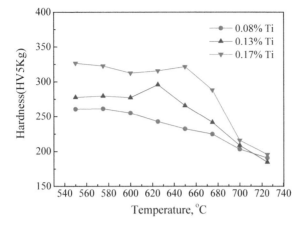

Fig. 4.26 Hardness of the microstructure isothermally transformed from deformed austenite in microalloyed steels with different Ti contents [6]

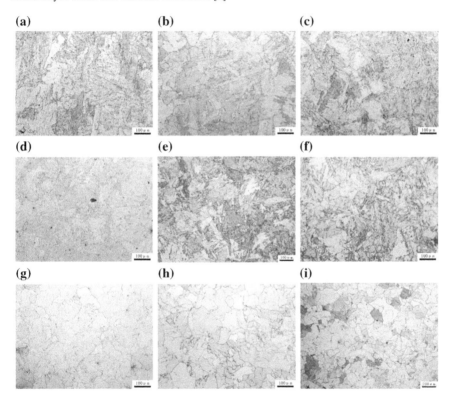

Fig. 4.27 Microstructure isothermally transformed at different temperatures from undeformed austenite of the 0.044%C–0.5%Mn–0.1%Ti microalloyed steel: **a** 550 °C, **b** 575 °C, **c** 600 °C, **d** 625 °C, **e** 650 °C, **f** 675 °C, **g** 700 °C, **h** 725 °C, **i** 750 °C

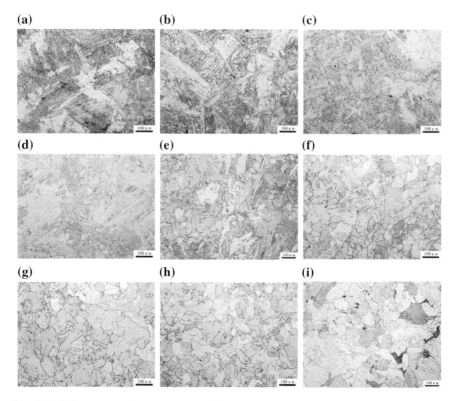

Fig. 4.28 Microstructure isothermally transformed at different temperatures from undeformed austenite of the 0.044%C–1.0%Mn–0.1%Ti microalloyed steel: **a** 550 °C, **b** 575 °C, **c** 600 °C, **d** 625 °C, **e** 650 °C, **f** 675 °C, **g** 700 °C, **h** 725 °C, **i** 750 °C

Figure 4.28 shows the type of the microstructure transformed from low temperature to high temperature of the 1.0% Mn steel: bainite (550–575 °C) → bainite + massive ferrite (600–650 °C) → polygonal ferrite (675–750 °C). It can be seen that the microstructural evolution inthe Ti microalloyed steels with different manganese contentsare basically the same, the difference is that with the increase of manganese content, the volume fraction of the polygonal ferrite isothermally transformed for the same time at high temperatures (such as 750 °C) is decreased, and the lowest temperature at which polygonal ferrite can be fully obtained decreases from 700 °C for the 0.5% Mn steel to 675 °C for the 1.0% Mn steel, and to 650 °C for the 1.5% Mn steel, and the grain size of polygonal ferrite is also increased. This is mainly due to the fact that Mn, an austenite forming element, reduces the equilibrium temperature of proeutectoid ferrite transformation and driving force, thus inhibiting the ferrite nucleation and growth.

Figure 4.29 shows the microhardness of the three steels with the isothermal temperature. Hardness at higher temperatures corresponds to the hardness of the proeutectoid ferrite or the massive ferrite, and the hardness at lower temperatures at which

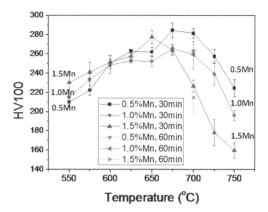

Fig. 4.29 Influence of Mn content on the hardness of the isothermal transformation microstructure of Ti-microalloyed steel

no ferrite is formed corresponds to the hardness of bainite. It can be seen that all the three steels have peak hardness, which corresponds to the polygonal ferrite. With the increase of manganese content, the hardness of the proeutectoid ferrite transformed at high temperatures is decreased, and the hardness of bainite/massive transformation ferrite formed at low temperatures is increased. In addition, with the increase in the isothermal time at high temperatures, the hardness of the polygonal ferrite is decreased.

Figure 4.30 shows the TiC precipitates in ferrite isothermally transformed at 725 °C of the three steels. It can be seen that with the increase of manganese content, a part of the precipitates are distributed in a chain-like manner, and the particle size is obviously coarsened. The results can explain the decrease in hardness of the polygonal ferrite isothermally formed at high temperatures as the manganese content increases. The phenomenon of chain-like precipitates with large size at high temperatures in the steel with higher Mn content can be explained as follows: Firstly, the element diffusion is fast at high temperatures, but the interface migration rate is not very fast. Secondly, increasing manganese content reduces the driving force for ferrite growth and the migration rate of ferrite grains into austenite at high temperatures, which further promotes the nucleation and growth of TiC precipitates along the interface. However, the actual production process of steel is continuous cooling process, increasing the manganese content significantly reduces the actual ferrite transformation temperature, which significantly reduces the diffusion capacity of the elements and is conducive to obtain tiny TiC particles.

The isothermal transformation characteristics of undeformed austenite of 0.044%C–1.5%Mn–0.10%Ti–0.20%Mo steel were studied by using the same Gleeble process as 0.046%C–1.5%Mn–0.10%Ti steel. Figure 4.31 shows the microstructure of isothermal transformation of Ti-Mo steel. Polygonal ferrite microstructure can be fully obtained at temperatures in the range of 700–725 °C. When the isothermal temperature is reduced to 700 °C, the volume fraction of polygonal ferrite is obviously reduced and the austenite transforms incompletely. This is different from the microstructural evolution in 1.5%Mn–0.1%Ti steel, which may be related to

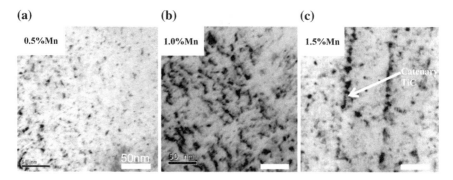

Fig. 4.30 Influence of Mn content on interphase precipitation at 725 °C for isothermal transformation in Ti-microallyed steel

the effect of molybdenum on precipitation. Firstly, molybdenum is a strong carbide formation element, which can increase the solubility of carbon in the austenite and ferrite and also can be precipitated in the MC phase to fix a portion of carbon. So it is more likely to form fully ferritic microstructure in Mo-added ultra-low carbon steel than that in Mo-free steel. Secondly, Mo plays a role in promoting the precipitation of MC during phase transformation [7], so when the ferrite transformation once occurs at higher temperatures, the MC phase could be precipitated largely due to the effect of Mo and the high element diffusibility at high temperatures, thus fixing the carbon in the steel, which leads to a rapid decrease in the carbon content in austenite matrix, thereby promoting the ferrite transformation. The interphase precipitates are nucleated repeatedly, and the precipitation and phase transformation promote each other, resulting in ferrite transformation at high temperatures. When the temperature is decreased to a certain value, the decrease in the diffusion capacity of the elements slows the precipitation of the MC phase, thereby suppressing the ferrite transformation. When the temperature is dropped to the range of 650–675 °C, only a small amount of ferrite is formed, and the rest of the austenite will be retained after the isothermal transformation. As the temperature is further reduced, the bainite and massive transformation ferrite are formed.

Figure 4.32 shows the microhardness of ferrite or bainite for 1.5%Mn–0.1%Ti microalloyed steel (without Mo) and Ti–Mo microalloyed steel. It can be seen that the hardness of Ti–Mo steel is higher than that of Ti microalloyed steel in the isothermal transformation temperature range, and the addition of molybdenum shows better precipitation strengthening effect.

Figure 4.33 gives TEM images of interphase precipitates in 1.5%Mn–0.1%Ti microalloyed steel and Ti–Mo microalloyed steel annealed at 750 °C. It can be seen that the precipitation size in Ti–Mo microalloyed steel is smaller than that in Ti microalloyed steel. Figure 4.34 shows the morphology of nano-sized precipitates in ferrite isothermally transformed at different temperatures of Ti–Mo microalloyed steel. With the decrease of temperature, the size of the precipitates decreases, which is consistent with the results observed by Wang et al. [5] in a Ti microalloyed steel

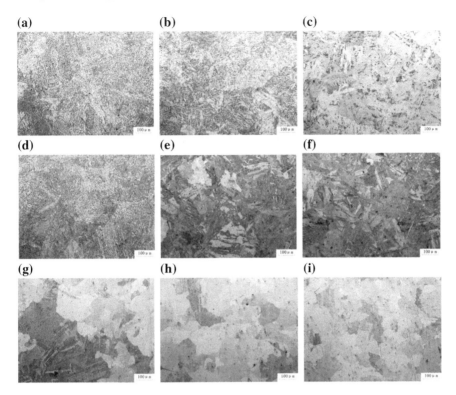

Fig. 4.31 Isothermal transformation microstructures of Ti–Mo steel at different temperatures: **a** 550 °C, **b** 575 °C, **c** 600 °C, **d** 625 °C, **e** 650 °C, **f** 675 °C, **g** 700 °C, **h** 725 °C, **i** 750 °C

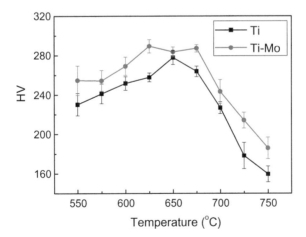

Fig. 4.32 Influence of Mo on the hardness of the microstructure obtained by isothermal transformation of Ti microalloyed steel. (Ti represents the 1.5%Mn–0.1%Ti steel)

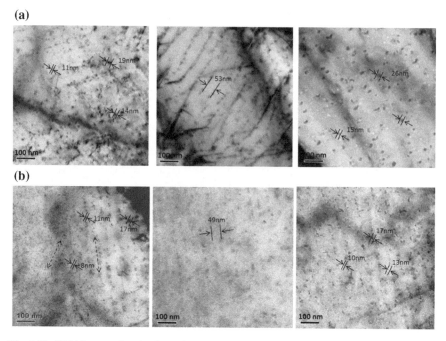

Fig. 4.33 TEM images showing interphase precipitates at 750 °C in the **a** Ti microalloyed steel and **b** Ti–Mo microalloyed steel

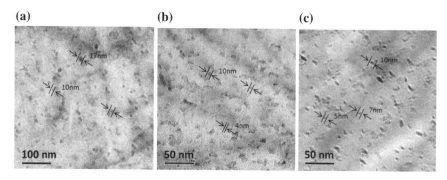

Fig. 4.34 Influence of isothermal temperature on the size of interphase precipitates in Ti–Mo microalloyed steel: **a** 750 °C, **b** 725 °C, **c** 700 °C

(Fig. 4.24). The two statistical diameters of precipitates in ferritic matrix of the two studied steels at different temperatures are shown in Fig. 4.35. It can be seen that the average diameters of precipitates in the Ti–Mo microalloying steel is smaller than that of Ti microalloyed steel at any studied temperature. It is the main reason that the higher hardness of ferrite is obtained in Ti–Mo microalloyed steel.

4 Physical Metallurgy of Titanium Microalloyed Steel …

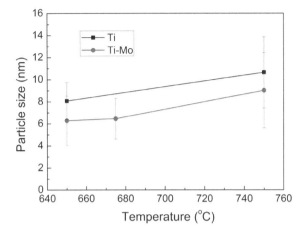

Fig. 4.35 Mean particle size in the Ti–Mo microalloyed steel and Ti microalloyed steel (1.5%Mn–0.1%Ti)

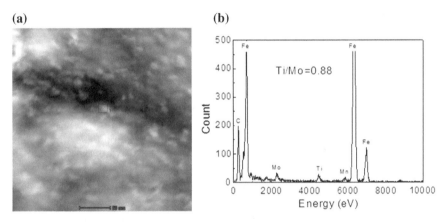

Fig. 4.36 **a** HAAADF image showing nano-sized (Ti, Mo)C particles and **b** EDS analysis

Figure 4.36 shows a high-angle annular dark field image (HAADF-STEM) and energy dispersive spectroscopy (EDS) of typical nano-sized (Ti, Mo)C particles. The contrast of HAADF-STEM image is proportional to the square of the atomic number, so the white precipitates should have a higher atomic number of elements than the Fe matrix. The atomic number of Fe is 26, Mo is 42, Ti is 22, C is 12, thus the contrast indicates that the precipitates should contain a higher Mo content. The EDS shows that the Mo content in the (Ti, Mo) C phase is higher than Ti and Ti/Mo = 0.88 (see Fig. 4.36b). Mo involved in the MC phase composition is directly confirmed, therefore, adding Mo also increases the total amount of MC precipitates. It is possible to explain the phenomenon that the ferrite has higher hardness for Ti–Mo microalloyed steel than that for pure Ti microalloyed steel by the effect of Mo on the refinement of MC particle size.

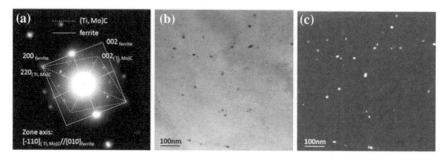

Fig. 4.37 TEM images showing the crystal structure of (Ti, Mo)C and the orientation relationship with respect to the ferrite matrix: **a** electron diffraction (SAD), **b** bright field and **c** dark field

In addition, the addition of Mo does not change the Baker-Nutting relation between the MC phase and the ferrite matrix, as shown in Fig. 4.37, $(001)_{(Ti, Mo) C} // (001)_\alpha$, $[-110]_{(Ti, Mo) C} // [010]_\alpha$.

4.2.2 Continuous Cooling Transformation

The 0.055%C–1.53%Mn–0.11%Ti steel samples were austenitized at 1150 °C for 3 min using a Gleeble simulator, then cooled to 1050 °C for a deformation with 50% reduction at a strain rate of 1 s^{-1} and cooled to 920 °C for a deformation with 30% reduction at a strain rate of 5 s^{-1}, respectively, after deformation the samples were cooled to room temperature at different cooling rates or first cooled to 550–600 °C at a cooling rate of 20 °C/s, and then slowly cooled to 450 °C at a cooling rate of 0.2 °C/s, followed by air cooling to room temperature, the latter is chosen for simulation of the laminar and coiling cooling processes.

During continuous cooling to room temperature, it can be seen from Fig. 4.38 that the starting and finishing transformation temperatures decrease as the cooling rate increases. It can be seen from Fig. 4.39 that when the cooling rate is not higher than 1 °C/s, the steel shows a typical ferrite/pearlite microstructure. When the cooling rate is increased to 3 °C/s, the microstructure is composed of granular bainite (bainitic ferrite matrix with dispersively distributed M/A islands); when the cooling rate is increased to 10 °C/s, lath bainite is formed, with the further increase in cooling rate, the fraction of lath bainite gradually increases, when the cooling rate is increased to 20 °C/s, the microstructure is basically composed of lath bainite.

Figure 4.40 shows the curves of transformation fraction with temperature of the studied steel during the simulated coiling. After deformation at a cooling rate of 20 °C/s, the starting transformation temperature is 650 °C, which is not higher than the final cooling temperature of simulated laminar cooling. The finishing temperature is about 450 °C, which is lower than the final cooling temperature of simulated laminar cooling. Therefore, phase transformation will partially occur during the

Fig. 4.38 Starting and finishing transformation temperatures of the studied steel at different cooling rates

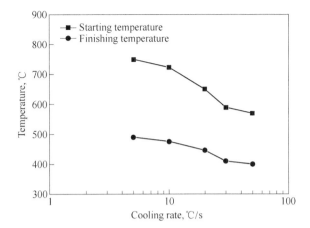

coiling process, and the higher the coiling temperature is, the larger the transformation fraction is. As the final cooling temperature decreases from 650 to 550 °C, the finishing transformation temperature of decreases first and then increases.

As shown in Fig. 4.41, the microstructure obtained by simulated coiling after final cooling to 620 and 650 °C is basically composed of full ferrite, almost no pearlite, which is due to the decrease in dissolved carbon content in austenite caused by sufficient TiC precipitates during high temperature coiling. When the simulated coiling temperature is below 600 °C, the microstructure is composed of granular bainite and TiC particles are not fully precipitated.

As shown in Fig. 4.42, when the steel is continuously cooled to room temperature, the hardness increases with the increase of the cooling rate, which is due to the gradual refinement of the microstructure and the increased fraction of the low temperature transformation microstructure. But in the lower cooling rate range, there is a small hardness peak, which is due to the increased precipitation strengthening effect caused by the increasing amount of TiC precipitates at lower cooling rates. As shown in Fig. 4.43, the studied steel shows a three-stage variation characteristic of hardness with the decrease in simulated coiling temperature, which increases first, then decreases and finally increases again. This is the result of the combined effects of microstructural changes and TiC precipitates on microhardness. The temperature of the minimum incubation time for TiC precipitation is 620 °C, which has the greatest precipitation strengthening effect, leading to the hardness peak at 620 °C. At lower temperatures, the fraction of bainitic microstructure increases and the effect of phase transformation strengthening is enhanced. So the hardness increases again.

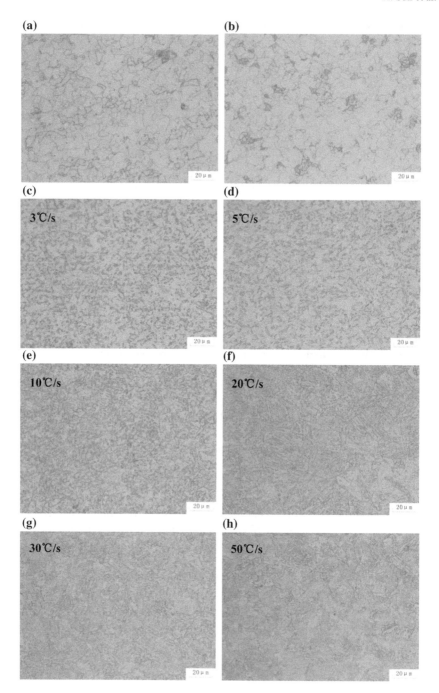

Fig. 4.39 Microstructures of the studied steel transformed at different cooling rates: **a** 0.5 °C/s, **b** 1 °C/s, **c** 3 °C/s, **d** 5 °C/s, **e** 10 °C/s, **f** 20 °C/s, **g** 30 °C/s, **h** 50 °C/s

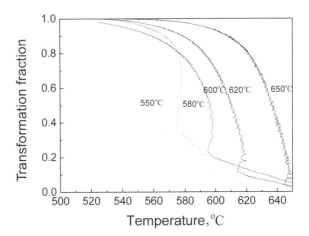

Fig. 4.40 The curves of phase transformation fraction with temperature of the studied steel during the simulated coiling

4.3 Recrystallization of Cold Rolled Ferrite

4.3.1 Recrystallization Thermodynamics

Recrystallization annealing is a process that heating a cold deformed metal to above the recrystallization temperature and below the A_{c1} temperature, maintaining and then cooling. The determination of the recrystallization temperature is the key for the recrystallization annealing, which can be measured by the means of isothermal annealing for half an hour or one hour, and continuous heating of simulated large-scale production. For the half an hour isothermal experiment, hardness test and metallography method are used for analyzing the variations of hardness and microstructure of the metal at various temperatures for 30 min, then the recrystallization temperature, is defined as the temperature at which the material is 50% softened or recrystallized.

From the hardness test (Fig. 4.44) and microstructural observation (Fig. 4.45), the annealing process of Ti microalloyed cold-rolled high-strength steel can be divided into three stages: recovery, recrystallization and grain growth:

(1) At temperatures below 640 °C, recovery is the dominant process, where the deformed microstructure remains mainly unchanged and the hardness decreases slowly with fluctuating sometimes.
(2) At temperatures in the range of 640–720 °C, recrystallization grains are largely nucleated, hardness drops dramatically, the Vickers hardness decreases by about HV90.
(3) At temperatures higher than 720 °C, the process is dominated by recrystallization and grain growth, the hardness decreases slowly.

In Fig. 4.44, the recrystallization temperature of the Ti microalloyed cold rolled high strength steel is 715 °C, which is significantly higher than that of the con-

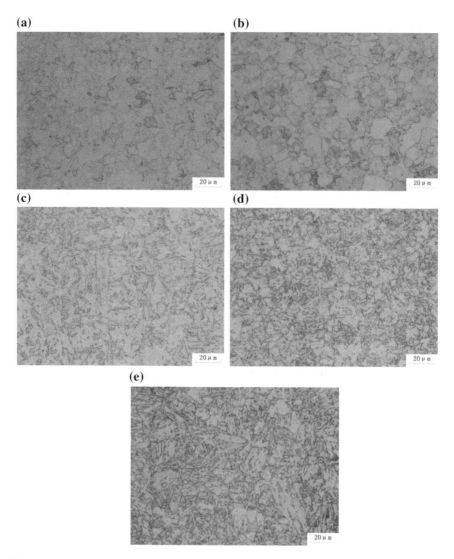

Fig. 4.41 Microstructure of the studied steel obtained by simulated coiling: **a** 650 °C; **b** 620 °C; **c** 600 °C; **d** 580 °C; **e** 550 °C

ventional cold rolled sheet. Therefore, in order to obtain cold-rolled high-strength steels with good combination of strength and ductility, it is necessary to improve the annealing temperature of the steel. In order to understand the effect of annealing time on recrystallization behavior at different temperatures, study on recrystallization kinetics is needed.

4 Physical Metallurgy of Titanium Microalloyed Steel …

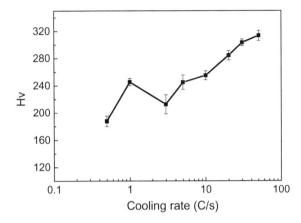

Fig. 4.42 Microhardness of the studied steel after continuous cooling transformation

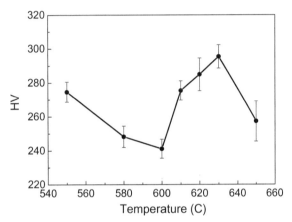

Fig. 4.43 Microhardness of the studied steel after the simulation of the laminar and coiling cooling processes

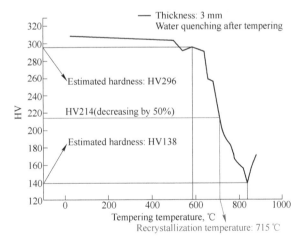

Fig. 4.44 Variation of hardness of the steel plate with annealing temperature by half an hour of isothermal annealing

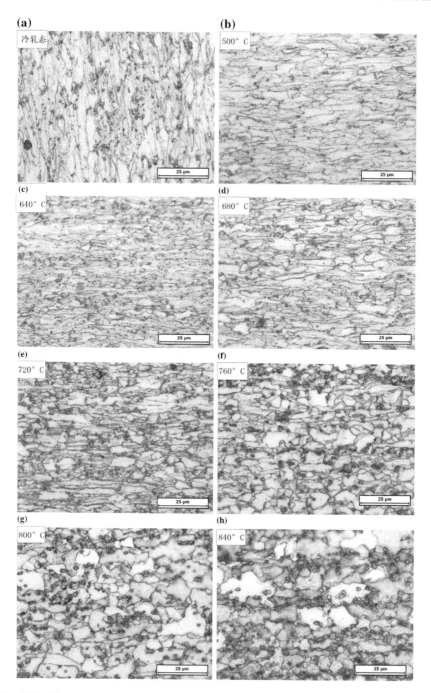

Fig. 4.45 Microstructures quenched at different temperatures after annealing for 30 min: **a** cold rolled, **b** 500 °C, **c** 640 °C, **d** 680 °C, **e** 720 °C, **f** 760 °C, **g** 800 °C, **h** 840 °C

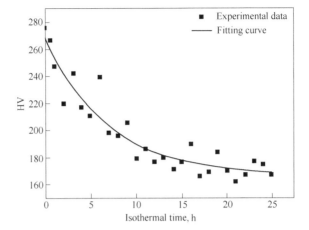

Fig. 4.46 Hardness of the steel samples annealed at 630 °C for different time

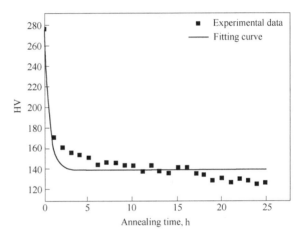

Fig. 4.47 Hardness of the steel samples annealed at 715 °C for different time

4.3.2 Recrystallization Kinetics

In order to understand the recrystallization process of Ti microalloyed high strength steels, the isothermal recrystallization kinetics curves are measured. The isothermal temperatures are the industrial annealing temperature (630 °C) and the recrystallization temperature (715 °C) measured experimentally, respectively. The holding time is 1–25 h. Figures 4.46 and 4.47 show the hardness of the steel samples isothermally annealed at 630 and 715 °C for different time, respectively.

It can be seen from Fig. 4.46 that the hardness of the samples decreases with the increasing annealing time at 630 °C, and the value is fluctuant. The hardness is HV 179 after annealing for 10 h, and is still HV 167 after 25 h, indicating that the hardness does not change after annealing for 10 h. From the metallographic images in Fig. 4.48, it is found that the aspect ratio of the elongated grains is decreased slightly with increasing annealing time, but even when the annealing time reaches 25 h, the grains

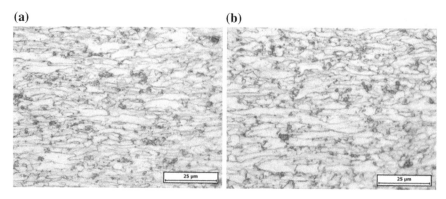

Fig. 4.48 Microstructure of the samples annealed at 630 °C for different time: **a** 10 h, **b** 25 h

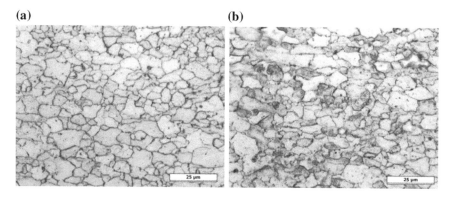

Fig. 4.49 Microstructure of the samples annealed at 715 °C for different time: **a** 10 h, **b** 25 h

are still elongated along the rolling direction. Recrystallization microstructure that is characterized by equiaxed grains cannot be observed. This indicates that complete recrystallization does not occur when the annealing temperature is low, even at 630 °C for 25 h, the microstructural changes are still affected by the recovery process. This also explains the reason why the cold rolled sheet has high strength, low elongation and poor formability after annealing at 630 °C.

It can be seen from Fig. 4.47 that the hardness of the sample decreases sharply from HV 276 to HV 180.7 after annealing at 715 °C for 0.5 h, then decreases slowly, and drops to HV 160.5 after 2 h, which is lower than that (HV 167) at 630 °C isothermal treatment for 25 h. After that, the hardness of the sample slowly drops to HV 120-130. From the metallographic images in Fig. 4.49, it is shown that fully equiaxed grains are appeared after 10 h annealing, indicating that complete recrystallization has occurred. Until 25 h, the grains do not appear to grow abnormally, indicating the determination of recrystallization temperature is reasonable.

4.3.3 Influence of Precipitation on Recrystallization

Annealing temperature on complete recrystallization has a greater influence than annealing time. The recrystallization behavior of Ti microalloyed cold rolled high strength steel shows that the recrystallization temperature increases obviously to 715 °C compared to conventional cold rolled steel. At 630 °C, even for 25 h annealing, recrystallization microstructure that is characterized by equiaxed grains cannot be observed. The mechanism of recrystallization in the annealing process needs to be studied in depth.

The precipitation and strengthening of nano-sized TiC are important physical-metallurgical characteristics of Ti microalloyed high strength steel, but the solid solution and precipitation of nano-sized TiC in Ti microalloyed cold rolled steel are still lack of study. The previously reported studies are mainly focused on the effect of precipitates on recrystallization behavior in titanium stabilized IF steel.

Shi and Wang [8] studied the microstructural evolution in titanium stabilized IF steels. The results showed that there is no significant change in size and morphology of TiN, TiS, and Ti_2CS between hot rolled, cold rolled, and batch annealed/continuously annealed samples. Because annealing temperature plays a more important role than annealing time on TiC evolution, the size of TiC after continuous annealing is larger than that after batch annealing. In addition, the TiC particles in continuously annealed samples apparently distribute on grain boundaries, while in batch annealed samples distribute more randomly. The authors also considered that TiC and other precipitates have no obvious effects on recrystallization and texture evolution before cold rolling, and precipitates such as FeTiP formed during annealing have an effect on grain boundary pining.

Choi et al. [9] studied the precipitation and recrystallization processes of two different titanium-containing ultra-low carbon steels. The results showed that the recrystallization temperature of high titanium steel is higher than that of low titanium steel. Goodenow and Held [10] studied the recrystallization behavior of titanium stabilized IF steel and concluded that the recrystallization time of titanium-containing steels is considerably longer than that of rimmed steel and aluminum killed steel at the same temperature. These studies show that the microalloying element titanium does have an effect on the recrystallization behavior of the cold rolled steel, but there are differences in the mechanism. One point of view [11] is that the recrystallization temperature of titanium-containing ultra-low carbon steels is more closely related to the solute titanium, rather than the TiC precipitates. This is due to the solute drag effect of titanium. However, there are no literatures that systematically evaluated the solute drag effect of titanium on recrystallization of cold rolled steels; another point of view is that [12], the pinning effect of the particles on the grain boundary mobility during recrystallization and grain growth is shown to be a significant factor for controlling the recrystallization texture in an IF steel, which has been confirmed by more research work.

Toroghinejad and Dini [13] added titanium to a ST14 steel and found that the recrystallization temperature in titanium-containing steels increases with increasing

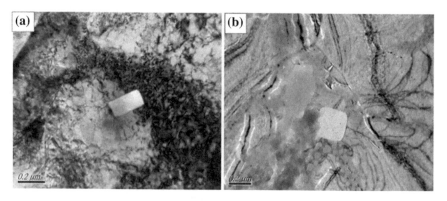

Fig. 4.50 TEM images showing the morphology of square particles in steel: **a** cold rolled plate and **b** after 880 °C annealing

titanium content. Laboratory annealing shows that the industrial annealing temperature (670 °C) of titanium-containing cold-rolled steel is not sufficient to produce fully recrystallized ferrite grains, while higher temperature annealing will lead to a decrease in strength and an increase in elongation.

The precipitates of Ti microalloyed cold rolled high strength steels were studied in our work. Through the TEM observation, many square particles with hundred nanometers in size are found in the cold rolled plate and the sample after half an hour isothermal annealing, as shown in Fig. 4.50, EDS spectrum shows that these particles are TiN. TiN particles are formed at higher temperatures, and the morphology and size remain almost unchanged in the subsequent processes, which can be found in the hot-rolled plate, cold rolled plate and cold-rolled annealing plate.

As can be seen from Fig. 4.50, the deformation introduces a large amount of dislocations in the cold-rolled hard plate. As the annealing temperature increases, the dislocation density is significantly reduced and almost no dislocations can be observed in the 880 °C annealed samples. Figure 4.51 also shows the similar phenomenon. Previous studies have shown that there are a large number of nano-sized TiC precipitates in Ti microalloyed hot rolled steels. As can be seen in Fig. 4.51, the size and distribution of nano-sized precipitates in hot rolled and cold rolled steels do not change significantly. As the annealing temperature increases to 880 °C, the average size of the precipitates increases, while the precipitation amount decreases gradually and the dislocation density reduces significantly.

The solubility of TiC in steel is higher than that of other titanium compounds, thus interphase precipitation or precipitation in ferrite of a large amount of nano-sized TiC will occur during austenite-ferrite transformation. The hindrance of these particles to dislocation movement leads to precipitation strengthening, which is also an important strengthening mechanism for Ti microalloyed steel. During the annealing process after cold rolling, when the annealing temperature is lower than 640 °C, nano-sized TiC particles will hinder the dislocation movement, and inhibit the formation of new grains, only the recovery process occurs. Therefore, compared with low-

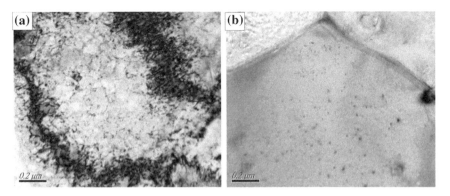

Fig. 4.51 The variation of nano-sized TiC precipitates at different stages: **a** cold rolled and **b** after 880 °C annealing

carbon titanium-free steel, titanium-containing steel has a higher recrystallization temperature. For higher temperature annealing, the precipitates will be coarsened by Ostwald and cannot effectively impede dislocation movement. New equiaxed grains containing low dislocation density begin to form and grow, therefore completed recrystallization occurs.

References

1. Vega M I, Medina S F, Quispc A, et al. Recrystallization driving forces against pinning forces in hot rolling of Ti-microalloyed steels [J]. Materials Science and Engineering A, 2006, 423: 253–261.
2. Medina S F, Chapa M, Valles P, et al. Influence of Ti and N Contents on Austenite Grain Control and Precipitate Size in Structural Steels [J]. ISIJ International, 1999, 39 (9): 930–936.
3. Medina S F, Mancilla J E. Determination of Static Recrystallization Critical Temperature of Austenite in Microalloyed Steels [J]. ISIJ International, 1993, 33 (12): 1257–1264.
4. Boratto F, et al. Effect of Chemical Composition on Critical Temperature of Microalloyed Steels [C] // THERMEC'88 Proceedings. Iron and Steel Institute of Japan, Tokyo, 1988: 383–390.
5. Wang T P, Kao F H, Wang S H, et al. Isothermal treatment influence on nanometer-size carbide precipitation of titanium-bearing low carbon steel [J]. Materials Letters, 2011, 65: 396–399.
6. Zhao Peilin. Study of microstructure evolution during hot rolling process of Ti microalloyed steel and development of high strength steel [D]. Central Iron and Steel Research Institute, Beijing, 2013.
7. Jang J H, Lee C H, Heo Y U, et al. Stability of (Ti, M)C (M = Nb, V, Mo and W) Carbide in Steels using First-Principles Calculations. Acta Mater., 2012, 60: 208–217.
8. Shi J, Wang X. Comparison of precipitate behaviors in ultra-low carbon, titanium-stabilized interstitial free steel sheets under different annealing processes [J]. Journal of Materials Engineering and Performance, 1999, 8(6): 641–648.
9. Choi Jae-Young, Seong Baek-Seok, Baik Seung-Chul, et al. Precipitation and Recrystallization Behavior in Extra Low Carbon Steels [J]. ISIJ International, 2002, 42(8): 889–893.
10. Goodenow R H, Held J F. Recrystallization of low-carbon titanium stabilized steel [J]. Metallurgical Transactions, 1970, 1: 2507–2515.

11. Rika Yoda. Ichiro Tsukatani, Tsuyoshi Inoue, et al. Effect of chemical composition on recrystallization behavior and $\bar{r}$ value in Ti-added ultra low carbon sheet steel [J]. ISIJ International, 1994, 34 (1): 70–76.
12. Subramanian S V, Prikyrl M, Gaulin B D, et al. Effect of precipitate size and dispersion on lankford values of titanium stabilized interstitial-free steels [J]. ISIJ International, 1994, 34(1): 61–69.
13. Toroghinejad M R, Dini G. Effect of Ti-microalloy addition on the formability and mechanical properties of a low carbon (ST14) Steel [J]. International Journal of ISSI, 2006, 3 (2): 1–6.

Chapter 5
Production, Structure and Properties Control of Titanium Microalloyed Steel

Jixiang Gao

As a kind of microalloying element, titanium significantly improves the comprehensive properties of steel. However, when compared with niobium and vanadium microalloying technology, Ti-microalloying technology had not been extensively used in industry for a long time. It is because the properties of Ti-microalloyed steel fluctuate largely and the production process is not stable. Titanium is very active and tends to react with oxygen, nitrogen and sulfur to form large Ti-bearing phases which are harmful to the comprehensive properties of steel, such as TiO, TiS and Ti_2CS. More importantly, the formation of these phases consumes a portion of titanium. This consumption not only reduces the volume fraction of TiC precipitated at low temperatures, but also significantly changes the chemical free energy of TiC precipitation. Therefore, the precipitation behavior of TiC is remarkably changed and its strengthening effect is greatly affected. In addition, the precipitation of TiC is sensitive to temperatures and the variation of production parameters remarkably affects the properties of steel. As a result, the mechanical properties of Ti-microalloyed steel products of different batches with the same chemistry or even different locations of the same batch fluctuate greatly.

In recent years, due to the rapid development of steel production technology, the content of impurities in steel is significantly reduced and the recovery rate of titanium is effectively controlled. In addition, the research on the chemical and physical metallurgy principles of Ti-microalloyed steel promotes the development of production technology. First of all, this chapter introduces the smelting production control in the actual production process, particularly the technology of deep deoxidation, deep desulfurization and low nitrogen control. Then, the influence of the main parameters

J. Gao (✉)
Guangdong Polytechnic Normal University, Guangzhou, China
e-mail: 13602772322@163.com

© Metallurgical Industry Press, Beijing and Springer Nature Singapore Pte Ltd. 2019
X. Mao (ed.), *Titanium Microalloyed Steel: Fundamentals, Technology, and Products*,
https://doi.org/10.1007/978-981-13-3332-3_5

of continuous casting process on the quality of Ti-microalloyed steel slabs is introduced. Finally, the effect of the parameters of hot rolling process on the microstructure and properties of Ti-microalloyed steel is elaborated.

5.1 Key Smelting Process

5.1.1 Technology of Deep Deoxidation and Inclusions Control

5.1.1.1 Deep Deoxidation Technology

Oxygen content in molten steel is one of the key indexes that evaluate the quality of molten steel. It does not only determine the amount of inclusions but also affect the size, morphology and distribution of inclusions. In order to reduce the oxygen content, aluminum, which is the strong deoxidizer, is generally used to kill the oxygen in molten steel. After deoxidation during steel tapping, the element controlling oxygen is changed from carbon to aluminum. The reaction is as follows:

$$2[Al] + 3[O] \rightarrow (Al_2O_3) \tag{5.1}$$

According to the thermodynamical data, the relationship between oxygen and aluminum in molten steel can be expressed by Eq. (5.2) [1]:

$$\lg[a_{Al}]^2[a_O]^3 = -\frac{62780}{T} + 20.17 \tag{5.2}$$

At 1600 °C, $[a_{Al}]^2[a_O]^3 = 4.5 \times 10^{-14}$.

Figure 5.1 shows the relationship between the content of aluminum and the content of dissolved oxygen according to Eq. (5.2). The content of oxygen in molten steel obviously decreases with aluminum increasing when the content of aluminum is less than 0.02%, but does not remarkably further decrease with aluminum increasing when the content of aluminum is in the range of 0.02–0.035%. And the content of oxygen keeps stable with aluminum increasing when the content of aluminum is more than 0.035%. In practice, aluminum in molten steel can easily react with oxygen in slags and also reduce SiO_2 and MnO in slags with the result of increasing the amount of aggregated Al_2O_3 in molten steel. In addition, high content of aluminum in molten steel also aggravates the re-oxidation during casting and produce Al_2O_3 inclusions.

Aluminum or aluminum-bearing deoxidizer is generally used to kill the oxygen in molten steel during tapping in the titanium microalloyed steel production process. Meanwhile, the synthetic slags with strong capabilities of deoxidation and desulphurization and lime are also used for refining slags. The content of the synthetic slag is as follows: CaO 50–55%, SiO_2 3–5%, Al_2O_3 30–35%, MgO 7–12%,

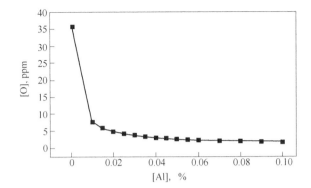

Fig. 5.1 Relationship between soluble aluminum content and soluble oxygen content in molten steel at 1600 °C

FeO + MnO < 1%. According to a large amount of production data, the matched amount of aluminum is 2–2.5 kg/t to the oxygen content in the range of 700–900 ppm.

It's very important to reserve some soluble aluminum in molten steel after deoxidization. It can be explained as follows. During the tapping process, iron and oxygen in molten steel reacts to form FeO, and manganese alloy reacts with oxygen in molten steel to form MnO. The content of FeO and MnO in refining slag in ladles is about 5% because of slagging off during tapping. The slag is black and the stability is lower than that of Al_2O_3. Thus, FeO and MnO in the slag will resolve and transfer oxygen, iron and manganese to molten steel, and aluminum in molten steel will kill the resolved oxygen further. The reaction continues until the content of FeO and MnO in the slag is less than 1%, and the slag becomes white. Finally, the oxygen in molten steel is controlled stably by aluminum.

Besides aluminum content, soft argon blowing process has a great effect on the oxygen content in molten steel. Figure 5.2 [2] shows the relationship between soft argon blowing time and content of oxygen, T[O]. T[O] decreases with the increase of argon blowing time. T[O] can be less than 15 ppm when the argon blowing time is more than 8 min, and the minimum is 10 ppm. In practice, soft argon blowing time is set to be 8–10 min in order to improve the purity of molten steel and take into account the production schedule.

5.1.1.2 Calcium Treatment Technology

Though oxygen can be controlled at low-level in aluminum killed steel, Al_2O_3 is inevitable. Al_2O_3 existing on steel plate surface can cause surface defects, and existing inside of the steel plate will deteriorate the mechanical performance. In addition, Al_2O_3 inclusions tend to gather and solidify in the inner of nozzle because the diameter of the submerged nozzle for thin slab casting is small. As a result, the nozzle is clogged, obstructing the flow of molten steel and leading to the fluctuation of liquid level. In severe case, the nozzle can be blocked and casting is interrupted. Figure 5.3 shows the morphology of a submerged nozzle and Fig. 5.4 presents the composition of the clogs analyzed by electron probe micro-analyzer (EPMA). The inner of nozzle is covered by a thick layer of clogs, which are composed of Al_2O_3 and calcium aluminates with high melting point.

In practice, modification of Al_2O_3 by calcium treatment is usually applied to improve the surface and internal quality of steels and to prevent nozzle clogging during the continuous casting. Figure 5.5 shows the equilibrium phase diagram of $CaO–Al_2O_3$ system. The reaction is as follows.

$$x[Ca] + yAl_2O_3 = x(CaO) \cdot (y - x/3)Al_2O_3 + 2/3x[Al] \qquad (5.3)$$

Alumina inclusions can be modified into $CaO \cdot Al_2O_3$ (CA) or $12CaO \cdot 7Al_2O_3$ ($C_{12}A_7$) according to Eq. (5.3). The melting temperature of $C_{12}A_7$ phase is 1455 °C, which is the lowest among all the phases. Therefore, $C_{12}A_7$ phase is liquid at the temperature of molten steel. $C_{12}A_7$ inclusions in molten steel tend to gather and grow up, then be removed through floating. So the purpose of calcium treatment is to modify Al_2O_3 into $C_{12}A_7$.

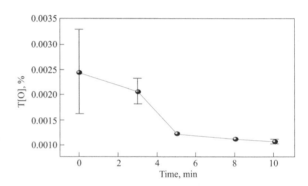

Fig. 5.2 Relationship between soft argon blowing time and oxygen content in molten steel

Fig. 5.3 Submerged nozzle adhered by clogs

Figures 5.6 and 5.7 show the changes of the compositions and morphologies of inclusions before and after Ca treatment, respectively [2]. The inclusions are scraps of Al_2O_3 and $MgO \cdot Al_2O_3$ in the initial stage of refining, as shown in Figs. 5.6 and 5.7. Then Al_2O_3 and $MgO \cdot Al_2O_3$ inclusions are modified into $CaO-Al_2O_3-MgO$ inclusions during the ladle furnace (LF) refining because of the reaction between the molten steel and the refining slag with high basicity. Most inclusions are $CaO-Al_2O_3-MgO$ with high melting points before calcium treatment. The morphologies of inclusions apparently change after calcium treatment because of the reaction between calcium and Al_2O_3. The compositions of $CaO-Al_2O_3-MgO$ move to liquid region from solid region in the ternary phase diagram of $SiO_2-CaO-Al_2O_3$ (shadow region in Fig. 5.6). The inclusions have been spheroidized or are being spheroidized after calcium treatment, and most are composite inclusions. Two types of inclusions have been observed. One is the homogeneous phase of the $CaO-Al_2O_3$ system. The other is the calcium aluminates containing MgO in core and surrounded

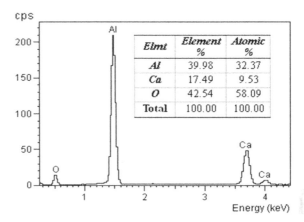

Fig. 5.4 Composition of clogs measured by EPMA

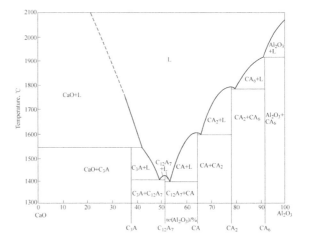

Fig. 5.5 Equilibrium phase diagram of $CaO-Al_2O_3$ system

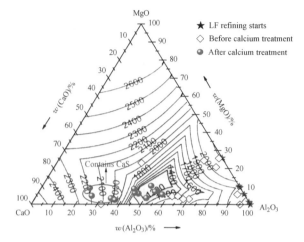

Fig. 5.6 Changes in the composition of inclusions in molten steel during LF refining process

by CaS. The size of the later is large and they can be removed during soft argon blowing. The proportion of inclusions with the size less than 5 μm rises remarkably to more than 80% after soft argon blowing.

In practice, Al_2O_3 can be modified into $12CaO·7Al_2O_3$ or calcium aluminates with compositions close to $12CaO·7Al_2O_3$ and low melting points by calcium treatment when the ratio of calcium to aluminum is above 0.09.

5.1.2 Deep Desulfurization Technology

Desulfurization reaction can be expressed by Eq. (5.4) for titanium microalloyed steel deoxidized by aluminum [1].

$$3[S] + 2[Al] + 3(CaO) = (Al_2O_3) + 3(CaS) \tag{5.4}$$

The key parameters of desulfurization during smelting are as follows.

(1) Aluminum content in molten steel. Aluminum content should be in the range of 0.04–0.05% before refining to create the conditions for rapid desulfurization. It should be in the range of 0.025–0.035% before titanium alloying after refining to improve the yield of titanium.
(2) Reducing slag for deep desulfurization. The content of Al_2O_3 in slag increases because of aluminum oxidation during desulfurization. Therefore, lime is necessary in the slag. The ratio of $w(\%CaO)/w(\%Al_2O_3)$ in slag should be about 1.8 to speed up the desulfurization reaction if the content of sulfur is high. Desulfurization can be significantly improved by the slag with high CaO content and good fluidity.

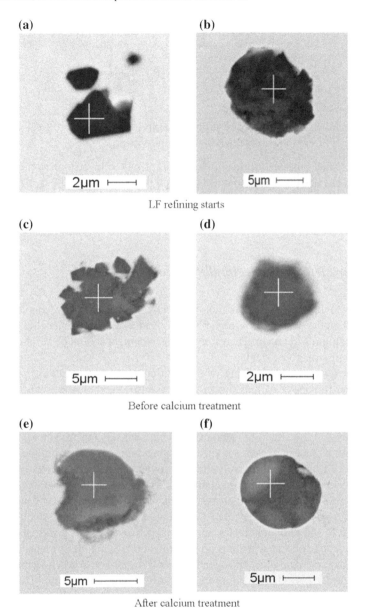

Fig. 5.7 Morphologies of the typical inclusions in molten steel during refining process

(3) Temperature. The equilibrium constant of desulfurization in Eq. (5.4) increases with temperature increasing, which means high temperature can promote the desulfurization reaction. So the temperature is generally set to be above 1600 °C.

Table 5.1 Desulfurization for titanium microalloyed high strength weathering steel

Before refining (%)	After refining (%)	Average desulphurization rate (%)	Max rate (%)	Min rate (%)
0.0193	0.004	79.27	92.59	75.2

(4) Argon blowing process. The argon pressure should be more than 0.75 MPa during refining to promote desulfurization reaction.

Table 5.1 shows the desulfurization for titanium microalloyed high strength weathering steel. The desulphurization rate during refining is 75.2–92.59% by the methods above, the average rate is 79.27%, and the sulfur content is less than 40 ppm at the end of refining.

5.1.3 Low Nitrogen Steelmaking Technology

Coarse TiN tends to precipitate in molten steel by the reaction between nitrogen and titanium when the content of nitrogen is high. And as a consequence, the beneficial effect of titanium is reduced. Therefore, nitrogen content should be strict controlled in titanium microalloyed steel.

Pillild investigated the change of nitrogen content in the smelting process by electrical arc furnace (EAF), as shown in Fig. 5.8 [3]. Symbol A presents the period of electrode insertion, during which the nitrogen content keeps stable. Symbol B presents the first charge and the formation of small molten bath. Because the surface area of scraps during smelting is large and there is no slag covered on molten steel, the nitrogen content increases obviously by absorbing nitrogen from the electrode zone. Symbol C presents the development of molten bath. In the period, the amount of molten steel in molten bath and volume of slag increase, while the amount of nitrogen picking up reduces. So the nitrogen content decreases because of the dilution by increasing melting steel. Symbol D presents the period of scrap melting down and heating to the temperature for carbon-oxygen reaction. The nitrogen content remains unchanged because there is no denitrification. Symbol E presents the period of decarburization. And much nitrogen is removed by carbon-oxygen reaction. Symbol F presents the period of shielding arc heating to tapping temperature. The nitrogen content doesn't change because of the terminal reaction between carbon and oxygen. Symbol G presents the period of tapping. Nitrogen content increases because of the nitrogen picking up from air. Symbol H presents the period of steel holding in ladle, during which the nitrogen content keeps unchanged. Then, the nitrogen content increases a little during casting (symbol I).

The nitrogen in scraps has a significant influence on the nitrogen content in molten steel after scrap melting down, as shown in Fig. 5.9 [4]. So proper charging is very important to reduce the nitrogen content introduced by scraps.

Fig. 5.8 Change in nitrogen content during EAF smelting process

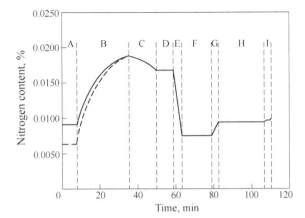

Due to more amounts of microalloying elements added and the nitrogen solubility increased by microalloying elements, denitrification for titanium microalloyed steel is even more difficult. Therefore, the key of nitrogen controlling for titanium microalloyed steel is to reduce nitrogen picking up in each process. The methods are as follows:

(1) Remove the nitrogen in converter during steelmaking. The nitrogen is mainly removed in converter because of the good kinetics conditions. The technologies for low nitrogen steelmaking are presented as follows.

   (1) Technology of controlling bottom blowing process in converter. Types and flow rate of bottom blowing gas influence the nitrogen content in molten steel at the end of steelmaking. The switch from nitrogen to argon is usually carried out during the intermediate stage of blowing, and the intensity

Fig. 5.9 Influence of nitrogen content in steel scraps on nitrogen content after scraps melting down

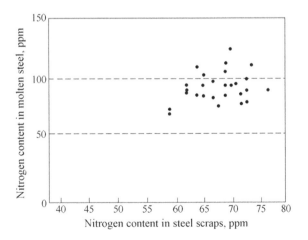

of bottom blowing gas increases gradually. The intensity of argon keeps maximum during the intermediate and the final stage of blowing. And the argon blowing time is more than 1 min before tapping. In addition, inspection and maintenance on the bottom blowing system are necessary to ensure the effect of blowing for all purging plugs.

(2) Technology of end-point carbon and temperature controlling in converter. The end-point carbon and temperature influence the operation of LF refining, such as deoxidization, temperature controlling and so on. Therefore, it is in favor of the nitrogen controlling in LF refining to have the reasonable end-point carbon and temperature.

The molten steel tends to pick up nitrogen from air during the contacting because of the high content of nitrogen in air. The methods to reduce nitrogen picking up in practice are shown as follows. Clean up the slag remained outside of the tap hole before steelmaking. Change the tap hole in the case of long tapping time or bad slag damming. Maintain the tap hole and reasonably control the tapping time. Blow argon to the ladle before tapping to reduce the nitrogen pickup. Cover the surface of molten steel with synthetic slag to insulate molten steel from air, by which to reduce the nitrogen pickup during tapping. The nitrogen content is less than 25 ppm after tapping by the methods above.

During EAF steelmaking, nitrogen pickup of molten steel is mainly from the nitrogen in scraps, the ionized nitrogen in the arc zone and the absorbed nitrogen from air during tapping. Accordingly, the methods to reduce nitrogen pickup are presented as follows. Charge reasonably to reduce the nitrogen introduced by scraps. Foam slag during EAF steelmaking to prevent the exposure of molten steel to the open air and nitrogen absorption. In addition, cover the surface of molten steel with synthetic slag to insulate molten steel from air, by which to reduce the nitrogen pickup during tapping.

Nitrogen can be removed by the large amount of CO bubbles produced during the decarburization of EAF steelmaking process because of the strong stirring effect of CO bubbles on the molten steel. The amount of denitrification increases with the amount of decarburization or the CO bubbles produced during decarburization. In order to ensure the high capacity of denitrification during EAF steelmaking, charging with high carbon (by addition of pig iron and HBI) and increasing the oxygen blowing intensity are necessary.

(2) Reduce nitrogen pickup during LF refining. Prevention of nitrogen pickup is the main measure during LF fining because LF furnace doesn't have the function of denitrification. The introduced nitrogen from scraps and the absorbed nitrogen from air are the main causes of nitrogen pickup during LF refining. Accordingly, the methods to reduce nitrogen pickup are presented as follows. Charge reasonably to reduce the nitrogen introduced from scraps. Heat the molten steel rapidly to the target temperature by the technology of slagging with short-arc and submerged-arc, which can reduce the heating time and prevent nitrogen absorption. In addition, controlling the stirring intensity during

refining and maintaining the vapor pressure over air pressure are also important to reduce nitrogen absorption. The nitrogen pickup can be controlled to be less than 5–10 ppm by the methods above.

(3) Reduce nitrogen pickup during continuous casting. The exposure of molten steel to air is the main source of nitrogen pickup during casting. Therefore, casting with whole-process protection is adopted to reduce or prevent reoxidation of molten steel. For example, ladle covering flux is put on the surface of molten steel after refining to prevent reoxidation. Ladle with the heat insulating cover is used to reduce the temperature drop. Long nozzle casting with shielding argon is also used to prevent reoxidation of molten steel by air. Tundish covering flux is put on the surface of molten steel to prevent reoxidation. According the methods above, the nitrogen content of molten steel by converter steelmaking is about 40 ppm, while it's about 60 ppm by EAF steelmaking.

5.1.4 Control of the Yield of Titanium

Because Ti is very active, Ti is prone to react with oxygen, nitrogen and sulfur in molten steel, leading to the low and unstable yield of titanium. Therefore, increasing and stabilizing the yield of Ti is the key during steelmaking for Ti microalloyed steel. The technologies of deep deoxidization, desulphurization and low nitrogen steelmaking are developed and applied in practice, which can keep the contents of oxygen, sulfur and nitrogen at low levels. Meanwhile, the appropriate addition process for Fe–Ti alloy should be developed. The addition of Fe–Ti alloy should be prior to the end of refining. The molten steel should be well deoxidized, of which the oxygen content is less than 0.003% and the aluminum content is in the range of 0.025–0.035%. It is beneficial to raising and stabilizing the yield of titanium [5]. Table 5.2 shows the industrial test data by the methods above. The yield of Ti during refining is 70–80% for the molten steel by EAF steelmaking, while it is 75–85% by converter steelmaking because of the lower nitrogen content.

5.2 Key Technologies for Continuous Casting

5.2.1 Key Process Parameters

In order to ensure the Production of continuous casting, no steel leaking and slab quality for titanium microalloyed steel, the casting process should be designed by the following principals besides the composition, cleanliness and inclusion morphology of molten steel. Increase the cooling rate of ingot as much as possible to refine the precipitates. Protect the whole casting process to reduce and prevent re-oxidation of molten steel. Solve the problems relevant to casting temperature, settings of the vibra-

Table 5.2 Yield of titanium for titanium microalloyed steel by EAF steelmaking

No.	Weight of molten steel (t)	Amount of Ti–Fe (kg)	Residual Ti (%)	Ti at the end of refining (%)	Yield of Ti (%)
104044760	151.1	170	0.002	0.0375	78.9
104063890	152.7	270	0.001	0.056	79.2
104080770	151.2	330	0	0.063	72.2
204113760	154	350	0.001	0.069	74.8
204113770	155	360	0	0.065	70.0
205031770	155		0.001	0.088	74.9
205031780	152.5	670	0	0.126	71.7
205031790	155	700	0	0.133	73.6
Average	153.8				73.9

tion curves and setting of the secondary cooling system. Optimize the mould flux, and match the speed and the temperature. Increase the charging temperature of ingot. The methods all above can improve the surface and internal quality of slabs.

(1) Casting temperature system. The reasonable casting temperature system is very important to the casting process and the improvement of the slab quality. High casting temperatures will aggravate the center segregation and the thickness inhomogeneity of the initially solidified shell at the meniscus of mould, leading to the thinning of the shell leaving mould and slab cracks and even breakout. Low casting temperatures will decrease the fluidity of molten steel and influence the smelting of mould flux, which will lead to longitudinal cracks. The liquidus temperature of titanium microalloyed steel is 1523–1524 °C when the content of titanium is in the range of 0.038–0.14%. The production data show that it can prevent the breakout and improve the slab quality by controlling the superheat in the range of 20–30 °C.

(2) Protecting the whole casting process. In order to prevent the re-oxidation of molten steel and improve the slab quality, the casting process is protected in the whole process. The methods include protection for molten steel in ladles, long nozzle casting with shielding argon gas, protection for molten steel in tundish and application of submerged nozzle. Tundish covering fluxes are put on the surface of molten steel to prevent re-oxidation.

(3) Optimizing the mould flux. A few issues need to be taken into account for selecting the mould flux, such as the matching of the melting point of mould flux to the liquidus temperature of steel, the matching of the melting rate (or powder consumption) of mould flux to the casting speed, the matching of the bulk density and moisture content of mould flux to the superheat of molten steel, the matching of the basicity and viscosity of mould flux to the Ti content of molten steel and so on. The surface quality of slabs is acceptable and can meet the requirements of the industrial use only when the properties of mould flux are stable and match with the casting speed and superheat.

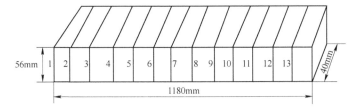

Fig. 5.10 Schematic diagram of sampling from the cast slab

(4) Controlling the casting speed. The casting speed should not only take into account the matching of the cracking sensitivity of steel to the temperature of molten steel, but also the connection between each process. In practice, the casting speed is usually limited by the processing time during refining for sequential continuous casting.

(5) Choosing the oscillation types of caster mould. Different oscillation types are chosen during the casting process for titanium microalloyed steel, which is significantly important to the surface quality of slabs and the casting process without breakout and sticking. The oscillation type is sinusoidal oscillation for thin slab casting. The amplitude increases with the casting speed and the negative strip rate is about 21%. Generally, the amplitude varies for different casters.

(6) Controlling the heat flux density in the mould. The heat flux density in the mould and the ratio of the heat flux on the narrow side to the wide side are the very important casting parameters, which can reflect the thickness of initially solidified shell. They are significant for estimating the change of the properties of mould flux, the quality of molten steel and the condition of casting. Maintaining the stable and suitable ratio of heat flux on the narrow side to the wide side can improve the slab quality. In addition, it is effective to improve the slab quality on the narrow side (e.g. cracks by hot-shortness) by adjusting the mould taper and maintaining the suitable cooling intensity.

5.2.2 Control of Slab Quality

5.2.2.1 Analysis of Slab Quality

Analysis of Composition Segregation of Slab [6]

The segregations of carbon, silicon, manganese, phosphorus, sulfur and titanium in the cross section of slab of titanium microalloyed steel produced by thin slab casting and direct rolling (TSCR) are investigated by in situ metal analyzer. The composition is 0.05%C–0.45%Si–0.45%Mn–0.08%P–0.005%S–0.07%Ti. Figure 5.10 shows the schematic of sampling. The widths of No. 1 and No. 13 are 95 mm, and the widths of No. 2 to No. 13 are 90 mm.

Figure 5.11 shows the two dimensional contour maps for each element in the cross section of slab by stitching together the two-dimensional contour maps for each element of the 13 samples. The segregation band caused by carbon segregation is obvious in the center of slab and it's the most severe segregation. Segregation bands caused by phosphorous and sulfur are also observed in the center of slab, but the segregations are less severe than that of carbon. The segregations of silicon, manganese and titanium are not distinct and there are no significant segregation bands in the center of slab.

Analysis of Slab Density

The distribution of density in the cross section of slab is shown in Fig. 5.12. The density is high and within the range of 0.924–0.955. Little difference is observed among the densities of all samples. In addition, there is no sharp reduction of density in the center of the cross section of slab. The result of in situ analysis shows that there are no central porosity and line shrinkage.

Analysis of Inclusions

Figure 5.13 shows the distribution of the content of alumina inclusions in the samples. The distribution of inclusions is homogeneous in the cross section of slab, and the content of inclusions fluctuates in the range of 0.0014–0.0022%. The results are consistent with the characteristics of inclusions exiting in slab and the content fluctuation is within the normal range.

5.2.2.2 Effect of Liquid Core Reduction on Slab Quality [7]

Liquid core reduction refers to the following process. Press the solidified shell in the mould exit and keep the liquid core in the slab, then the liquid core keeps shrinking until it completely solidifies by passing the segments of secondary cooling. In practice, liquid core reduction affects the quality of slab. This section will introduce the effect of liquid core reduction on the composition segregation and density of slab.

The slab with the thickness of 60 mm is pressed until the thickness is reduced to 55 mm by liquid core reduction equipment. The sample with liquid core reduction is labeled No. 1. Meanwhile, the sample without liquid core reduction on the same slab is labeled No. 2. The segregations of the elements and densities of the cross sections of the two samples are measured by in situ metal analyzer. Figure 5.14 shows the schematic diagram for sampling at 1/2, 1/4 and the right edge locations along the width of slab. The samples from the three different locations on No. 1 slab are labeled No. 11, No. 12 and No. 13 respectively. And the reference samples from No. 2 slab are labeled No. 21, No. 22 and No. 23 respectively.

Effect of Reduction on Segregation of Slab

Table 5.3 shows the effect of reduction on the segregation of slab. The segregation of carbon in the center of slab is obviously alleviated by liquid core reduction, and less

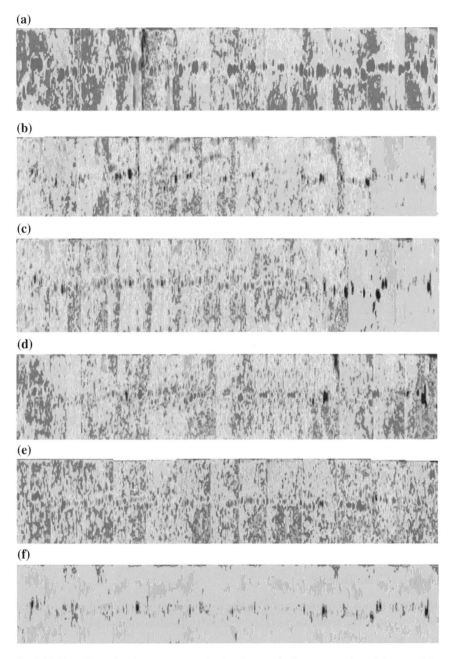

Fig. 5.11 Two dimensional contour maps for the elements in the cross section of the cast slab **a** Carbon, **b** Silicon, **c** Manganese, **d** Phosphorous, **e** Sulfur, **f** Titanium

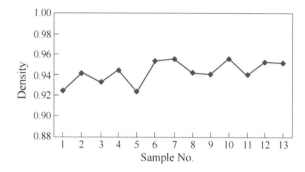

Fig. 5.12 Distribution of density across the cross section of the cast slab

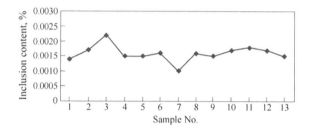

Fig. 5.13 Distribution of alumina inclusion content across the cross section of the cast slab

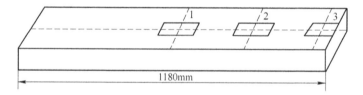

Fig. 5.14 Schematic diagram of sampling from the cast slab

obviously improved at 1/4 location along the width of slab. However, the segregation aggravates at the edge of slab. The segregations of other elements are improved, but not significantly.

Analysis of Slab Density

Figure 5.15 shows the densities of different locations of slabs with and without liquid core reduction. The result shows that the density of slab is remarkably improved by liquid core reduction.

Table 5.3 Effect of liquid core reduction on the segregation of slab

Items		With liquid core reduction			Without liquid core reduction		
		No. 11	No. 12	No. 13	No. 21	No. 22	No. 23
C	Average content (%)	0.040	0.040	0.039	0.038	0.039	0.039
	Maximum segregation	4.405	1.943	11.856	29.42	3.893	3.546
Si	Average content (%)	0.428	0.428	0.423	0.434	0.433	0.435
	Maximum segregation	1.137	1.068	1.155	1.114	1.119	1.180
Mn	Average content (%)	0.434	0.436	0.432	0.431	0.430	0.429
	Maximum segregation	1.067	1.078	1.117	1.045	1.150	1.152
P	Average content (%)	0.077	0.079	0.078	0.081	0.080	0.080
	Maximum segregation	1.342	1.219	1.235	1.391	1.349	1.478
S	Average content (%)	0.007	0.007	0.007	0.008	0.008	0.008
	Maximum segregation	1.992	1.345	2.195	1.503	1.443	1.908
Ti	Average content (%)	0.063	0.064	0.066	0.066	0.064	0.064
	Maximum segregation	1.157	1.123	1.087	1.262	1.199	1.382

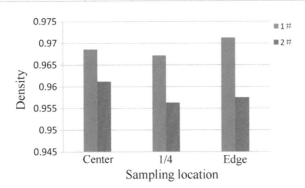

Fig. 5.15 Comparison of densities of the cast slabs with and without liquid core reduction

5.3 Key Hot Rolling Processes

The size, morphology and volume fraction of precipitates are important factors to determine the strengthening effect of Ti-microalloyed steel. The temperature and reduction schedules of hot rolling process affect the precipitation behavior of titanium, and thus have an important effect on the microstructure and properties of final products. According to the precipitation behavior of Ti(C, N), it is important to develop a reasonable hot rolling schedule so as to improve the overall properties of Ti-microalloyed steel.

5.3.1 Temperature Schedule

Temperature is one of the most important factors affecting the precipitation behavior of Ti(C, N). The temperature schedule of hot rolling process mainly includes the discharge temperature, the finishing temperature and the coiling temperature.

5.3.1.1 Discharge Temperature

The discharge temperature of TSCR process is generally in the range of 1100–1200 °C. According to the thermodynamic condition, TiN precipitates in this temperature range. TiN precipitates have been observed by scanning electron microscopy (SEM), but the difference among the precipitates precipitated at different temperatures is not obvious. In order to further analyze the effect of discharge temperature on the properties of Ti-microalloyed steel, a large number of production data are analyzed statistically. The results are shown in Figs. 5.16 and 5.17.

According to the results, there is no evident regularity to describe the effect of discharge temperature on the mechanical properties of Ti-microalloyed steel. Low discharge temperature should be applied, which has two benefits. One is to reduce the growth and coarsening tendency of TiN precipitates during the reheating process. The other is to minimize the precipitation of TiN at high temperatures which leads to

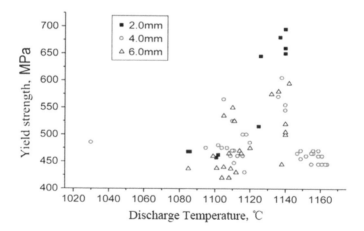

Fig. 5.16 Relationship between discharge temperature and yield strength

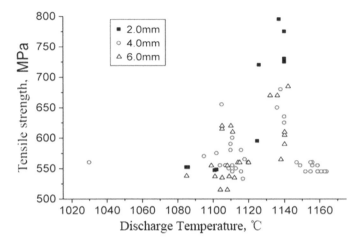

Fig. 5.17 Relationship between discharge temperature and tensile strength

precipitation of more fine TiN precipitates at lower temperatures. Taking into account the load capacity and rolling stability of current mills, the discharge temperature is selected to be above 1120 °C.

5.3.1.2 Finishing Temperature

The effect of finishing temperature on the properties of Ti-microalloyed steel is complicated. Low finishing temperature refines ferrite grains, thereby improving the grain refinement strengthening. However, low finishing temperature induces the deformation induced precipitation of carbonitrides. These carbonitride precipitates inhibit the growth of austenite grains and contribute to a certain effect of grain refinement strengthening and precipitation strengthening. However, they are relatively large compared to the nano-scale particles precipitated in ferrite, reducing the precipitation strengthening effect. Therefore, high finishing temperature suppresses the deformation induced precipitation of Ti(C, N) in high temperature austenite and thus promotes the dispersive precipitation of TiC in ferrite, which improves the precipitation strengthening effect. But high finishing temperature is not advantageous to refinement of ferrite grains, and reduces the grain refinement strengthening effect to some extent. Figures 5.18 and 5.19 show the statistical results of production data. Different from plain carbon steel, decrease in the finishing temperature cannot improve the strength of Ti-microalloyed steel, on the contrary, if the finishing temperature increases from 860–880 °C to 890–900 °C, the strength is significantly improved.

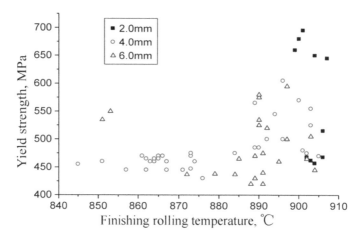

Fig. 5.18 Relationship between finishing temperature and yield strength

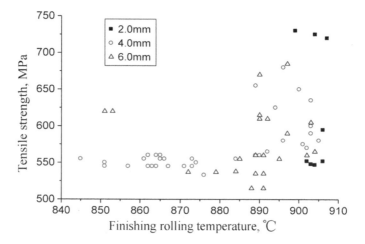

Fig. 5.19 Relationship between finishing temperature and tensile strength

5.3.1.3 Coiling Temperature

The coiling temperature is a key factor affecting the precipitation behavior of precipitates of Ti-microalloyed steel. Production data shown in Figs. 5.20 and 5.21 indicate that the coiling temperature has a significant impact on the strength of steel. The strength reaches the maximum when the coiling temperature is in the range of 580–610 °C.

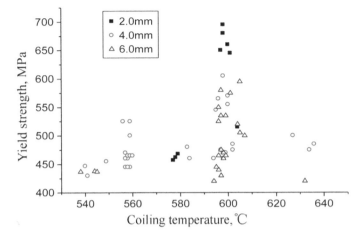

Fig. 5.20 Relationship between coiling temperature and yield strength

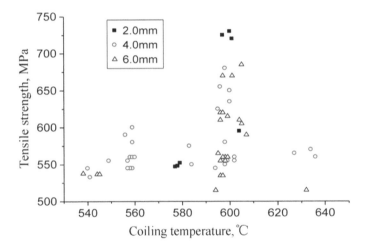

Fig. 5.21 Relationship between coiling temperature and tensile strength

5.3.2 Reduction Schedule and Reduction Ratio

The results of the comparison tests using different reduction schedules are shown in Table 5.4. The slabs are from the same batch. The effect of reduction schedules on the strength is not evident according to the results. In the production process, F1 and F2 stands should apply large reduction ratio of more than 50% as much as possible so as to avoid the occurrence of mixed grains.

Table 5.4 Production data showing the relationship between reduction schedule and strength

Heat No.	Coil No.	Ti content in production (%)	Thickness (mm)	Reduction rate (%)						Yield strength (MPa)	Tensile strength (MPa)
				F1	F2	F3	F4	F5	F6		
1	1	0.062	6.0	41.58	44.23	31.85	25.65	21.23	14.62	470	560
	2	0.062	6.0	55.39	53.03	0	34.18	0	19.48	465	560
	3	0.062	6.0	52.78	56.46	0	32.58	0	20.04	465	555
	4	0.060	6.0	52.78	56.22	0	32.72	0	20.32	475	560
	5	0.063	6.0	53.20	56.39	0	32.06	0	20.06	525	610
2	1	0.068	6.0	41.60	43.37	31.87	25.67	21.24	15.81	535	615
	2	0.068	6.0	48.57	47.60	34.97	0	23.31	17.35	550	620
	3	0.068	6.0	48.05	48.20	35.14	0	23.16	17.33	535	620

In addition, the reduction ratio is one of the main factors affecting the strength. As shown in Figs. 5.22 and 5.23, suppose equal thickness of the initial cast slabs, the thinner the product gauge, i.e. the greater the reduction ratio, the finer the grains of the product.

5.4 Comprehensive Controlling Technology for Microstructure and Properties

5.4.1 Effect of Titanium Content on Yield Strength

Figure 5.24 shows the relationship between yield strength and titanium content. The effect of titanium content on the strength of steel is divided into three regions. When the titanium content is lower than 0.045%, the yield strength increases slowly with the titanium content. When the titanium content is between 0.045 and 0.095%, the yield strength increases linearly with the titanium content. When the titanium content is more than 0.095%, the yield strength remains basically unchanged. The increase in the titanium content significantly improves the yield strength, and the yield strength can reach 750 MPa and more [8].

5.4.2 Controlled Rolling Patterns of Ti-Microalloyed Steel

There are two methods to refine the grains of microalloyed steel produced by controlled rolling. One is recrystallization controlled rolling and the other is non-recrystallization controlled rolling. The former method refines austenite grains by repeated recrystallization of austenite in the hot rolling process, and ultimately refines ferrite grains. The effect of microalloying elements is to control the coarsening of recrystallized austenite grains during rolling and after rolling. Representative steel are V–N and V–Ti–N microalloyed steel. For the non-recrystallization controlled rolling, austenite does not experience recrystallization during the whole process or the last several passes of finishing rolling, and transforms into pan-caked austenite with a high density of defects. Therefore, the nucleation rate of ferrite is improved and thus the final ferrite grains are refined. Almost all steel produced by non-recrystallization controlled rolling contain niobium, because niobium has a strong inhibitory effect on the austenite recrystallization. The inhibitory effect of titanium on recrystallization is between that of niobium and vanadium. The coarse microstructure of original austenite in Ti-microalloyed steel produced by TSCR can achieve complete static recrystallization through a significant high temperature deformation exerted by F1 stand. TiN particles precipitated in cast slabs effectively prevent the coarsening of recrystallized austenite grains, realizing the recrystallization controlled rolling. The dragging effect of titanium solute and the strain-induced precipitates of TiC inhibit

Fig. 5.22 Microstructure of titanium microalloyed steel plates with different thicknesses: **a** 6.0 mm, **b** 4.0 mm, **c** 1.6 mm

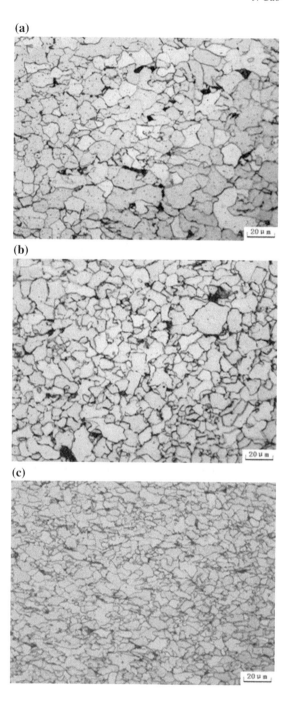

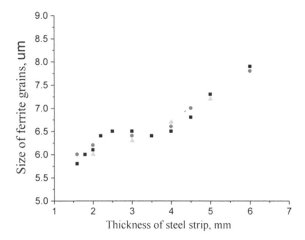

Fig. 5.23 Relationship between the size of ferrite grains and the thickness of titanium microalloyed steel (0.0558% Ti)

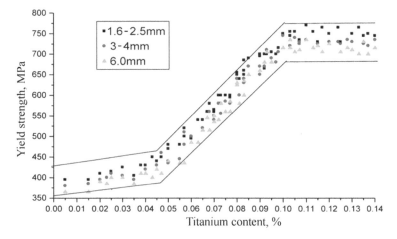

Fig. 5.24 Relationship between yield strength and titanium content

the recrystallization of austenite to some extent, realizing the non-recrystallization controlled rolling. By effectively controlling the parameters of hot rolling process, this joint controlled rolling pattern of recrystallization controlled rolling and non-recrystallization controlled rolling for producing Ti-microalloyed steel can be realized in the TSCR process [9].

5.4.3 Strengthening Mechanism of Ti-Microalloyed High Strength Steel

The main strengthening mechanisms of steel shown in Fig. 5.25 include solute strengthening, dislocation strengthening, grain refinement strengthening and precipitation strengthening.

The yield strength of steel can be described according to the extended Hall-Petch formula:

$$\sigma_y = \sigma_i + \sigma_{ss} + \sigma_p + \sigma_d + \sigma_{gs} \tag{5.5}$$

where σ_i is the internal lattice strengthening, $\sigma_i = 48$ MPa for low carbon steel
σ_{ss} is the solute strengthening
σ_p is the precipitation strengthening
σ_d is the dislocation strengthening
σ_{gs} is the grain refinement strengthening, $\sigma_{gs} = Kd^{-1/2}$.

Solute Strengthening The main microscopic mechanism of solute strengthening is the elastic interaction. The entrance of solute atoms into the crystal lattice of matrix distorts the lattice. The distortion produces elastic stress fields, which interact with the elastic stress fields around the dislocation. The solute strengthening effect is related to the amount of solute atoms. It is generally regarded that the solute strengthening effect in a certain composition range is proportional to the amount of solute. The proportional coefficient, i.e. the yield strength increment k_M produced by 1 mass

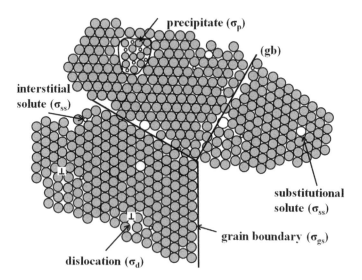

Fig. 5.25 Main strengthening mechanisms of steel

percent of solute, can be measured by experiments. Yong et al. [10] summarized the commonly used strengthening coefficients, as shown in Table 5.5.

The amount of element dissolved in steel is equal to the element content of steel if the element exists only in the form of solute in steel. For the element that exists either as solute in steel or in secondary phase, the respective amount of element dissolved in steel and in secondary phase must be theoretically calculated or experimentally measured based on the thermal history and equilibrium solubility at different temperatures. The solute strengthening increment is only related to the amount of element dissolved in steel [M].

A large number of experiments have confirmed that the increase in the yield strength of general dilute solid solution due to solute strengthening can be expressed by Eq. (5.6).

$$\Delta\sigma_s = 37[Mn] + 83[Si] + 59[Al] + 38[Cu] + 11[Mo] + 33[Ni] - 30[Cr] \\ + 680[P] + 2918[N] \quad (5.6)$$

where [M] is the mass percent of solute.

Nitrogen is mainly fixed by titanium in the form of TiN in Ti-microalloyed steel. Even in the case of low titanium content, nitrogen is fixed by aluminum to form AlN. Therefore, the content of nitrogen dissolved in steel is very low and can be ignored. Titanium in steel mainly combines with carbon, nitrogen and sulfur to form a variety of compounds so the content of titanium dissolved in steel is also very low and can be ignored. A part of carbon combines with titanium to form Ti(C, N) and another part forms cementite Fe_3C. However, there is still a considerable amount of carbon dissolved in ferrite so the strengthening effect should be taken into account. According to the chemical phase analysis of Ti-microalloyed steel, the content of carbon dissolved in ferrite [C] is taken as 0.01 wt%. Elements, such as manganese, copper, silicon, phosphorus, chromium and nickel, are present in steel in the form of solute, and their amounts dissolved in steel are directly obtained from the steel composition. The calculated solute strengthening increment is 120–140 MPa.

Dislocation strengthening Dislocation strengthening is one of the effective methods to strengthen metal materials. The relationship between flow stress and dislocation density ρ is described by the following equation:

$$\sigma_d = M\alpha\mu b\rho^{1/2} \quad (5.7)$$

where M is the orientation factor and α is the proportional coefficient.

Dislocation density mainly depends on the reduction ratio. In general, the larger the reduction ratio and the lower the finishing temperature there are, the higher the dislocation density and the greater the dislocation strengthening contribution to the yield strength there will be. σ_d is very small and does not exceed 45 MPa in the temperature range of the formation of polygonal ferrite. The dislocation density in ZJ330B steel sheet with a thickness of 1.0 mm is 2.8×10^{13} m/m³, and the calculated dislocation strengthening contribution to the yield strength is 46.1 MPa [11]. The dislocation strengthening contribution is different among Ti-microalloyed high

Table 5.5 Yield strength increment k_M produced by 1 mass percent of solute in ferrite (MPa)

C (solubility < 0.2%)	N (solubility < 0.2%)	P	Si	Ti	Cu	Mn	Mo	V	Cr	Ni	Sn	Comment
		247	82		96	70	8			33	113	
	354.2		83			32						
	2918		83			37						
		677	59			40						
			84			33						
4370	3750	350	86		39	50	22					
4570	4570	67.6	84	80	38	32	11	3	−30	0		
		468										
5000	5000	680	84		38	32	11		−30	33		
	5197											
4570	4570	470	83	80	38	37	11	3	−30	0	113	Recommended Value

5 Production, Structure and Properties Control of Titanium ...

strength steel sheets with different gauges. The lower the finishing temperature and the thinner the sheets, the greater the strengthening effect. The dislocation strengthening contribution is taken as 20–40 MPa for Ti-microalloyed high strength steel sheets with different gauges.

Grain refinement strengthening Researchers are always devoted to research on grain refinement because it is the only way to simultaneously improve the strength and toughness of steels. The grain refinement strengthening can be described by the Hall-Petch formula.

$$\sigma_g = k_y d^{-1/2} \qquad (5.8)$$

where d is the effective grain size and k_y is the proportional coefficient.

Effective grain size refers to the size of the smallest grain composed of the boundaries which impede the dislocations slip and result in the pilling up of dislocations. The sub-grain boundaries cannot become the effective grains because there is generally no dislocation pilling up nearby. Ti-microalloyed steel is generally ferrite-pearlite steel and the effective grain size is the size of ferrite grains. Theoretical calculations show that the proportional coefficient k_y is about 24.7 MPa mm$^{1/2}$. A large number of experiments have confirmed the Hall-Petch formula, and the proportional coefficient k_y can be obtained according to these experimental results. The results indicate that when the strain rate is between 6×10^{-4} and 1 s^{-1} and the grain size is in the range of 3 μm to several millimeters, the proportional coefficient k_y is 14.0–23.4 MPa mm$^{1/2}$. The proportional coefficient k_y is usually set as 17.4 MPa mm$^{1/2}$ for low carbon steel.

Figure 5.26 shows the relationship between grain refinement strengthening effect and titanium content of steel sheets with the thickness of 4.00 mm. The grain refinement strengthening increment increases with the increase in the titanium content in the beginning. Then, the increment gradually reaches a stable value of about 210 MPa when the titanium content is larger than 0.045%. The grain refinement strengthening effect increases with the decrease in the thickness, as shown in Fig. 5.27.

Precipitation strengthening Precipitation strengthening increment can be obtained by subtracting the contribution of solute strengthening, grain refinement strengthening and lattice force of pure iron from the yield strength according to Eq. (5.5). Figure 5.26 shows the relationship between precipitation strengthening effect and titanium content of steel sheets with the thickness of 4.00 mm. The precipitation strengthening increment increases slowly with the increase in the titanium content initially. Then, it increases rapidly when the titanium content exceeds 0.045%, but the increase slows down when the titanium content exceeds 0.095%.

When the titanium content is less than 0.045%, titanium mainly combines with nitrogen and sulfur to form TiN and $Ti_4C_2S_2$ particles with the size ranging from several dozens to several hundred of nanometers. Among those precipitates, the smaller ones refine the grains by suppressing the growth of recrystallized austenite grains, thereby playing the role of grain refinement strengthening. In addition, they have small precipitation strengthening effect.

When the titanium content continues to increase, the excess titanium is consumed by precipitation in austenite, interphase precipitation and precipitation in

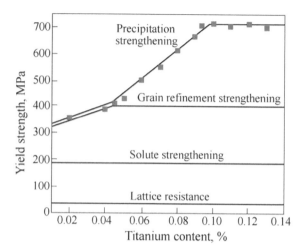

Fig. 5.26 Relationship between yield strength and titanium content of steel plates with the thickness of 4.0 mm

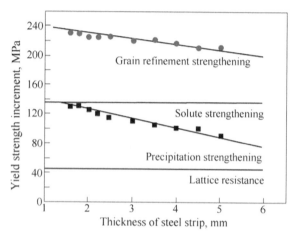

Fig. 5.27 Relationship between strengthening increment and thickness of steel strip containing 0.058% Ti

ferrite successively. These TiC precipitates are spherical and their sizes are ranging from several nanometers to several dozens of nanometers. They are massively and dispersively distributed in ferrite matrix and dislocation lines, so the precipitation strengthening effect is remarkable. In particular, the TiC particles precipitated in ferrite are the smallest because of the lowest precipitation temperature, so their precipitation strengthening effect is the strongest. Therefore, the yield strength is significantly enhanced with the increase in the titanium content when the titanium content exceeds 0.045%. The amount of TiC precipitated in ferrite is limited because the carbon content of ferrite is low. When the titanium content increases to a certain value, the carbon in ferrite completely combines with titanium to form TiC and the strengthening effect tends to saturate. When the titanium content further increases, the further strengthening effect comes from TiC particles precipitated in austenite and inter-phase precipitation. However, TiC particles tend to grow and coarsen because

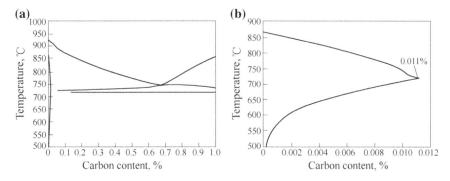

Fig. 5.28 **a** Fe (Mn, Cr, Si, P, S, Cu, P)–C phase diagram, and **b** enlarged diagram of single phase region of ferrite

the precipitation temperature is high due to the high titanium content. Therefore, their strengthening effect is not as significant as those precipitated in ferrite. The strengthening increment slows down when the titanium content exceeds 0.095%.

The maximum solubility of carbon in ferrite is 0.0218% at the eutectoid temperature according to the Fe-C binary phase diagram. The addition of alloying elements usually decreases the carbon solubility of ferrite. Figure 5.28a shows the phase diagram of Fe (Mn, Cr, Si, P, S, Cu, P)–C system of Ti-microalloyed high strength steel. Figure 5.28b is the amplification of the single phase region of ferrite, indicating the maximum solubility of carbon in ferrite is 0.011%. However, the maximum solubility of carbon in ferrite is slightly larger under the actual non-equilibrium cooling conditions. In addition, the carbon content of ferrite should be lower than the actual maximum solubility. Therefore, it can be roughly estimated that the carbon content of ferrite is 0.01%, and the carbon can combine with 0.04% titanium to form TiC. If the titanium consumptions including combination with nitrogen and sulfur, precipitation in austenite and inter-phase precipitation are taken into account, the value of the inflection point reflecting the slowdown of the strengthening increment, i.e. the second inflection point, should be no less than 0.085% Ti, which is almost consistent with the production data (0.095%).

In addition, the decrease in the steel product thickness enhances the precipitation strengthening if the titanium content is constant, as shown in Fig. 5.27. This is because when the thickness decreases, the cooling rate increases, thereby refining Ti(C, N) precipitates.

Titanium combines with carbon to form TiC in Ti-microalloyed high strength steel. TiC particles precipitated during the cooling and coiling processes are very fine and their sizes reach the magnitude of 10 nm, resulting in a significant precipitation strengthening effect. According to Gladman's theory, the precipitation strengthening effect can be quantitatively calculated.

According to the analysis above, the strengthening components of thin gauge high strength steel are estimated, as shown in Fig. 5.26 [12]. The maximum contribution of precipitation strengthening is about 250 MPa. The cooling rates of steel strips with different thicknesses are different. As a result, the sizes and mass fractions of precipitate particles are different, so the precipitation strengthening effects are different.

Figure 5.24 shows that there are two inflection points at 0.045% and 0.095% Ti. This is because only the grain refinement strengthening and precipitation strengthening are related to the titanium content among the hardening components. When the titanium content is less than 0.045%, the yield strength increases slowly with the increase in the titanium content. There are two reasons. One is that the grain refinement strengthening effect increases the yield strength by about 20 MPa with the increase in the titanium content. The other is that the precipitation strengthening effect of TiN and $Ti_4C_2S_2$ particles is small because these particles precipitate at high temperatures and they are large (about several dozens of nanometers to several hundred of nanometers). When the titanium content is above 0.045%, the grains are no longer refined with the increase in the titanium content. The increase in the strength is mainly due to the precipitation strengthening effect of TiC precipitates. The strength increases rapidly with the titanium content because the strengthening effect from the dispersive and fine TiC particles is large. The yield strength does not increase anymore when the titanium content is higher than 0.095%, because higher titanium content leads to precipitation of TiC in austenite before rolling, and the mass fraction of TiC particles from strain induced precipitation and precipitation in ferrite remains almost unchanged.

References

1. Qu Y. Principles of steelmaking [M]. Beijing: Metallurgical Industry Press, 1994.
2. Zhu W J, Ou T, Li G Q. Influence of LF refining on T[O] and inclusions of container steel [J]. Journal of University of Science and Technology Beijing, 2011, 33(S1), 137–140.
3. Pillild C F. Variables affecting the nitrogen content of carbon and low alloy acid electric arc furnace steel [C]. Electric Furnace Conference Proceedings, ISSI, 1988, 46, 107–110.
4. Thomas J, Scheid C, Geiger G. Nitrogen control during EAF steelmaking [J]. April 1993-January 1994, I & SM.
5. Mao X P, Lin Z Y, Li L J, et al. A method of improving and stabilizing the yield of titanium of titanium microalloyed high strength weathering steel during the smelting and continuous casting process [P]. Patent: 200510102239.0, China.
6. Mao X P. Microalloying technology on thin slab casting and direct rolling process [M]. Beijing: Metallurgical Industry Press, 2008.
7. Su L, Tian N Y, Mao X P, et al. Influence of liquid core reduction on the microstructure and mechanical properties of high strength weathering steel [C]. Society of Technical Exchange and Development of Thin Slab Casting and Direct Rolling. Proceedings of the Fourth Technical Exchange Conference on Thin Slab Casting and Direct Rolling, Ma Anshan, China, 2006: 538–543.
8. Huo X D, Mao X P, Li L J, et al. Strengthening mechanism of Ti micro-alloyed high strength steels produced by thin slab casting and rolling [J]. Iron and Steel Supplement, 2005, 40: 464–468.
9. Mao X P, SunX J, WangS Z. Control rolling technology of Ti-microalloyed strip produced by TSCR [J]. Iron & Steel, 2016, 51(1): 1–7.

10. Yong Q L, Ma M T, Wu B R. Physical and mechanical metallurgy of microalloyed steel [M]. Beijing: Machinery Industry Press, 1989.
11. Yu H. Research on the microstructure refinement and strengthening mechanism of hot strip of low carbon steel produced by compact strip production [D]. Beijing: University of Science and Technology Beijing, 2003.
12. Mao X P, SunX J, Kang Y L, et al. Physical metallurgy for the titanium microalloyed strip produced by thin slab casting and rolling process [J]. Acta Metallurgica Sinica, 2006, 42(10): 1091–1095.

Chapter 6
Design, Development and Application of Titanium Microalloyed Steel

Qilin Chen and ShuiZe Wang

In the 1920s, titanium was used as a microalloying element. Initially, it was used in the trace-titanium microalloying treatment to improve the welding performance of steel. With the deepening of the research on the effect of titanium in steel and continuous progress of technology, such as smelting and rolling, the role of titanium steel is further highlighted and the product range is been continuously broadened. Representative products mainly include German QStE series of steel (titanium content ≤ 0.16%), YS-T50 produced by Youngstown Sheet and Tube Company, automobile beam steel NSH52T (titanium content: 0.08–0.09%) produced by Nippon Steel and so on. The research on the titanium microalloying technology and product development in China launched late and the first titanium microalloyed steel was not developed until the 1960s. The representative product was 15MnTi (yield strength: 390 MPa). Then, a series of products were gradually developed, such as the hull structure steel 14MnVTiRE (titanium content: 0.07–0.16%), automobile beam steel 06TiL, 08TiL and 10TiL (titanium content: 0.07–0.20%) and weathering steel 09CuPTi. After 2000, many researches were carried out on thin slab casting and direct rolling (TSCR) process, titanium microalloying technology and high-strength steel, and a great leap forward development was achieved in China. A series of 450–700 MPa grade titanium microalloyed high strength steel products was developed, which is mainly applied for the container, automobile, construction machinery and other industry fields. This chapter mainly introduces the representative achievements of titanium microalloyed high strength steel in China.

Q. Chen (✉)
Guangzhou Automobile Group Co., Ltd, Guangzhou, China
e-mail: chenqilin@gaei.cn

S. Wang
Baosteel Central Research Institute, Wuhan, China
e-mail: wangshuze-316@163.com

© Metallurgical Industry Press, Beijing and Springer Nature Singapore Pte Ltd. 2019
X. Mao (ed.), *Titanium Microalloyed Steel: Fundamentals, Technology, and Products*,
https://doi.org/10.1007/978-981-13-3332-3_6

6.1 Container Steel

6.1.1 New Generation of Container Steel

More than 90% of international trade goods are transported through the ocean shipping. As the most important transport carrier for the ocean shipping, containers play an indispensable role. Currently, the global quantity of container possessive is expected to exceed 36 million twenty-foot equivalent units (TEU). With the increasing use of containers as the shipping carriers, containers will play a more significant role in international trade and transportation. China is the dominant country, which manufactures the containers and currently contributes to annual output and sales with more than 3 million TEU, accounting for more than 90% of the world.

According to statistics, more than 85% of the container steel applies the steel SPA-H with the yield strength of only 355 MPa in the field of container manufacturing. The weight of each TEU box is about 1.6 tons, which significantly increases the consumption of steel and the transportation cost. In the wake of the changes in the demand of international trade and the pursuit of reducing the transportation cost, lighter and more solid containers have drawn great attentions and lightweighting has become one of the most important trends in the future development of the containers. The lightweighting of the containers requires that the stiffness and strength should still meet the requirements of the users with the reduction of the weight. As a result, it provides a way to reduce the transportation cost. Increasing the strength of the container steel and reducing the thickness of the steel sheet are an important way to lightweight the containers. The practice proves that the weight of the container can be reduced by 12.7–14.4% if the new generation 550 MPa grade container steel is used to replace the conventional steel SPA-H. As a result, nearly 500,000 tons of steel can be economized annually, reducing the consumption of fuel for container transportation by about 1.2 million tons [1].

6.1.1.1 Standards and Performance Requirements

JIS G3125, a Japanese standard, is the most widely adopted standard for container steel currently, but the product grades prescribed in the standard only include SPA-H. Tables 6.1 and 6.2 show the chemical composition and the requirements for the mechanical properties of SPA-H, respectively. The container industry proposed the requirements for the performance of the new generation of lightweight container steel by a comprehensive evaluation, as shown in Table 6.3.

Production statistics show that the yield strength of SPA-H based on the JIS G3125 standard is 400–450 MPa, which is lower than the requirement for the new generation of container steel by 100–150 MPa. Therefore, the composition and production process of the steel need to be re-designed to get the new steel, which meets the requirements for the container steel.

Table 6.1 Chemical composition of container steel SPA-H prescribed in the standard JIS G3125-1987 (wt%)

C	Si	Mn	P	S	Cu	Ni	Cr
≤0.12	0.25–0.75	0.20–0.5	0.07–0.15	≤0.040	0.25–0.60	≤0.65	0.30–1.25

Table 6.2 Mechanical properties of container steel SPA-H prescribed in the standard JIS G3125

Yield strength (MPa)	Tensile strength (MPa)	Elongation (%)	Cold bending (180°)	
≥345	≥480	≥22	≤6 mm, $d=a$	>6 mm, $d=1.5a$

Table 6.3 Requirements for the mechanical properties of the new generation of lighweight container steel

Grade	Yield strength (MPa)	Tensile strength (MPa)	Elongation (%)	Cold bending (180°)
ZJ550 W	≥550	≥620	≥16 ($h \leq 6$ mm)	$d=1.5a$ Qualified

6.1.1.2 Composition and Process Design

Composition design is the prerequisite and basis for product development. The new generation of container steel is developed according to the requirements for mechanical properties, weather resistance, formability and weldability. The Cu–P–Cr–Ni component system of the ordinary container steel SPA-H is used as the basis, and titanium microalloying technology is applied to improve the strength by grain refinement strengthening and precipitation strengthening [2]. The typical microstructure of the ordinary container steel SPA-H is shown in Fig. 6.1. The size of the ferrite grains is 8–10 μm. If titanium microalloying is applied, the grain sizes can be refined to be 6–7 μm. According to the Hall-Petch formula, the contribution of the grain refinement strengthening is 20–30 MPa.

Titanium microalloying can provide significant precipitation strengthening effect in addition to grain refinement strengthening. According to the Orowan mechanism, the precipitation strengthening effect of the secondary phase particles is $\Delta\sigma_p$ [see Eq. (1.5)]. Suppose that the total content of titanium in steel is "Ti%", the titanium providing the precipitation strengthening effect is "effective Ti", the content of nitrogen in steel is 70 ppm and the content of sulfur is 50 ppm. According to the ideal chemical ratios, "Effective Ti" = Ti% − 3.4 × N% − 3 × S% = Ti% − 3.4 × 0.007 − 3 × 0.005 = Ti% − 0.03889. Suppose that all the "effective Ti" exists in the form of TiC, the weight percent of TiC can be calculated from the weight percent of "effective Ti":

$$\text{TiC\%} = (Ti\% - 0.03889) \cdot \frac{A_{TiC}}{A_{Ti}} = (Ti\% - 0.03889) \cdot \frac{59.9}{47.9} \quad (6.1)$$

Fig. 6.1 Metallographic microstructure of steel SPA-H (thickness: 1.6 mm)

where A is the atomic weight or molecular weight. The weight percent can be converted to volume fraction by plugging the densities of TiC and Fe (4.944 g/cm³ and 7.87 g/cm³, respectively) into Eq. (6.1).

$$f = (Ti\% - 0.03889) \cdot \frac{59.9}{47.9} \times \frac{\rho_{Fe}}{\rho_{TiC}} \quad (6.2)$$

Suppose that the diameters of TiC precipitate particles are 5, 10 and 15 nm respectively. The relationship between the precipitation strengthening effect and the content of titanium can be obtained by plugging Eq. (6.2) and the diameter d into Eq. (1.5), as shown in Fig. 6.2.

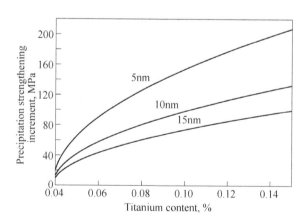

Fig. 6.2 Relationship between precipitation strengthening increment and titanium content

The yield strength of the ordinary container steel SPA-H is 400–450 MPa. The strength should be increased by 100–150 MPa to produce the 550 MPa grade high strength container steel. Subtracting the contribution from grain refinement strengthening (20–30 MPa), the precipitation strengthening should be 80–120 MPa. Generally, the diameter of TiC particles precipitated by deformation induced precipitation or precipitation in ferrite is 5–10 nm. According to Fig. 6.2, the addition of titanium in steel should be 0.07–0.09%.

Based on the analysis above, the chemical composition of the new generation of the 550 MPa grade lightweight container steel is shown in Table 6.4.

The deformation induced precipitation of Ti(C, N) should be prevented in the rolling process, and the precipitation of TiC should be promoted in the cooling and coiling processes, because the strength of steel is improved mainly by the precipitation strengthening of TiC particles. The finishing rolling temperature and coiling temperature are very important to the production process of titanium microalloyed steel. The low finishing rolling temperature is advantageous to the refinement of grain size of ferrite and the improvement of grain refinement strengthening effect. However, meanwhile, the low finishing rolling temperature leads to the deformation induced precipitation of carbonitrides. These precipitates can inhibit the grain growth of austenite, providing a certain grain refinement strengthening and precipitation strengthening effect. But they are larger compared to the nano-scale precipitates in ferrite, so their precipitation strengthening effect is reduced. Through increasing the finishing rolling temperatures, the deformation induced precipitation of Ti (C, N) in the high temperature austenite can be avoided, and the precipitation of TiC in ferrite can be promoted. In addition, increasing the finishing rolling temperature is beneficial to the large deformation during the first two passes, which promotes grain refinement and ensures the plate shape and dimensional accuracy. The finishing rolling temperature is promoted to be 880–900 °C according to the analysis above and the precipitation-temperature-time (PTT) curves for TiC precipitation in austenite. The coiling temperature is another important factor affecting the precipitation of TiC. In a certain temperature range, the secondary phase particles gradually precipitate with the decrease of temperature. The lower the precipitation temperature, the smaller the critical size of the nuclei and the finer the precipitates. On the other hand, according to the kinetic analysis, the precipitation of TiC is the result of long range diffusion. The precipitation of TiC is suppressed when the coiling temperature is too low, so the volume fraction of TiC precipitates is reduced and the precipitation strengthening effect is decreased. Therefore, the selection of the coiling temperature should take the PTT curve for the precipitation of TiC in ferrite into account so as to get the desirable temperature range.

Table 6.4 Chemical composition of the new generation of container steel ZJ550 W (wt%)

C	Si	Mn	P	S	Ti	Ni+Cu+Cr
≤0.07	≤0.30	≤1.0	≤0.075	≤0.01	≤0.10	0.60–0.90

6.1.1.3 Microstructure and Properties

The metallographic samples were cut from the steel plate, sanded with sandpaper, polished, and etched with 4% nitric acid alcohol solution. The microstructures of the samples were observed with an optical microscope and the average grain size was measured by image analysis software. Figure 6.3 shows the microstructure of a typical new generation of lightweight container steel. The microstructure is mainly composed of ferrite with the grain size of about 5.8 μm. There is a small amount of pearlite at the grain boundaries of ferrite. The thickness of the strip has a significant effect on the grain size and uniformity of the product. Figure 6.4 shows the calculated average grain sizes of strips with different thicknesses. The average grain diameter of the new generation of container steel with the thickness ranging from 1.5 to 5.0 mm is in the range of 5.8–9.8 μm.

The type, morphology and size of precipitates in the new generation of container steel were analyzed by scanning electron microscope (SEM), transmission electron

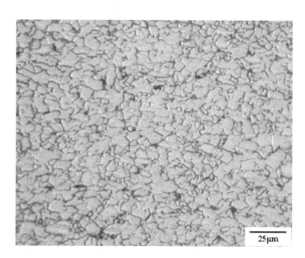

Fig. 6.3 Microstructure of typical product (thickness: 1.5 mm)

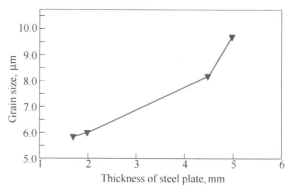

Fig. 6.4 Relationship between grain size and thickness of steel plate

6 Design, Development and Application of Titanium … 225

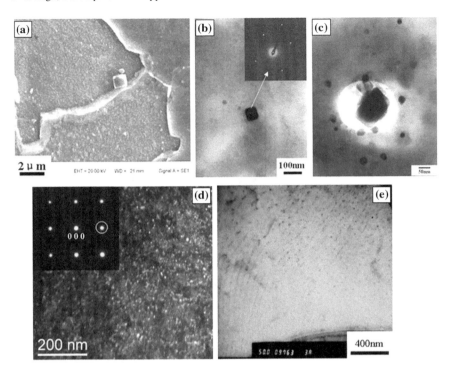

Fig. 6.5 Precipitates in the new generation of container steel: **a** TiN precipitated in liquid, **b** TiN precipitated in solid solution, **c** TiC precipitated in austenite and $Ti_4C_2S_2$, **d** TiC precipitated in super-saturated ferrite, **e** TiC formed by inter-phase precipitation

microscope (TEM) and energy dispersive spectrometer (EDS), and the results are shown in Fig. 6.5. The titanium-bearing precipitates mainly include TiN precipitated in liquid, TiN precipitated in solid solution, $Ti_4C_2S_2$ and TiC particles. Among all these precipitates, the TiC particles, which are spherical or flaky, provide the main precipitation strengthening effect. The TiC particles can be classified into two types according to their sizes. One is the larger particle with the diameter of about 20 nm, as shown in Fig. 6.5c. The other is the smaller particle with the diameter below 10 nm, as shown in Fig. 6.5d. They are uniformly dispersed in the matrix. Analysis by TEM diffraction indicates that there is no definite orientation relationship between the larger particles and the ferrite, while the smaller particles maintain the Bake-Nutting orientation relationship with the ferrite. According to the sizes and the orientation relationships of the particles with the ferrite, it is clarified that the two types of TiC particles correspond to the two different precipitation stages. The larger particles precipitate in austenite through deformation induced precipitation, while the smaller particles precipitate in the supersaturated ferrite. In addition, the TiC particles formed by inter-phase precipitation during the phase transformation process were observed, as shown in Fig. 6.5e. Their sizes are less than 10 nm, and the spacings of the precipitate arrays are in the range of 50–100 nm.

Table 6.5 Mechanical properties of the new generation of container steel

Grade	R_{eL} (MPa)	R_m (MPa)	δ (%)	Yield ratio	Comment
ZJ550 W	550	620	21.0	0.83	Minimum
	595	715	34.0	0.90	Maximum
	571	656	26.1	0.87	Average

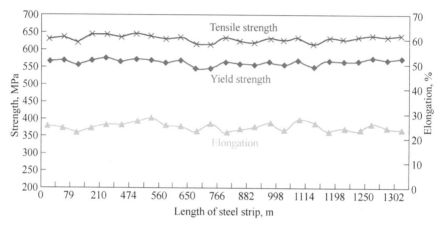

Fig. 6.6 Mechanical properties over the coil of the new generation of container steel ZJ550 W

Table 6.5 shows the results of the mechanical tests and statistics analysis on the new generation of container steel (typical grade: ZJ550 W). The yield strength and tensile strength are 550–595 MPa and 620–715 MPa respectively. The total elongation is 21–34%, and the yield ratio is 0.83–0.90.

The mechanical properties at different locations of the whole coil were measured, and the results are shown in Fig. 6.6. The differences in the yield strength and tensile strength are within 30 and 25 MPa, respectively, indicating that the mechanical properties are stable over the whole coil.

6.1.1.4 Service Performance

Weather Resistance

The new generation of container steel is mainly used for manufacturing containers for ocean shipping. A high standard of corrosion resistance is required for the containers because of the maritime climate and erosion by the waves during the long distance transport. The weather resistance is an important index to evaluate the service performance of container steel. According to the standard TB/T2375 and the national standard GB/T 10125 in China, the cyclic infiltration test and salt spray test were used to qualitatively and quantitatively measure the corrosion resistance of the new generation of container steel. The results were compared to those of the conventional container steel. Three samples (A, B, C) were obtained from the new generation of container steel ZJ550 W(1#) and conventional container steel SPA-H(2#), respectively. Cyclic infiltration tests were carried out on these samples simultaneously, and the results are shown in Table 6.6.

The rate of weight loss of the new generation of container steel ZJ550 W is in the range of 1.224–1.296 g m^{-2} h^{-1}, with an average value of 1.268 g m^{-2} h^{-1}. The rate of weight loss of the conventional container steel SPA-H is in the range of 1.217–1.354 g m^{-2} h^{-1}, with an average value of 1.295 g m^{-2} h^{-1}. The comparison shows that the weather resistance of the new generation of container steel ZJ550 W can rival that of the conventional container steel SPA-H, indicating that ZJ550 W can meet the requirements for container steel.

Weldability

The container steel plates need to be stitched by welding. The welding test was carried out to evaluate the weldability and the microstructure evolution of the new generation of container steel ZJ550 W. The welding rod was chosen according to the strength grade and the chemical composition of steel. In this experiment, the welding rod was HTW-50 and the diameter of the solder wire was 1.0 mm. CO_2 was used as the shielding gas. After welding, the transverse samples were taken from the welded steel plates according to the national standard GB 2649-1989, and tensile tests were carried out on these samples according to the national standard GB 2651-2008. All the samples fractured at the base metal during the tensile tests. The tensile properties before and after welding are shown in Table 6.7. The comparison shows that the steel plates still have good tensile properties after welding, indicating the weldability is qualified.

The microstructures were investigated by metallographic observation, as shown in Fig. 6.7. The microstructures in the weld and heat affected zone are homogeneous and fine, and the grain size after welding does not change greatly, so the weld quality is evaluated to be good.

Cold Bending and Springback Properties

Container steel needs to bear the bending deformation during the service period. Cold bending test is one method to evaluate the plastic deformation property of the steel at room temperature. In order to evaluate the formability of the new generation of container steel, cold bending tests were carried out on the longitudinal samples taken from the container steel plates with 4 different thickness specifications. The widths of the samples were 20 mm and 40 mm, respectively. The radius d was 0, and the bending angle was 180°. The results shown in Fig. 6.8 state that all samples have good cold bending property, and they are qualified for stamping.

The springback property was tested by V-shape bending mold. The results shown in Table 6.8 indicate that the springback of the new generation of container steel is small.

Table 6.6 Results of the cyclic infiltration tests on the new generation of container steel ZJ550 W and conventional container steel SPA-H

No.	Length (mm)	Width (mm)	Thickness (mm)	Surface area (mm^2)	Original weight (g)	Weight after corrosion (g)	Weight loss (g)	Rate of weight loss (g m^{-2} h^{-1})	Average (g m^{-2} h^{-1})
1A	50.04	40.12	3.24	4599.446	49.5474	48.6013	0.9461	1.224	1.268
1B	50.08	40.06	3.22	4592.911	49.432	48.4319	1.0001	1.296	
1C	50.04	40	3.20	4579.456	49.4438	48.4555	0.9883	1.284	
2A	50.20	40.18	1.30	4269.06	19.9545	19.013	0.9415	1.312	1.295
2B	50.24	40.18	1.52	4312.163	23.2305	22.3485	0.882	1.217	
2C	50.30	40.06	1.56	4311.959	23.9786	22.9977	0.9809	1.354	

Table 6.7 Tensile properties of container steel ZJ550 W before and after welding

Thickness (mm)	Before welding			After welding		
	Yield strength (MPa)	Tensile strength (MPa)	Elongation (%)	Yield strength (MPa)	Tensile strength (MPa)	Fraction location
4.0	605	680	23	585	670	Base metal

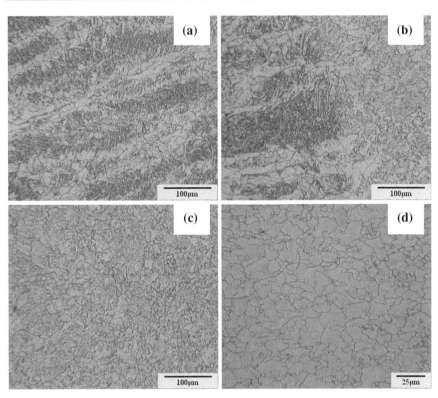

Fig. 6.7 Microstructures of different regions of the weld joint of container steel ZJ550 W: **a** weld, **b** weld and transition region, **c** transition region, **d** base metal

6.1.1.5 Applications

The new generation of container steel is mainly used to manufacture the lightweight containers. The 20-foot and 40-foot standard containers shown in Fig. 6.9 account for more than 90% of the total containers. The new generation of container steel ZJ550 W was used to replace the conventional container steel SPA-H after comprehensively evaluating the effect of thickness reduction, safety and other factors. The thicknesses of various container parts are reduced with varying degrees, as shown

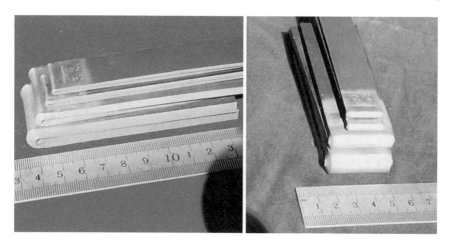

Fig. 6.8 Samples of the new generation of container steel for cold bending test ($d=0$)

Table 6.8 Springback of the new generation of container steel

Thickness (mm)	Radius (mm)	Springback (°)
1.6	2	4
2.0	2	5
4.0	5	2.5
5.0	5	1.5

Fig. 6.9 Lightweight containers. **a** 20-foot DV container, **b** 40-foot HC container

in Table 6.9. Table 6.10 shows the total weight reduction of the container. The new generation of 20-foot container weighs 1900 kg, while the conventional container weighs 2220 kg. Therefore, the weight is reduced by 14.41% and the steel consumption is reduced by 15.2%. The new generation of 40-foot container weighs 3350 kg, while the conventional container weighs 3840 kg. As a result, the weight is reduced by 12.76% and the steel consumption is reduced by 14.75%.

Table 6.9 Lightweighting design of container parts

Parts	20-foot standard container			40-foot standard container		
	Thickness (mm)		Ratio of weight reduction (%)	Thickness (mm)		Ratio of weight reduction (%)
	ZJ550 W	SPA-H		ZJ550 W	SPA-H	
Main side panel	1.5	1.6	6.25	1.5	1.6	6.25
Front wall panel, roof panel, side wall panel	1.7	2.0	15	1.7	2.0	15
Door rail	2.5	3.0	16.67	2.5	3.0	16.67
Front cross member, door header	3.0	4.0	25	3.0	4.0	25
Door sill, bottom side rail	4.0	4.5	11.11	4.0	4.5	11.11
Tunnel beam, gooseneck tunnel	–	–	–	3.0	4.0	25
Front corner post	4.5	6.0	25.0	5.0	6.0	16.67
Fork lift pocket bottom plate	5.0	6.0	16.67	–	–	–
Rear corner post	5.0	6.0	16.67	5.0	6.0	16.67

Table 6.10 Lightweighting of the new generation of containers

Type	20-foot standard container		40-foot standard container	
	Steel consumption (kg)	Weight (kg)	Steel consumption (kg)	Weight (kg)
New generation	1450	1900	2600	3350
Conventional	1710	2220	3053	3840
Weight reduction	260	320	453	490
Ratio of weight reduction (%)	15.2	14.41	14.75	12.76

The new generation of lightweight containers has passed 15 standard tests, including the stacking, ceiling, fork, vertical bolt, end wall strength, side panel strength, floor strength, roof strength, transverse rigidity, longitudinal rigidity, watertight and so on. The containers have been certified by Bureau Veritas (BV) and American Bureau of Shipping (ABS), and have been used in the international ocean shipping.

6.1.2 Special Container Steel

The sizes of containers for ocean shipping are generally 20 feet or 40 feet. However, because the highway and railway systems in the inland of USA are very developed, larger containers with the size of 53 feet are used in USA to save transportation cost and improve transportation efficiency. The goods contained in the 20-foot or 40-foot containers shipped to USA should be unloaded at the port of USA and loaded into the 53-foot containers, which is time-consuming and laborious. Therefore, the container industry proposed the demand of 53-foot special containers for seamlessly connecting ocean shipping and inland transport.

53-foot special containers were originally manufactured by domestic manufacturers in USA. Most of them were made of riveted aluminum alloy or composite plates. The production process is complicated, labor-intensive and costly, and the maintenance cost is high. In addition, the strength is low and the structure is not durable. Therefore, it is unable to meet the container standards of ISO and cannot be used for maritime transport. According to the comparative analysis on the properties, advantages and disadvantages, and prices of aluminum alloy, composite and steel plates, the container industry in China proposed the use of steel to replace the aluminum alloy and composite materials.

The conventional container steel is SPA-H, which has good resistance to the corrosion of marine climate. However, the use of conventional container steel increases the weight of the container. The consumption of steel for each TEU is 4625 kg, and the dead load of each TEU is 5630 kg. Because the maximum rated weight of each TEU (sum of load capacity and dead load) is constant, an increase in the dead load reduces the load capacity. If the dead load is reduced and the strength is maintained simultaneously, the load capacity will increase and the consumption of materials will decrease, which saves the energy and reduces the environmental pollution. Therefore, the container industry put forward the demand for ultra-high strength weathering steel.

6.1.2.1 Standards and Performance Requirements

Ultra-high strength weathering steel refers to its yield strength above 700 MPa. In addition to high corrosion resistance, the properties such as higher strength, better formability and weldability are demanded. Thus, the requirements for the metallurgical process and equipment control are very high. Only a few countries can produce this kind of steel. The major product is the DOMEX weathering steel produced by SSAB. The steel grade is DOMEX700 W, and its chemical composition and mechanical properties are shown in Tables 6.11, 6.12 and 6.13 [3]. However, up to now, there are no mature and complete industry standards and national standards.

Table 6.11 Chemical composition of steel DOMEX700 W produced by SSAB (wt%)

C	Si	Mn	P	S	Cu	Cr	Ni	Mo	Nb,V,Ti
≤ 0.12	≤ 0.6	≤ 2.1	≤ 0.03	≤ 0.015	0.25–0.55	0.3–1.25	≤ 0.65	≤ 0.3	Nb+V+Ti ≤ 0.22

Table 6.12 Mechanical properties of steel DOMEX700 W produced by SSAB

Yield Strength (MPa)	Tensile Strength (MPa)	Elongation (%)		Cold bending (90°)		
		<3 mm	$\geq$3 mm	<3 mm	3–6 mm	>6 mm
$\geq$700	$\geq$750	$\geq$12	$\geq$12	1.5a	2.0a	2.0a

Table 6.13 Low temperature toughness of steel DOMEX700 W produced by SSAB

Grade	V-notch impact test		
	Test direction	Temperature (°C)	Impact energy (J)
DOMEX700 W	Longitudinal	–20	$\geq$40

6.1.2.2 Composition and Process Design

Baosteel Co., Benxi Steel Co. and Taiyuan Iron and Steel Co. have studied the 700 MPa grade ultra-high strength steel [4–6]. The chemical compositions are shown in Table 6.14. The design of Baosteel is multiple-microalloying low carbon steel with niobium, titanium and molybdenum. The composition system of Benxi Steel is the low carbon high manganese steel with the addition of molybdenum, chromium, niobium and titanium. Taiyuan Iron and Steel Co. developed the steel by multiple niobium and titanium microalloying and adding a proper amount of molybdenum.

Zhujiang Steel Co. developed the low carbon and high manganese composition system microalloyed by single titanium based on the TSCR process. The performance requirements for the ultra-high strength weathering steel and the economy of production were taken into account in the design. The steel was developed by adding a certain amount of manganese and titanium into the composition system of the conventional container steel. The ferrite grains are refined and the precipitation strengthening effect is improved due to the synergistic effect of manganese and titanium. As a result, in addition to the weather resistance, the requirements for strength, formability and weldability are satisfied. The typical grade is ZJ700 W, and

Table 6.14 Chemical compositions of 700 MPa grade ultra-high strength steel developed by steel manufacturers in China (wt%)

Company	C	Si	Mn	P	S	Cr	Nb	Ti	Mo	V
Baosteel	0.07	0.25	1.8	0.025	0.015	–	S	S	S	–
Benxi	0.05	0.15	2.0	0.008	0.003	0.41	0.05	0.12	0.2	0.15
Taiyuan	0.084	0.14	1.82	0.011	0.001	–	S	S	S	–

S means a small amount

Table 6.15 Chemical composition of the ultra-high strength weathering steel produced by Zhujiang Steel (wt%)

C	Si	Mn	P	S	Cu	Ni	Cr	Ti
≤0.07	≤0.60	≤2.0	≤0.03	≤0.01	0.55	≤0.3	≤0.6	≤0.15

the chemical composition is shown in Table 6.15. The alloying cost is significantly reduced. The design of the hot rolling process parameters, such as the finishing rolling temperature and coiling temperature, is mainly based on the PTT curves for the precipitation of TiC particles in austenite and ferrite, so as to maximize the precipitation of TiC in ferrite.

6.1.2.3 Microstructure and Properties

The microstructure of the steel is shown in Fig. 6.10. The size of the grains with high angle grain boundaries (misorientation angle greater than 15°) is about 3.3 μm. Compared with the new generation of the container steel ZJ550 W with the equal thickness (Fig. 6.3), the grains of ZJ700 W are greatly refined, indicating that the synergistic effect of manganese and titanium is significant. According to the Hall-Petch formula, the contribution from grain refinement strengthening is about 80 MPa. Moreover, as shown in Fig. 6.11, the precipitates in ZJ700 W are significantly refined because of the synergistic effect of manganese and titanium. Compared with ZJ550 W, the proportion of precipitates with the size below 5 nm in ZJ700 W rises from 5 to 23%, as shown in Fig. 6.12. Therefore, the precipitation strengthening effect increases from about 130 MPa to about 200 MPa.

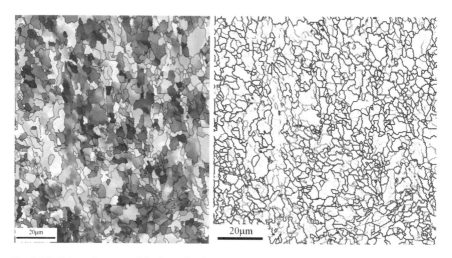

Fig. 6.10 Orientation map of ferrite grains in steel ZJ700 W (thickness: 2.0 mm)

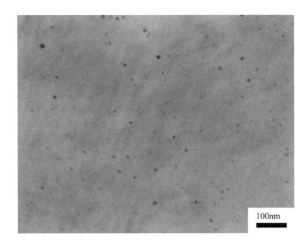

Fig. 6.11 Distribution of precipitates in steel ZJ700 W

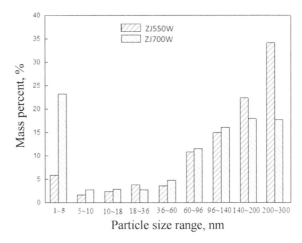

Fig. 6.12 Comparison of particle size distribution of precipitates in steel ZJ550 W and ZJ700 W

Table 6.16 Tensile properties of titanium microalloyed steel ZJ700 W

Grade	R_{el} (MPa)	R_m (MPa)	δ (%)	Yield ratio	Comment
ZJ700 W	715	760	17	0.86	Minimum
	770	850	25	0.91	Maximum
	740	820	20	0.89	Average

Table 6.16 shows the mechanical properties of the ultra-high strength steel ZJ700 W used for special containers. The yield strength is 715–770 MPa and the tensile strength is 760–850 MPa. The total elongation is 17–25% and the yield ratio is 0.86–0.91.

6.1.2.4 Service Performance

Corrosion Resistance

The corrosion rates of ZJ700 W, SPA-H and ZJ550 W were measured by cyclic infiltration corrosion tests to evaluate the corrosion resistance. The standard is TB/T 2375-1993. The sample specifications are shown in Table 6.17. The test time was 72 h. The results shown in Table 6.18 indicate that the weather resistance of ultra-high strength weathering steel is basically the same as that of the conventional container steel.

Weldability

The welding test was carried out to evaluate the weldability of the steel and the microstructure evolution due to welding. The samples were obtained along the rolling direction of the steel plates with thickness of 6.0, 4.0 and 3.5 mm. Then they were cut from the transverse center. The welding rod was HTW-70 and the diameter of the solder wire was 1.0 mm. CO_2 was used as the shielding gas. The tensile properties of the samples after welding are shown in Table 6.19. The steel plates maintain good tensile properties after welding, indicating the weldability is qualified.

The metallographic samples were obtained at the weld region, sanded with sandpaper, polished and then etched with nitric acid alcohol solution. The microstructures of the samples were observed under an optical microscope, as shown in Fig. 6.13. The microstructure of the base metal after welding is composed of ferrite and pearlite. Ferrite grains are uniform and fine. The microstructure of the weld is composed of massive eutectoid ferrite, a large amount of Widmanstanten structure, and pearlite distributed between ferrite. The microstructure of the transitional area between the base metal and the weld consists of ferrite, Widmanstanten structure and pearlite.

Table 6.17 Sample specifications

Sample number	Grade	Company	Thickness (mm)	Dimension (mm × mm × mm)	Number of samples
1#	SPA-H	–	6.0	60 × 40 × 5.5	4
2#	SPA-H	–	4.5	60 × 40 × 3.8	3
3#	SPA-H	Zhujiang Steel	4.5	60 × 40 × 3.4	3
4#	ZJ550 W	Zhujiang Steel	4.5	60 × 40 × 3.8	4
5#	ZJ550 W	Zhujiang Steel	4.8	60 × 40 × 4.5	3
6#	ZJ700 W	Zhujiang Steel	3.2	60 × 40 × 2.6	4
7#	ZJ700 W	Zhujiang Steel	4.5	60 × 40 × 4.5	3

Table 6.18 Results of the cyclic infiltration corrosion tests

Sample Number	1#	2#	3#	4#	5#	6#	7#
Corrosion rate ($g\ m^{-2}\ h^{-1}$)	1.933	1.54	1.448	1.559	1.572	1.554	1.874

Table 6.19 Tensile properties of the samples after welding

Sample number	Thickness (mm)	Yield strength (MPa)	Tensile strength (MPa)	Fracture location
1#	3.5	715	770	Base metal
2#	5.0	725	785	Base metal
3#	6.0	705	765	Base metal

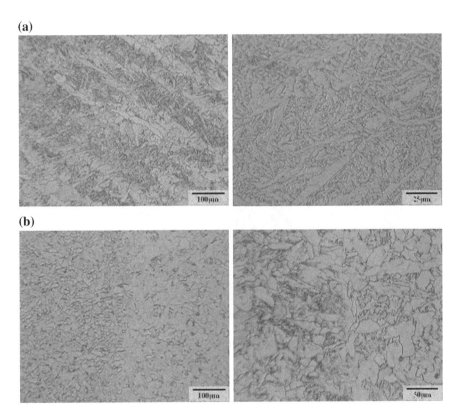

Fig. 6.13 Microstructures of different regions of the weld joint of ultra-high strength weathering steel ZJ700 W: **a** weld, **b** transition region between weld and base metal

Impact Toughness

A series of impact tests in the temperature range from room temperature to −100 °C were carried out (20, 0, −20, −40, −60, −80, −100 °C) to evaluate the low temperature toughness of ultra-high strength weathering steel ZJ700 W. According to the national standard GB/T 229-1994, small samples were obtained along the direction

Table 6.20 Impact toughness of ZJ700 W at different temperatures (J/cm^2)

Sample number	Thickness (mm)	Room temperature	0 °C	−20 °C	−40 °C	−60 °C	−80 °C	−100 °C
1#	3.5	93.7	91.2	91.8	90.0	81.5	28.7	13.7
2#	5.0	86.9	84.4	95	81.3	73.8	23.4	10.3
3#	6.0	89.6	89.8	82.5	82.1	72.1	7.9	5.2

Table 6.21 Ductile-brittle transition temperature of ZJ700 W

Number	Thickness (mm)	FTE (°C)
1#	3.5	<−70
2#	5	<−60
3#	6	<−60

Fig. 6.14 Impact fracture of ultra-high strength weathering steel (Test temperature: −60 °C, thickness: 6 mm)

perpendicular to the rolling direction. The depth of V-notch is 2 mm. The impact toughness at different temperatures and the ductile-brittle transition temperature of ZJ700 W steel are shown in Tables 6.20 and 6.21.

Table 6.20 shows that the impact toughness of ZJ700 W at −60 °C exceeds 70 J/cm^2, indicating that its low temperature toughness is good. Table 6.21 shows that the ductile-brittle transition temperature of ZJ700 W is below −60 °C, indicating that the toughness of ZJ700 W maintains good at the low temperature. The impact fracture of steel 3# at −60 °C presented in Fig. 6.14 shows that there is still a certain percent of ductile fracture when the test temperature is −60 °C.

6.1.2.5 Applications

The high strength weathering steel ZJ700 W is mainly used for the production of special 53-foot containers. The main structure is shown in Fig. 6.15. Because of the successful development of ultra-high strength weathering steel, the container industry has used this steel to replace the conventional container steel SPA-H and carried out the lightweighting design for the special 53-foot container, as shown in Table 6.22. With the use of ultra-high strength weathering steel, the steel consumption for each TEU steel can be reduced by 330.65 kg, resulting in a weight loss rate of 34.5%.

It is particularly worth mentioning that the thickness of the steel plates used for the roof and door panels of the special containers is usually designed to be in the range of 1.1–1.2 mm in order to reduce the weight of the containers. Generally, the cold rolled products such as DOCOL700 W are used. In recent years, Wuhan Iron and Steel successfully developed hot rolled ultra-high strength weathering steel by using titanium microalloying technology on the TSCR process. The grade is WJX750-NH and the thinnest thickness of the steel reaches 1.1–1.2 mm [7]. The performance indicators meet the requirements for ultra-high strength weathering steel. At present, the product has replaced the cold rolled products in the manufacturing of the roof and door panels of the special containers, as shown in Fig. 6.16. The goal to replace the cold rolled with the hot rolled is realized and the cost of manufacturing cost is reduced significantly.

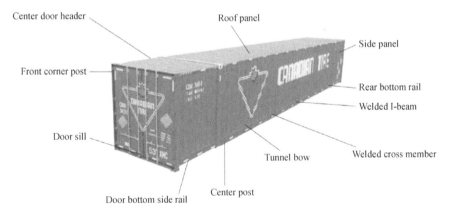

Fig. 6.15 Special 53-foot container

Table 6.22 Design of container parts by using ultra-high strength weathering steel

No.	Part	Number	Without ultra-high strength weathering steel			With ultra-high strength weathering steel		
			Materials	Thickness (mm)	Weight (kg)	Materials	Thickness (mm)	Weight (kg)
1	Front bottom side rail	2	SPA-H	4.0	64.07	ZJ700 W	3.0	48.05
2	Rear bottom side rail	2	SPA-H	4.0	63.29	ZJ700 W	3.0	47.47
3	Center corner post (external)	4	SPA-H	6.0	113.42	ZJ700 W	4.0	75.61
4	Center corner post (internal)	4	SPA-H	6.0	159.12	ZJ700 W	4.0	106.08
5	Center corner post reinforcement	4	SPA-H	4.5	74.15			
6	Tunnel side rail	2	SPA-H	6.0	66.44	ZJ700 W	4.0	44.29
7	Tunnel bolster (lower)	1	SPA-H	6.0	30.09	ZJ700 W	4.5	22.57
8	Tunnel bolster (upper)	1	SPA-H	6.0	35.95	ZJ700 W	4.5	26.96
9	Short tunnel bolster	2	SPA-H	6.0	15.81	ZJ700 W	4.5	11.86
10	Internal door sill	1	SPA-H	6.0	38.20	ZJ700 W	4.5	28.65
11	Door header	1	SPA-H	4.5	27.56	ZJ700 W	4.5	19.97
12	Door sill	1	SPA-H	6.0	35.67	ZJ700 W	6.0	25.85
13	Door sill bottom plate	1	SPA-H	6.0	21.90	ZJ700 W	6.0	15.87
14	Gooseneck tunnel panel	1	SPA-H	6.0	211.39	ZJ700 W	6.0	153.18
	Total consumption				957.06			626.41

6.2 Automotive Structure Steel

6.2.1 Automotive Body Structure Steel

The automotive body structure steel referred in this section is mainly used for the production of truck carriage, including the fence, floor, beams and other structural parts. It is also used for manufacturing rectangular tubes, which are used for body and chassis supports, luggage racks, roofs of passenger cars. The yield strength of the steel used for automotive body structure was generally below 600 MPa. In recent years, the

Fig. 6.16 Use of thin gauge WJX750-NH to manufacture containers

demand of high strength steel with the yield strength of 700 MPa exhibits an explosive growth because of the increasing demand for lightweighting in truck industry [8–10] and especially the release and implementation of the national standard GB 1589-2016, which limits the size, shaft load and weight limit of automobiles, trailers and trains. Due to the development of new energy cars and the strict requirements for the weight of some car models for export, the passenger car manufacturing industry in China has began to choose the thin gauge steel with the yield strength of 700 MPa from 2011.

6.2.1.1 Standards and Performance Requirements

European standard EN 10149 is currently widely applied to steel for cold forming. The steel prescribed in the standard is not only suitable for cold bending, roll forming, cold rolling and other forming methods, but also has good welding adaptability and galvanizing performance. Therefore, the steel is widely adopted by automotive body manufacturers. The chemical compositions and properties requirements of the typical ordinary strength steel S420MC and ultra-high strength steel S700MC in this standard are shown in Tables 6.23 and 6.24, respectively.

Table 6.23 Chemical compositions of S420MC and S700MC in the standard EN 10149 (wt%)

Grade	C	Si	Mn	P	S	Nb	V	Ti	Mo
S420MC	≤0.12	≤0.50	≤1.5	≤0.025	≤0.02	≤0.09	≤0.20	≤0.15	–
S700MC	≤0.12	≤0.60	≤2.1	≤0.025	≤0.015	≤0.09	≤0.20	≤0.22	≤0.50

Table 6.24 Mechanical properties of S420MC and S700MC in the standard EN 10149

Grade	Yield strength R_{eL} (MPa)	Tensile strength R_m (MPa)	Elongation (%)		Bending test (180°) d: diameter (mm) a: thickness (mm)
			$h<3$ mm $L_0 = 80$ mm	$h \geq 3$ mm $L_0 = 80$ mm	
S420MC	≥ 420	480–620	≥ 16	≥ 19	$d=0.5a$
S700MC	≥ 700	750–950	≥ 10	≥ 12	$d=2a$

Table 6.25 Chemical compositions of S420MC and S700MC developed by Wuhan Iron and Steel

Grade	C	Si	Mn	P	S	Nb	Ti	Mo
S420MC	0.05–0.09	≤ 0.03	0.40–0.70	≤ 0.02	≤ 0.010	–	0.05–0.09	–
S700MC	0.05–0.09	≤ 0.15	1.2–1.9	≤ 0.012	≤ 0.005	0.03–0.06	0.10–0.14	0.1–0.2

6.2.1.2 Composition and Process Design

Ordinary strength steel used for automotive body structure, such as S420MC, is mainly based on the C-Mn-Nb composition system. Wuhan Iron and Steel developed S420MC by microalloying C-Mn system with single titanium. The special rolling process is implemented to solve the problem that the fluctuation of the strength of the steel microalloyed by single titanium is large. The strengthening ways, such as precipitation strengthening and grain refinement strengthening, must be fully exploited to develop the ultra-high strength steel with the yield strength of more than 700 MPa. Currently, the composition system is based mainly on titanium microalloying, which makes full use of the precipitation of fine and dispersive TiC particles to achieve significant precipitation strengthening. Addition of a small amount of niobium raises the austenite recrystallization temperature and expands the austenite non-recrystallization zone. Therefore, the workpiece can be completely deformed at high temperatures, resulting in a fine microstructure. In addition, it is beneficial for producing thin gauge products. Adding a small amount of molybdenum and vanadium influences the precipitation behavior of TiC. Therefore, the steel strength can be further improved and the property fluctuation can be reduced. Wuhan Iron and Steel developed the S420MC and S700MC, and the chemical compositions are shown in Table 6.25.

For S420MC microalloyed with single titanium, the fine TiC particles precipitated during the rolling and coiling processes provide the main strengthening effect. According to Chap. 3, effective Ti = Ti% − 3.4 × N% − 3 × S%. Therefore, the fluctuations of nitrogen and sulfur contents lead to the fluctuation of effective titanium content, which has impact on the strength of S420MC. Figure 6.17 shows the influence of the fluctuation of effective titanium content on the tensile strength of S420MC with the same thickness and rolling process. The coiling temperature is dynamically set according to the effective titanium content of the smelting composition so as to reduce the fluctuation of the strength of S420MC.

The nitrogen in the molten steel is liable to react with titanium to form coarse TiN. The higher the nitrogen content, the higher the precipitation temperature of TiN. The

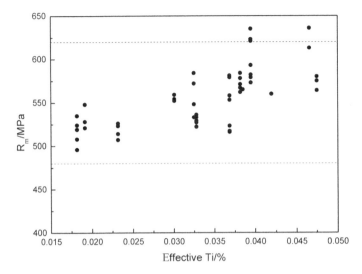

Fig. 6.17 Influence of effective titanium content on the tensile strength of S420MC

size of TiN can reach 5 μm. The size of the critical crack in the low strength steel is close to millimeter in magnitude, so the severe brittle fracture will not occur as long as the maximum size of inclusions is controlled to be below millimeter in magnitude. However, the size of the critical crack in the ultra-high strength steel is about 10 μm, so the precipitation of large TiC particles in steel with the yield strength of more than 700 MPa must be avoided. Therefore, the nitrogen content is generally controlled to be below 50 ppm.

Two stages of cooling are used in the production of ultra-high strength steel with the yield strength of more than 700 MPa. The intermediate cooling temperature has great influence on the final microstructure of S700MC. As shown in Fig. 6.18, the high intermediate cooling temperature results in the formation of polygon ferrite, in contrast, the low temperature leads to the formation of bainite and a larger amount of Fe_3C.

The effect of cooling mode and coiling temperature on the microstructure and properties of ultra-high strength steel with the yield strength above 700 MPa was studied [11]. The results are shown in Fig. 6.19. The yield strength and tensile strength are above 700 and 800 MPa respectively due to high coiling temperature. If the low coiling temperature is used, the yield strength and tensile strength are lower, indicating that the high temperature coiling can significantly improve the yield strength and tensile strength of titanium microalloyed steel.

The effect of coiling temperatures on the microstructure is shown in Fig. 6.20. The microstructure resulted from the high temperature coiling is mainly composed of quasi-polygonal ferrite. In contrast, the low temperature coiling leads to the formation of a large number of dispersive and fine particles, most of which are distributed along the grain boundaries.

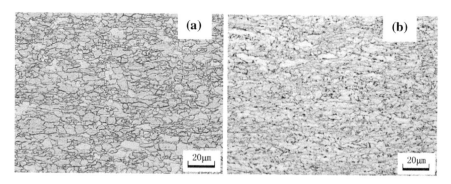

Fig. 6.18 Influence of intermediate cooling temperature on the microstructure of S700MC: **a** high intermediate cooling temperature, **b** low intermediate cooling temperature

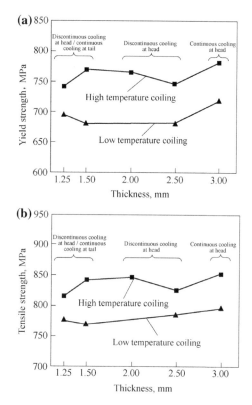

Fig. 6.19 Influence of cooling mode and coiling temperature on the strength: **a** tensile strength, **b** yield strength

The fine particles precipitated during the low temperature coiling process is shown in Fig. 6.21. The EDS analysis shows that the precipitates are Fe_3C. During the low temperature coiling process, a large number of cementite particles with the

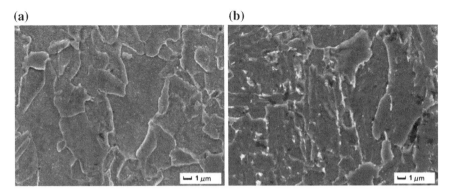

Fig. 6.20 Influence of coiling temperature on the microstructure of S700MC: **a** high temperature coiling, **b** low temperature coiling (thickness: 1.5 mm)

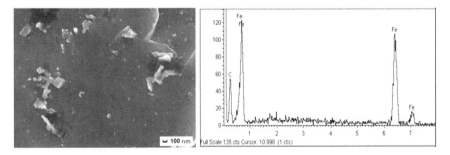

Fig. 6.21 Particles precipitated during the low temperature coiling

size in the range of 200–300 nm precipitate. The main strengthening mechanism of titanium microalloyed high strength steel is the precipitation strengthening of TiC particles. However, there is no sufficient time for completing the precipitation of TiC during the hot rolling and cooling process. Therefore, the coiling process is an important stage for the precipitation of TiC and the coiling temperature is an important parameter affecting the precipitation of TiC. If the coiling temperature is too low, the precipitation of TiC is suppressed, resulting in a significant reduction in the number of nano-size particles in steel. In addition, the excess carbon precipitates in the form of Fe_3C with the size of several hundred nanometers, so the precipitation strengthening effect is not obvious and the strength of steel is significantly decreased.

Two stages of cooling with high intermediate cooling temperature and high temperature coiling are used in the production of S700MC. As a result, the precipitation of TiC is promoted to achieve the high strength. In addition, the microstructure is composed of polygonal ferrite, which is beneficial for the formability of S700MC.

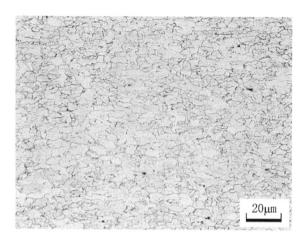

Fig. 6.22 Metallographic microstructure of S420MC (thickness: 4 mm)

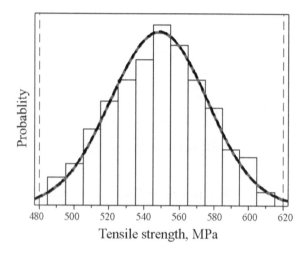

Fig. 6.23 Tensile strength distribution of S420MC

6.2.1.3 Microstructure and Properties

The metallographic microstructure of S420MC is shown in Fig. 6.22. As shown in Fig. 6.23, the problem of large fluctuation in the strength is solved when the coiling temperature is dynamically set according to the effective titanium content of the smelting composition.

The metallographic microstructure of S700MC is shown in Fig. 6.18a. The effect of thickness on the grain size is shown in Fig. 6.24. The grain size of S700MC is small and in the range of 2.4–3.4 μm. Moreover, the grains are obviously refined

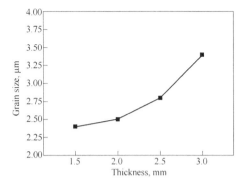

Fig. 6.24 Influence of thickness on grain size

Table 6.26 Typical mechanical properties of S700MC with different thickness specifications

Thickness (mm)	$R_{p0.2}$ (MPa)	R_m (MPa)	$R_{p0.2}$ (R_m)	A_{50} (%)
3.0	720	800	0.90	19.9
2.0	744	805	0.92	18.5
1.2	750	826	0.91	20.3

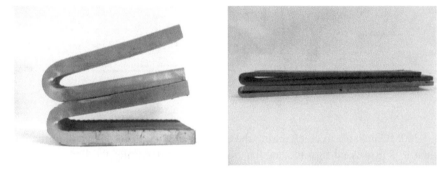

Fig. 6.25 Samples of S420MC for cold bending test

with the decrease of the thickness. The typical mechanical properties of S700MC with different thickness specifications are shown in Table 6.26.

6.2.1.4 Service Performance

Formability

The chemical composition of S420MC is based on low carbon and low manganese composition system, so the cold bending formability is good, as shown in Fig. 6.25. It can meet the requirements for roll forming, bending and other processes.

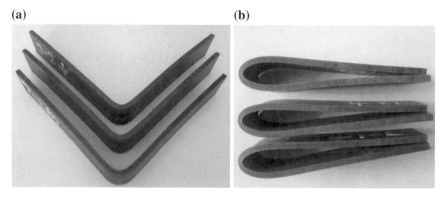

Fig. 6.26 Samples of S700MC for cold bending test: **a** 90 °C, **b** 180 °C

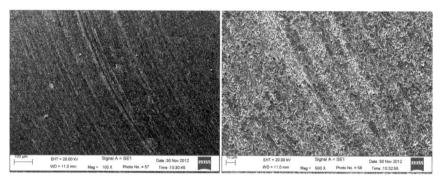

Fig. 6.27 Microstructure of the deformed region

S700MC is liable to crack during the cold bending because of large deformation when it is processed into rectangular tubes for manufacturing the body frame of bus [12]. Therefore, there is strict requirement for the cold formability. Figure 6.26 shows the results of the cold bending test on the ultra-high strength steel S700MC developed by Wuhan Iron and Steel. No cracks and lamination phenomenon were observed. Figure 6.27 shows the microstructure of the samples after cold bending. No micro-cracks were generated, indicating that the cold bending formability is good. The sample was further processed into rectangular tube, and flattened along the diagonal direction, as shown in Fig. 6.28. No cracks were produced, indicating that the service requirement is satisfied.

Fatigue Property

The axial fatigue test on S700MC with the thickness of 2.5 mm was carried out and the results are shown in Fig. 6.29.

Fig. 6.28 Rectangular tube made of S700MC

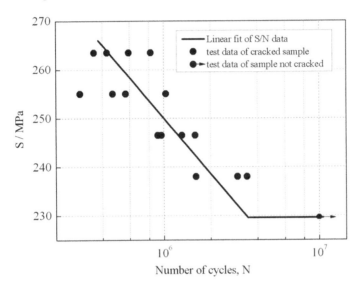

Fig. 6.29 Fatigue curve of S700MC

Table 6.27 Welding parameters

Grade	Materials	Current (A)	Voltage (V)	Speed (cm/min)	Energy input (kJ/cm)
S420MC	WER50	230	25	30	9
S700MC	WER70	140	28	50	4

Weldability

According to the national standards GB/T 2651-2008 and GB/T 2653-2008, the LincolnPower Wave 455 welder was used to test the weldability of S420MC with the thickness of 8 mm and S700MC with the thickness of 2 mm. The welding parameters are shown in Table 6.27.

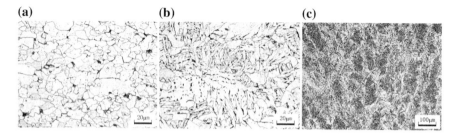

Fig. 6.30 Microstructures of different locations of the weld joint of S420MC: **a** base metal, **b** transition zone, **c** weld

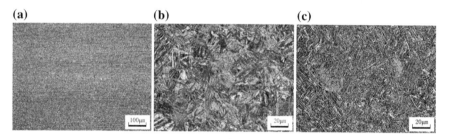

Fig. 6.31 Microstructures of different locations of the weld joint of S700MC: **a** base metal, **b** transition zone, **c** weld

Table 6.28 Results of the tensile test and bending test

Grade	Tensile test		Bending test		
	Tensile strength (MPa)	Fracture location	Bending diameter	Bending angle	Evaluation
S420MC	576	Base metal	$D=2a$	120°	Qualified
	591	Base metal			Qualified
S700MC	758	Base metal	$D=2a$	120°	Qualified
	781	Base metal			Qualified

The microstructures of different locations of S420MC sample are shown in Fig. 6.30. The microstructure of the base metal is composed of ferrite and pearlite. The microstructure of the transition zone is composed of ferrite, bainite and pearlite. The microstructure of the weld consists of acicular ferrite and pre-eutectoid ferrite. Figure 6.31 shows the microstructures of different locations of S700MC sample. The microstructure of the base metal is ferrite. The microstructure of the transition zone is lath martensite, and that of the weld is needle-like martensite.

The tensile test and bending test (120 °C) were carried out on the welded samples. The results are shown in Table 6.28. All the results of the bending test (120 °C) are qualified, indicating that S420MC and S700MC have good weldability to meet the service requirements.

Fig. 6.32 Truck parts made of S420MC

6.2.1.5 Applications

S420MC has good machinability and can be used to make all kinds of structural parts of trucks, as shown in Fig. 6.32. The manufacturing cost is significantly reduced by microalloying C-Mn system with single titanium. As shown in Fig. 6.33, S700MC can be used to make carriage cross beams, front and rear bumper beams and other parts. It can also be rolled into square rectangular tubes and used in body and chassis supports, luggage rack and roof of buses, as shown in Fig. 6.34. Especially, significant lightweighting can be achieved when it is applied to the manufacturing of the new energy buses. The thickness of the steel required is generally in the range of 1.2–3.0 mm. The strips with the thickness of 2.0 mm or more are mainly produced by hot rolling, and those with the thickness of 1.2–2.0 mm are mainly produced by cold rolling. It is worth mentioning that Wuhan Iron and Steel has successfully developed the ultra-thin specifications of ultra-high strength automotive structure steel on TSCR process titanium microalloying technology. The maximum strength reaches 700 MPa level and the minimum thickness reaches 1.2 mm. It partly replaces the cold rolled products with the thickness in the range of 1.2–2.0 mm, realizing the goal to replace the cold rolled with the hot rolled.

6.2.2 Automotive Beam Steel

High strength is an important goal in the development of automotive steel. Commercial vehicle frames are made of beam steel. Frame components include the longitudinal beams, lining beams, cross beams, reinforcement parts, etc. The weight reduction can reach 30% if they are made of high strength steel.

Initially, beam steel was produced by single-plate rolling. Its composition was based on the C-Mn steel system. In 1978, the first continuous hot rolling production line in China was put into operation in Wuhan Iron and Steel. The study on titanium-bearing steel was also started. In the early stage, the study on the characteristics

Fig. 6.33 Truck bumper beam made of S700MC

Fig. 6.34 Bus frame parts made of S700MC

of titanium in steelmaking and rolling processes was very insufficient. As a result, the titanium content in the 500 MPa grade beam steel was up to 0.14%, and the mechanical properties of the steel plates were very unstable.

During the Ninth Five-Year Plan (1996–2000) period in China, the Ministry of Metallurgical Industry in China organized the research on automotive steel, which was completed by Wuhan Iron and Steel and Baosteel. During this period, the development of QStE460TM was completed, and the experience of the development of titanium-bearing beam steel was accumulated.

After the year 2000, the requirement for the strength of beam steel is increasingly getting higher due to the continuous update of trucks, passenger cars and the requirement for lightweighting. The development of titanium-bearing beam steel has stepped into a period of rapid development. QStE550TM and QStE650TM were used by Dongfeng Commercial Vehicle in 2002 and 2009, respectively. QStE700TM was adopted by Liuzhou Motors in 2012.

In 2016, the national standard GB1589, which limits the frame size, shaft load and weight limit of automobiles, trailers and trains, was released to replace the 2004 version. The new national standard establishes definite provisions and restrictions on the weight and size of various types of commercial vehicles in market. The commercial vehicles in China have made extensive use of QStE420TM, QStE460TM and QStE500TM. With the implementation of the new national standard, the low strength beam steel is gradually replaced by high strength beam steel. The annual consumption of QStE650TM and QStE700TM has reached up to 100,000 tons.

6.2.2.1 Standards and Performance Requirements

The national standard GB/T3273, which is about hot rolled steel plate and steel strip for automotive beams, was released in 2015 to replace the 2005 edition. The grades prescribed in the new national standard are named after the tensile strength. Five grades are added, which are 600L, 650L, 700L, 750L and 800L. The specifications are shown in Table 6.29. Many automobile manufacturers in China adopt the grades prescribed in the German standard SEW092, in which the grades are named after the yield strength. The mechanical properties are shown in Table 6.30.

Table 6.29 Specifications of beam steel prescribed in the national standard GB/T3273

No.	Grade	Lower yield strength R_{eL} (MPa) $\geq$	Tensile strength R_m (MPa)	Elongation(%), $\geq$		Bending test (180°), $b=35$ mm	
				Thickness <3 mm	Thickness $\geq$3 mm	Thickness $\leq$12.0 mm	Thickness >12.0 mm
				A_{80mm}	A		
1	370L	245	370–480	23	28	$D=0.5a$	$D=a$
2	420L	305	420–540	21	26	$D=0.5a$	$D=a$
3	440L	330	440–570	21	26	$D=0.5a$	$D=a$
4	510L	355	510–650	20	24	$D=a$	$D=2a$
5	550L	400	550–700	19	23	$D=a$	$D=2a$
6	600L	500	600–760	15	18	$D=1.5a$	$D=2a$
7	650L	550	650–820	13	16	$D=1.5a$	$D=2a$
8	700L	600	700–880	12	14	$D=2a$	$D=2.5a$
9	750L	650	750–950	11	13	$D=2a$	$D=2.5a$
10	800L	700	800–1000	10	12	$D=2a$	$D=2.5a$

Table 6.30 Specifications of QStE series of beam steel

Grade	Thickness (mm)	Yield strength R_{eL} (MPa)	Tensile strength R_m (MPa)	Elongation (%) $A_{80\,mm}$, $b=20$ mm Nominal thickness (mm) <3.0	A ≥3.0	Bending test (180°) d: diameter a: thickness
QStE260TM	2.0–12.0	≥260	340–460	≥23	≥29	$d=0a$
QStE300TM	2.0–12.0	≥300	380–500	≥21	≥27	$d=0.5a$
QStE340TM	2.0–12.0	≥340	420–540	≥19	≥25	
QStE380TM	2.0–12.0	≥380	450–590	≥18	≥23	
QStE420TM	2.0–12.0	≥420	480–620	≥16	≥21	
QStE460TM	2.0–12.0	≥460	520–670	≥14	≥19	$d=1a$
QStE500TM	2.5–12.0	≥500	550–700	≥12	≥17	
QStE550TM	3.0–12.0	≥550	600–760	–	≥15	$d=1.5a$
QStE600TM	3.0–12.0	≥600	650–820	–	≥13	
QStE650TM	3.0–12.0	≥650	700–880	–	≥12	$d=2a$
QStE700TM	3.0–12.0	≥700	750–950	–	≥12	

6.2.2.2 Composition and Process Design

The high strength beam steel is developed by adding one or several microalloying elements such as titanium, niobium, vanadium and molybdenum into the C–Mn steel [13–15]. Because titanium has the cost advantage, it is preferred among the microalloying elements. Currently, the high strength beam steel is mainly developed based on the C–Mn–Ti–Nb system. The main strengthening mechanisms are grain refinement strengthening and precipitation strengthening. The beam steel with the yield strength of 700 MPa is developed by adding molybdenum to strengthen the microstructure. The compositions of 600–700 MPa grade beam steel developed by Wuhan Iron and Steel are listed in Table 6.31.

The composition design of beam steel should take into account the manufacturing process and service conditions. The manufacturing process of automotive beam steel

Table 6.31 Chemical compositions of titanium microalloyed beam steel developed by Wuhan Iron and Steel (wt%)

Grade	C	Si	Mn	P	S	Al	Ti	Nb	Mo	N
QStE600TM	0.070	0.10	AA	≤0.015	≤0.005	0.20–0.70	0.95	AA	–	≤0.006
QStE650TM	0.065	0.10	AA	≤0.015	≤0.005	0.20–0.70	1.15	AA	–	≤0.006
QStE700TM	0.060	0.10	AA	≤0.015	≤0.005	0.20–0.70	1.15	AA	AA	≤0.006

AA: appropriate amount

includes bending, stamping, roll-forming, welding, painting and other processes. The beams need to bear harsh working conditions, such as heavy load and a variety of impact, reverse and other complex forces.

The carbon content is controlled to be below 0.1%. It should decrease with the increase in the strength grade, so as to guarantee the toughness meet the requirement for formability. In addition, the banded structures are reduced, so do the sources of cracking during the forming process and service period. The content of silicon is controlled properly to reduce the formation of Fe_2SiO_4 with a strong adhesion to the matrix in the furnace, improving the descaling effect, the surface quality and the paintability. The inclusions formed by sulfur and phosphorus are very detrimental to the fatigue life and impact resistance. Therefore, these two elements must be strictly controlled to improve the purity of molten steel. In addition, the nitrogen element should be controlled to increase the content of effective titanium and stabilize the strengthening effect.

6.2.2.3 Microstructure and Properties

Influence of Heating System on the Precipitates and Properties

The high strength beam steel is developed mainly by titanium microalloying. The content of titanium in QStE600TM-QStE700TM is in the range of 0.095–0.115%. This steel is particularly sensitive to the heating system. The fluctuations of heating temperature and heating time cause the fluctuations of the tensile properties. The tensile properties of the two QStE650TM samples (1# and 2#) with different heating systems are shown in Table 6.32. The heating system of 1# is stronger than that of 2#. The yield strength and tensile strength of 1# are higher than those of 2# by 160 MPa and 150 MPa, respectively, and the elongation is equal to that of 2#.

The samples for TEM analysis were prepared by extraction replica technique. The morphology, distribution, size, quantity, compositions of the precipitates were investigated by JEM-2100F TEM equipped with EDS. As shown in Figs. 6.35, 6.36, 6.37, 6.38, when the heating temperature is high and the heating time is long, the large particles precipitated during the smelting process are dissolved, and then particles precipitate with the decrease in the temperature during the rolling process. Therefore, the particles are fine and their sizes are in the range of 10–80 nm. Moreover, they are dispersively distributed. When the heating temperature is low and the heating

Table 6.32 Mechanical properties of two QStE650TM samples with different heating systems

No.	Thickness (mm)	Heating temperature (°C)	Heating time (min)	R_{eL} (MPa)	R_m (MPa)	A (%)
1#	8	1300	220	725	795	19
2#	8	1200	180	565	645	19
Standard				≥650	700–880	≥12

Fig. 6.35 Morphology of precipitates in sample 1#

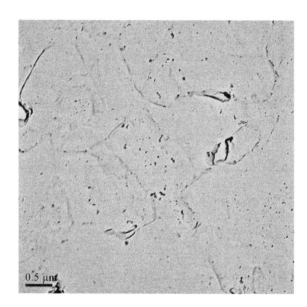

Fig. 6.36 Morphology of precipitates in sample 2#

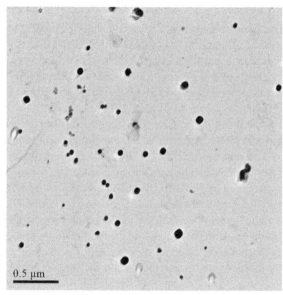

time is short, the large particles precipitated during the smelting process are not fully dissolved, and the precipitates are larger with the size ranging from 30 nm to 180 nm. These coarse precipitates contribute little to the performance of steel.

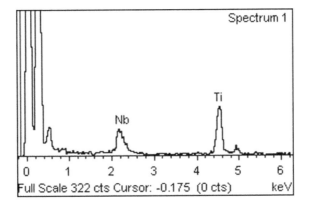

Fig. 6.37 EDS analysis of precipitates in sample 1#

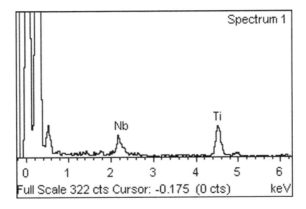

Fig. 6.38 EDS analysis of precipitates in sample 2#

Influence of Heating System on the Microstructure and Properties

The microstructures and sizes resulted from the two different heating systems are different, as shown in Fig. 6.39. The microstructure of 1# is bainite. In contrast, the microstructure of 2# consists of ferrite and pearlite, and the grade of the grain size is 12. The difference of the heating systems leads to the difference of the strength.

Effect of Finishing Rolling Temperature on the Microstructure and Properties

Table 6.33 shows the mechanical properties of QStE600TM samples produced by the same heating process, but two different finishing rolling temperatures. The yield strength and tensile strength are increased by 39 and 38 MPa respectively with the increase in the finish rolling temperature.

As shown in Fig. 6.40, the microstructure of 3# consists of bainite and ferrite, while that of 4# is composed of bainite and a small amount of ferrite. By increasing

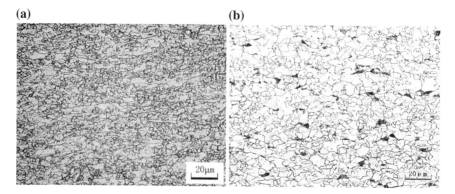

Fig. 6.39 Microstructures of steel produced by two different heating systems **a** sample 1#, **b** sample 2#

Table 6.33 Mechanical properties of QStE600TM

No.	Thickness (mm)	Finishing rolling temperature (°C)	R_{eL} (MPa)	R_m (MPa)	A (%)
3#	10	825	589	662	18
4#	10	869	628	700	19
Standard			≥600	650–820	≥13

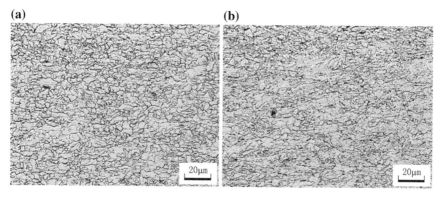

Fig. 6.40 Microstructures of QstE600TM produced by two different finishing rolling temperatures: **a** sample 3#, **b** sample 4#

the finishing rolling temperature, the amount of pre-eutectoid ferrite is reduced and the amount of bainite is increased, which enhances the strength.

Mechanical Properties

The tensile tests were carried out on different locations and directions of the samples by WE-60 universal tensile testing machine according to the national standard

GB228-2002. The results are shown in Table 6.34. The fluctuation of strength is within 50 MPa and the fluctuation of elongation is within 5%.

6.2.2.4 Service Performance

Cold Bending Property

The stamping formability is one of the main performance indexes of beam steel. The beam produced by stamping process is generally U-shaped and the bending angle is 90°. For the cold bending test in the lab, the beam bending angle is 180° and the bending diameter is 0. Thus, the test conditions in the lab are stricter than the actual stamping process. As shown in Table 6.35 and Fig. 6.41, the results of the bending tests on different locations and different directions of the samples indicate that the bending property is excellent and can meet the requirement for automotive beams produced by stamping.

Low Temperature Impact Property

Commercial vehicles have to bear a variety of impact forces in the service period. Therefore, the risk of impact fracture should be evaluated for service in the cold area. The low temperature toughness impact test is used to evaluate the service capacity of high titanium beam steel used in the cold area.

The QStE650TM plate was processed into V-notch impact test sample with the dimension of 7.5 × 10 × 55 mm. The impact test was carried out by using the pendulum testing machine according to the national standard GB/T 229. The longitudinal

Table 6.34 Mechanical properties of QStE650TM

Location	Direction	R_{eL} (MPa)	R_m (MPa)	A %	R_{eL}/R_m	Fluctuation (MPa)	
						Same location, different direction	Same direction, different location
Head	Longitudinal	645.0	722.5	20.8	0.89	$\Delta R_{eL} = 35.8$	$\Delta R_{eL} = 30$
	Transverse	680.8	745.0	16.8	0.91	$\Delta R_m = 32.5$	$\Delta R_{eL} = 30.8$
	45°	647.5	712.5	20.8	0.91	$\Delta A = 4\%$	$\Delta R_{eL} = 30.8$
Middle	Longitudinal	675.0	755.0	20.8	0.89	$\Delta R_{eL} = 40.0$	$\Delta R_m = 32.5$
	Transverse	710.0	780.0	17.3	0.91	$\Delta R_m = 45.0$	$\Delta R_m = 35.0$
	45°	670.0	735.0	21.3	0.91	$\Delta A = 4\%$	$\Delta R_m = 22.5$
Tail	Longitudinal	655.0	735.0	20.3	0.89	$\Delta R_{eL} = 47.5$	$\Delta A = 0.5\%$
	Transverse	695.0	760.0	16.8	0.91	$\Delta R_m = 45.0$	$\Delta A = 0.5\%$
	45°	647.5	715.0	21.3	0.91	$\Delta A = 4.5\%$	$\Delta A = 0.5\%$

Table 6.35 Cold bending properties of QStE650TM

Location	Direction	Thickness (mm)	Diameter	Evaluation
Head	0°	8	$d=0.5a, d=0$	Qualified
	90°	8	$d=0.5a, d=0$	Qualified
	45°	8	$d=0.5a, d=0$	Qualified
Middle	0°	8	$d=0.5a, d=0$	Qualified
	90°	8	$d=0.5a, d=0$	Qualified
	45°	8	$d=0.5a, d=0$	Qualified
Tail	0°	8	$d=0.5a, d=0$	Qualified
	90°	8	$d=0.5a, d=0$	Qualified
	45°	8	$d=0.5a, d=0$	Qualified

0° is for the rolling direction. All the bending angles are 180 °C

Fig. 6.41 QStE650TM samples for cold bending test

and transverse impact energies are shown in Figs. 6.42 and 6.43. The difference of the impact energies at different locations is not large. However, the impact energies of the longitudinal and transverse directions are different. The brittleness transition temperature is determined to be below −60 °C. The results indicate that the steel is qualified for the service at low temperatures.

Fatigue Property

Fatigue life is an important indicator of commercial vehicles, and it is relevant to the frame structure and the fatigue property of steel plate. According to the national standard GB/T 3075, the fatigue property of beam steel QStE650TM was measured by using tension-tension method. The results are shown in Fig. 6.44. The thickness

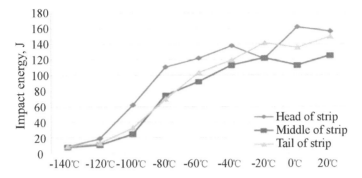

Fig. 6.42 Longitudinal impact energy of different locations of QStE650TM sample

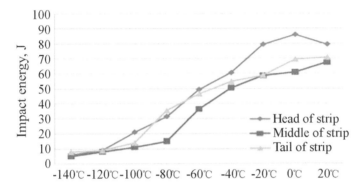

Fig. 6.43 Transverse impact energy of different locations of QStE650TM sample

of the sample is 8 mm, the stress ratio is 0.15, and the frequency is 160 Hz. The conditioned fatigue strength is 219.8 MPa. The fitting equation of the fatigue curve is $\lg N = 13.51 - 0.03 \times S$, and the fitting correlation coefficient r is 0.951.

6.2.2.5 Applications

In Europe and USA, the frames of Benz, Volvo, Mann, Scania and other heavy trucks commonly use single layer beam structure made of QStE650TM and QStE690TM plates with the thickness of 8–9 mm. With the rapid development of China's economy and the enforcement of relevant laws and regulations, the manufacturers such as FAW Group Corporation, Dongfeng Motor Co. Ltd., China National Heavy Duty Truck Group Co. Ltd. and many others have began to use high titanium and high strength steel QStE650TM, QStE700TM for manufacturing automotive beams.

The lightweighting design of the beam should take into account the improvement of the strength of steel, connections of beams and other structural parts, and sectional size of beams to meet the requirements for the stiffness of frames. The lightweighting

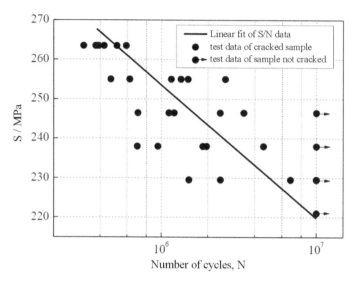

Fig. 6.44 Fatigue curve of QStE650TM

Table 6.36 Lightweighting design and weight reduction

Model	Without lightweighting			With lightweighting			Weight reduction (%)
	Single/Twin beam	Grade	Gauge	Single/Twin beam	Grade	Gauge	
Truck	Twin beams	QStE500TM	8+8 mm	Twin beams	External beam QStE650TM Inner beam QStE500TM	External beam 8 mm Inner beam 5 mm	18.75
Semi-trailer towing vehicle	Twin beams	QStE460TM	8+8 mm	Single beam	QStE650TM	8 mm	33
Semi-trailer towing vehicle	Twin beams	QStE500TM	8+7 mm	Single beam	QStE700TM	8 mm	28

scheme and the weight reduction by using high strength beam steel are shown in Table 6.36.

6.3 Engineering Machinery Steel

Engineering machinery refers to the equipment used for mining and all kinds of construction work, such as electric shovel, electric wheel dumpers, excavators, bulldozers, trucks, all kinds of cranes, scraper conveyors and hydraulic support used in coal mines. High titanium hot rolled welded high strength steel is generally used in the main structure of engineering machinery to withstand the complex and variable cyclic load. Therefore, it is required that the steel has high yield strength, high fatigue limit, good impact toughness, good cold formability and excellent weldability.

With the rapid development of global engineering machinery industry since the beginning of the 21st century, the market competition is becoming more intense. The requirements for the strength and toughness of steel from downstream customers are getting increasingly higher, and the cost control is becoming critical. A large amount of 700–800 MPa grade hot rolled welded high strength steel are demanded by the customers in China, such as the Sichuan Changjiang Engineering Crane Co., Ltd., Zoomlion Heavy Industry Science & Technology Co., Ltd., Xuzhou Construction Machinery Group Co., Ltd., Sany Heavy Industry Co. and so on. It is required that not only the steel plates have good balance between strength and toughness, but also they have good weldability to ensure the quality of the weld during service.

According to the technical standards, the performance indexes required by the construction machinery industry in China are at the same level as those prescribed in the European standards. The safety design requires that the tensile strength is not less than 685 MPa and the Charpy impact energy is not less than 27 J in the service environment. Moreover, the welding crack sensitivity index P_{cm} does not exceed 0.25%.

According to the specifications of steel plates, the thickness of the steel used for the jib and other parts of cranes is below 12 mm and the width is within 2000 mm. The steel plates with the specification (thickness: 4–10 mm, width: 1100–1600 mm, length: 6000–10,000 mm) account for about 90% of market share.

6.3.1 Standards and Performance Requirements

European standard EN10149 prescribes the requirements for the chemical composition and mechanical properties of the hot rolled welded high strength steel with the tensile strength of 700–800 MPa, as shown in Tables 6.37 and 6.38.

Table 6.39 shows the grades of the hot rolled welded high strength structure steel with the tensile strength of 700–800 MPa developed by Wuhan Iron and Steel and other manufacturers. The steel of Wuhan Iron and Steel is developed based on the European standards EN10025, EN10149 and the standards for manufacturing excavators and cranes from the manufacturers in China, such as Zoomlion Heavy Industry Science & Technology Co., Ltd., Sany Heavy Industry Co., Xuzhou Construction

Machinery Group Co., Ltd., LiuGong Group, Sichuan Changjiang Engineering Crane Co., Ltd. and so on. The mechanical properties are shown in Table 6.40.

6.3.2 Composition and Process Design

Table 6.41 shows the chemical composition of the hot rolled welded high strength steel with the strength of 700–800 MPa developed by Wuhan Iron and Steel. It is designed by adding titanium, niobium, chromium, molybdenum and other alloying elements into the C–Mn composition system. Considering the process cost and the influence of microstructure on the property, the target microstructure was designed to be ferrite and bainite for the 700–800 MPa grade hot rolled high strength structure steel. Based on the chemical composition, the continuous cooling transformation (CCT) curve was determined, as shown in Fig. 6.45. In the general cooling rate range of 5–50 °C/s, the bainite transition temperature is 550–600 °C. Therefore, the coiling temperature should be in this range so as to obtain the expected microstructure consisting of ferrite and bainite.

6.3.3 Microstructure and Properties

Figure 6.46 shows that the microstructure mainly consists of bainite and small ferrite. Table 6.42 shows that the yield strength is 741 MPa, the tensile strength is 817 MPa, and the elongation reaches 18%.

Table 6.37 Chemical compositions of the hot rolled welded high strength steel

Grade		C% Max	Mn% Max	Si% Max	P% Max	S% Max	Alt% Min	Nb% Max	V% Max	Ti% Max	Mo% Max	B% Max
Abbreviation	Material number											
S600MC	1.8969	0.12	1.90	0.50	0.025	0.015	0.015	0.09	0.20	0.22	0.50	0.005
S700MC	1.8974	0.12	2.10	0.60	0.025	0.015	0.015	0.09	0.20	0.22	0.50	0.005

Table 6.38 Mechanical properties of the hot rolled welded high strength steel

Grade		R_{eL} (MPa)	R_m (MPa)	A(%) Normal Thickness (mm)		Bending (180°)
Abbreviation	Material number			<3 $L_0 = 80$ mm	≥3 $L_0 = 5.65\sqrt{S_0}$	Diameter
S600MC	1.8969	≥600	650–820	11	13	1.5t
S700MC	1.8974	≥700	750–950	10	12	2t

Table 6.39 Grades of the hot rolled welded high strength structure steel developed by different manufacturers

Wuhan Iron and Steel	Standard (EU) EN10149	National standard (CN) GB/T 1591	SSAB	Baosteel
HG70	S600MC	Q620	DOMEX600	BS600
HG785	S700MC	Q690	DOMEX700	BS700

Table 6.40 Mechanical properties of the hot rolled welded high strength structure steel developed by Wuhan Iron and Steel

Grade	Thickness (mm)	Quality grade	R_{eL} (MPa)	R_m (MPa)	A_{50} (%)	Cold bending (180°)	Impact energy (J)	
							-20 °C	-40 °C
HG70	4.0–12.0	D	≥590	≥685	≥17	$d=2a$ qualified	≥40	/
		E					/	≥27
HG785	4.0–12.0	D	≥685	785–940	≥15	$d=3a$ qualified	≥40	/
		E					/	≥27

Table 6.41 Chemical composition (wt%)

Grade	C	Si	Mn	P	S	Mo	Cr	Nb	Ti	B
HG70	0.04–0.10	0.03–0.40	0.80–2.00	≤0.020	≤0.010	≤0.30	≤0.35	≤0.06	0.08–0.12	≤0.004
HG785	0.04–0.12	0.03–0.40	0.80–2.00	≤0.020	≤0.010	≤0.30	≤0.35	≤0.06	0.08–0.15	≤0.004

Table 6.42 Mechanical properties of HG785

Specification	R_{el} (MPa)	R_m (MPa)	A50 (%)	-20 °C KV_2(J)
Standard requirement	≥685	≥785	≥15	≥36
8 mm	741	817	18	62

6.3.4 Service Performance

Weldability

The butt welding of HG785 steel plate was carried out with argon gas protection under room temperature of 25 °C and humidity of 70%. The test conditions are shown as follows.

(1) The dimension of the sample was 680 mm × 150 mm × 8 mm, as shown in Fig. 6.47, with single V-shape groove and included angle of 60°. The root face and root gap were 1 and 2 mm, respectively.
(2) The welding power is LINCOLN Powerwave 455 Lincoln welder with direct current (DC) reversed polarity. The protection gas was a mixture of 80% Ar and 20% CO_2 with the flow rate of 18 L/min.
(3) WH70-G gas shielded welding wire with 1.2 mm diameter, manufactured by Wuhan Tiemiao Welding Materials Co., was used.

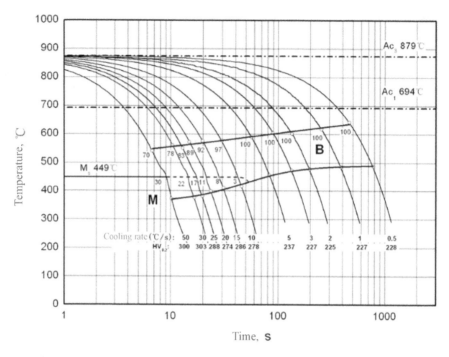

Fig. 6.45 CCT curve of steel sample

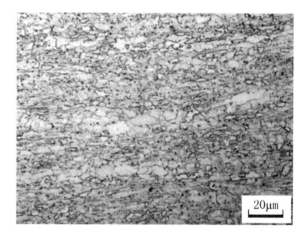

Fig. 6.46 Metallographic microstructure of steel HG785

(4) The welding parameters and the welding sequence are shown in Table 6.43 and Fig. 6.47, respectively. No preheating was applied before welding. The test was carried out with multi-layer and multi-bead. After the front side was welded, the reverse side was sanded to clear the root. The interlayer temperature was below

6 Design, Development and Application of Titanium …

Table 6.43 Welding parameters

Weld bead	Welding wire	Specification Φ (mm)	Gas	Flow rate (l/min)	Pre-heating temperature (°C)	Interlayer temperature (°C)	Current (A)	Voltage (V)	Welding speed (cm/min)	Energy input (kJ/cm)
1	WH70-G	1.2	80%Ar+ 20% CO_2	18	Room temperature	≤150	280–300	26	31	14.1–15.1
2							280–300	26	24	18.2–19.5
3							280–300	26	27	16.2–17.3

Table 6.44 Results of the tensile test and cold bending test on the welded joints

Specification (mm)	Tensile test		Side bending test (diameter of indenter $d=3a$)
	R_m (MPa)	Fracture location	
8	797	Base metal	Positive bending (2), 180°: Crack, 120°: No crack
	791	Base metal	Negative bending (2), 180°: Crack, 120°: No crack

Table 6.45 Results of the impact tests on the welded joints

Temperature (°C)	Impact energy KV_2 (J)		
	Weld	Fusion line	Heat affected zone
0	27 19 38 / 28	37 40 / 39	62 36 30 / 43
−20	16 23 20 / 20	35 16 31 / 27	31 30 65 / 42
−40	16 19 38 / 24	29 37 39 / 35	60 28 65 / 51

200 °C. No heat treatment was applied after welding. The test temperature was 25 °C and the humidity was 80%.

The results of the tensile test and cold bending test on the welded joints are shown in Table 6.44 and Fig. 6.48.

The results of the impact tests on the welded joints are shown in Table 6.45. The non-standard samples with the size of 5 mm were used in the tests.

The metallographic microstructure of the welded joint is shown in Fig. 6.49. There are no obvious pores, inclusions and other defects within the welded joint.

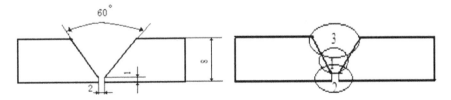

Fig. 6.47 Schematic diagrams of welding assembly and weld bead of HG785

Fig. 6.48 Samples for bending test

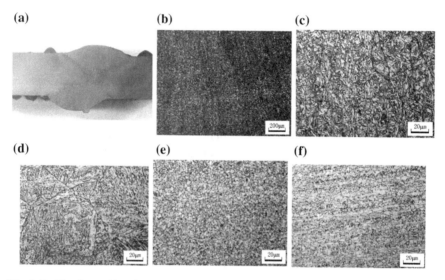

Fig. 6.49 Metallographic microstructures of different locations of the weld joint: **a** weld joint, **b** weld (100X), **c** weld (500X), **d** coarse grain zone, **e** fine grain zone, **f** critical heat affected zone

The microstructures are acicular ferrite and pre-eutectoid ferrite in the weld, bainite and ferrite in the critical heat affected zone, lath bainite in the coarse grain area of heat affected zone, and ferrite and pearlite in the fine grain area of heat affected zone.

In summary, the experimental results indicate the following conclusions:

(1) The tensile strength of the welded joint is greater than 791 MPa. The fracture location is at the base metal. The tensile strength of the welded joint meets the technical requirement for HG785 steel. The cold bending sample with $D = 3a$ cracked when the bending angle was 180°, while it did not crack when the bending angle was 120°, indicating that the plasticity of the joint is good.

(2) The results of the impact tests on the non-standard sample with the size of 5 mm show that the average impact energies of the weld, the fusion line and the heat

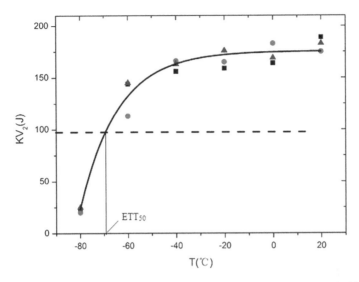

Fig. 6.50 Brittle-ductile transition temperature of HG785 sample with the thickness of 8 mm

affected zone at $-20\ °C$ are 20, 27 and 42 J respectively, indicating that the impact performance of the three zones meets the performance requirement for the steel plate at $-20\ °C$.

(3) The microstructures are acicular ferrite and pre-eutectoid ferrite in the weld, bainite and ferrite in the critical heat affected zone, lath bainite in the coarse grain area of heat affected zone, and ferrite and pearlite in the fine grain area of heat affected zone.

Ductile-Brittle Transition Temperature

The transition from brittleness to ductility of metal materials is completed in a temperature range rather than a temperature point. Therefore, the measured transition temperatures T_R are different according to different criteria. The current test adopts the following criteria to measure the T_R value. The standard Charpy V-notch impact test samples were used. Three samples were tested for each temperature. The impact energy W_t and the temperature were set as the ordinate and abscissa respectively, as shown in Fig. 6.50. The temperature positioned at the middle between the upper and lower platforms of impact absorbed energy-temperature is defined by ETT50.

The results show that HG785 with high titanium content has obvious ductile-brittle transition phenomenon. The range of ductile-brittle transition temperature is from -75 to $-60\ °C$, indicating that HG785 steel with high titanium content still has good ductility at low temperatures.

Axial High Cycle Fatigue

As shown in Fig. 6.51, the lower limit of the fatigue strength of HG785 steel with high titanium content is 146.7 MPa when the parameters are $R = 0.15$, 1.0×10^7

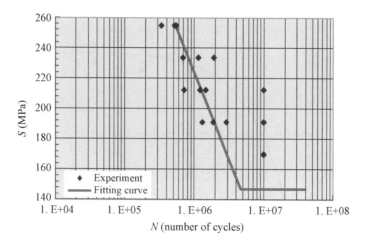

Fig. 6.51 S-N curve of HG785 sample with the thickness of 8 mm

cycles, confidence level of 95% and failure probability of 10%. The results indicate that the steel plate has high anti-fatigue strength. The fitting formula for the S-N curve is $\lg N = 7.945 - 0.00868 \times S$. S is the stress amplitude and r is -0.78.

6.3.5 Applications

At present, these series of steel products have been widely used in cranes and heavy-duty vehicles. As shown in Fig. 6.52, they are mainly used for important parts, such as truck crane, pump boom, chassis girder, fixed leg, side panel, lifting bracket, front rail and other key components. These products not only meet the requirements for reduction of the weight of engineering machinery, but also meet the needs of upgrading the products.

6.4 Wear Resistant Steel

Wear resistant steel is widely applied to manufacture machines bearing harsh working conditions, which are used in mining, construction, agriculture, cement production, port, power plant and metallurgical industry. These machines include scraper conveyor, stage loader conveyor, excavator, dump truck and a variety of mining machinery. Low alloyed martensite wear resistant steel is the most widely used wear resistant steel in coal industry in China. It has obvious advantages, such as simple production process, low cost and excellent comprehensive performance. It represents the development direction of wear resistant steel for coal mining. However, the conventional low alloyed martensite steel improves wear resistance simply by improving the hardness, resulting in deterioration of formability and significant tendency of cracking of the wear resistant parts. Through the strengthening of two-scale TiC particles, the wear resistance of the martensite wear resistant steel is promoted without increasing the hardness and decreasing the formability. Therefore, the development of high wear resistant steel for coal mining is greatly promoted.

6.4.1 Standards and Performance Requirements

The national standard GB/T 24186-2009 grades the low alloyed martensite steel based on the surface Brinell hardness. The grades are from NM360 to NM600, on behalf of different hardness levels. Word leading manufacturers of wear resistant steel, such as SSAB, JFE Steel and so on, also grade the steel according to the surface hardness, for example, HARDOX series of steel produced by SSAB. HARDOX450 represents the steel with the surface hardness in the range of HB425–HB475. The chemical composition and mechanical properties of the typical products are shown in Tables 6.46 and 6.47.

The higher the equivalent carbon content, the higher the hardness, but the lower the weldability and machinability. The structure of the equipment for coal mining is usually cast-weld structure, so the requirements for weldability and machinability are high. As a result, most coal companies only use the steel plate with the hardness not higher than HB450. The use of martensite wear resistant steel reinforced by two-

Fig. 6.52 Application of welded high strength steel for engineering machinery: **a** crane, **b** truck

Table 6.46 Chemical composition of the typical martensite wear resistant steel (wt%)

Grade	C	Si	Mn	P	S	Cr	B
HARDOX450	0.18	0.33	0.6	0.007	0.001	0.6	0.0014
JFE-EH400	0.15	0.33	0.6	0.006	0.001	0.5	0.0012

Table 6.47 Mechanical properties of the typical martensite wear resistant steel

Grade	Producer	Yield strength (MPa)	Tensile strength (MPa)	Elongation (A5/%)	Cold bending property (90°, $D=3a$)	Hardness (HB)
JFE-EH360	JFE	≥850	≥1050	≥12	Good	≥361
JFE-EH400	JFE	≥950	≥1200	≥10	Good	≥401
HARDOX400	SSAB	≥900	≥1100	≥10	Good	370–430
HARDOX450	SSAB	≥1000	≥1250	≥10	Good	425–475

scale TiC particles can improve the wear resistance of steel without increasing the hardness and decreasing the machinability.

6.4.2 Composition and Process Design

The microstructure is designed to be composed of ultra-fine martensite and two-scale TiC particles. The chemical composition, which is shown in Table 6.48, features high titanium content.

The hot rolling process is shown in Table 6.49. The hot rolling process should ensure that the micron-scale TiC particles in the core are broken as much as possible and evenly distributed. The quenching and tempering process is adjusted according to the thickness of the steel plate, as shown in Table 6.50. The cooling process of the quenching machine must ensure that the steel plate is hardened and the microstructure of the core is martensite. Meanwhile, the heating temperature for quenching cannot be too high, so as to maintain the ultra-fine microstructure of matrix during heating and obtain ultra-fine martensite in the quenching process. The main purpose of tempering process is elimination of quenching stress. The tempering process is longer than conventional tempering because of the quenching stress resulted from the difference of the thermal expansion coefficients of the matrix and the micron-scale TiC particles.

Table 6.48 Chemical composition of wear resistant steel ZM450 (wt%)

Process	C	Si	Mn	P	S	Cr	Ni	Mo	Cu	Ti	B
Steel in tundish	0.33	0.33	0.6	0.007	0.001	1.03	0.58	AA	AA	AA	0.0014
Cast slab	0.3	0.33	0.59	0.006	0.001	1.02	0.58	AA	AA	AA	0.0012

AA: appropriate amount

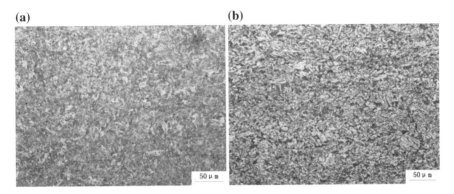

Fig. 6.53 Microstructure of steel ZM450: **a** hot rolled, **b** heat treated

6.4.3 Microstructure and Properties

Microstructure

Figure 6.53 show the microstructures of ZM450 after hot rolling and heat treatment respectively. The microstructure of ZM450 after hot rolling is bainite and the hardness is about HB300. The microstructure after heat treatment is tempered martensite and the hardness is about HB450. Figure 6.54 shows the microstructure after polishing and etching. The light gray area is the matrix and the dark gray particle is the micron-scale TiC. The morphology of TiC precipitate was observed by TEM, as shown in Fig. 6.55. The micron-scale TiC particles are blocky or short rod-shaped, while the nano-scale particles are spherical and their sizes are in the range of 10–50 nm.

Mechanical Properties

Table 6.49 Hot rolling process for producing ZM450

Heating	Rough rolling	Finishing rolling	Cooling	Final temperature
1200–1220 °C	Intermediate slab thickness $\geq 3H$	Initial rolling temperature ≤ 950 °C, End rolling temperature 860 ± 20 °C	ACC	650 ± 20 °C

H is the thickness of the plates

Table 6.50 Heat treatment process for producing ZM450

Heating before quenching	Time	Tempering temperature	Time
920 ± 10 °C	$H*3$ min	240 ± 5 °C	$H*10$ min

H is the thickness of the steel plates

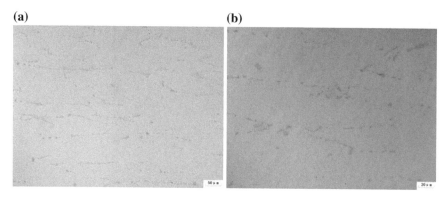

Fig. 6.54 Morphology of micron-scale TiC precipitates: **a** 500X, **b** 200X

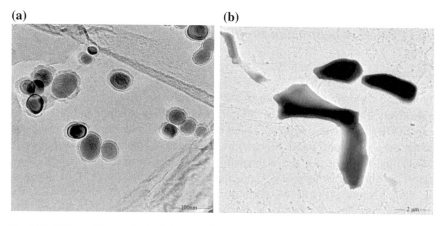

Fig. 6.55 TiC precipitates obtained by extraction replica technique: **a** micron-scale, **b** nano-scale

Table 6.51 shows the mechanical properties of ZM450 of different thicknesses after heat treatment. The strength is similar for the samples with different thicknesses. Compared with the conventional martensite wear resistant steel, the plasticity and toughness of wear resistant steel reinforced by TiC particles are slightly reduced. The plasticity and toughness increase with the decrease in the thickness of the steel plate (increase in the rolling compression ratio). The impact energy of the steel plate with the thickness of 30 mm at room temperature can reach up to 20 J.

6.4.4 Service Performance

The wearing mechanism of wear resistant steel for coal mining is mainly abrasive wear of low stress scratch. The stress exerted by the abrasives on the surface of the

Table 6.51 Mechanical properties of ZM450 of different thicknesses

Thickness (mm)	Tensile strength (MPa)	Yield strength (MPa)	Elongation (%)	Section shrinkage (%)	Impact energy at room temperature (J)
30	1531	1297	10.5	43	20
50	1533	1299	8	41	13
60	1522	1286	7	29	11

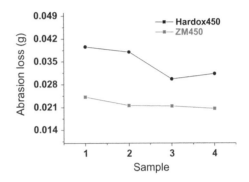

Fig. 6.56 Comparison of weight loss of ZM450 and HARDOX450 subjected to wear resistance test

part is lower than the crushing strength of the abrasives, and the surface of the part is scratched. Generally, the MLS-225 abrasive wear testing machine is used to test and evaluate the wear resistance of steel. The abrasive is quartz sand and the size is in the range of 40–80 mesh. The composition of the mixture is 1,500 g water and 1,500 g sand. The speed of the rubber wheel is 240 r/min. The number of the rounds of pre-grinding and finishing grinding are both 1000. Then, the weight loss is measured.

Figure 6.56 shows the comparison of the wear resistance of martensite steel reinforced by TiC particles and the conventional martensite steel. The wear resistance of ZM450 is about 1.5 times that of conventional martensite steel HARDOX450. Under the condition of abrasive wear, the wear resistance of martensite wear resistance steel reinforced by TiC particles is significantly higher than that of conventional martensite wear resistant steel. For the wear resistant steel used for coal mining, the micron-scale chips dominate the wear process in the service environment. The micro hardness of TiC phase is more than HV3000 and TiC particles can prevent the pass of furrow, or make the furrow narrow and shallow. Meanwhile, super-hard TiC particles can break the abrasives, blunt the sharp angles and reduce the plastic deformation during the wear process. The TiC particles can only function when the particle size is equal to the depth of the furrow, or greater than the width of the furrow.

Fig. 6.57 Floors of the chute of scraper conveyor made by ZM450 plates

6.4.5 Applications

ZM450 steel has been successfully used to manufacture the middle and bottom panels of the middle trough shown in Fig. 6.57. The service period of the scraper conveyor has exceeded 3 years, and the breakdown accident resulted from the quality of the middle and bottom plates of the middle trough never happened. The weight loss of the ZM450 steel plate is less than half of that of conventional martensite steel for per million tons of coal.

6.5 Magnetic Yoke Steel

Hydropower is a clean energy. The development of hydropower in China was started in the second half of the 20th century. Currently, about two hundred hydropower stations of different sizes have been built. The unit capacities of Wudongde and Baihetan hydropower stations under construction are 850,000 kW and 1 million kilowatts respectively. They will be the hydropower stations with the largest unit capacity in China. Survey data show that by the end of 2015, the installed capacity of China's hydropower reached 320 million kilowatts, accounting for 59% of the total exploitable capacity. There are great prospects to develop the water resources in the future. The rotor of the hydro-generator is mainly composed of magnetic pole, magnetic yoke, rotor bracket and so on. The magnetic yoke is the most important part of the rotor. Its main function is to produce the moment of inertia and mount the magnetic pole. It is also a part of the magnetic circuit, as shown in Figs. 6.58 and 6.59.

The development of magnetic yoke steel is broadly divided into four stages:

(1) Development of low strength magnetic yoke steel (1997–2000). The steel was developed based on the economic titanium-bearing steel composition system. The strength is in the range of 235–550 MPa.

Fig. 6.58 Schematic diagram of hydro-generator

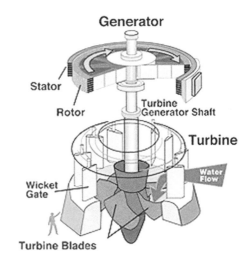

Fig. 6.59 Schematic diagram of rotor yoke

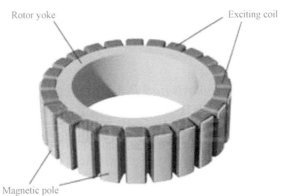

(2) Development of high strength magnetic yoke steel (2005–2016). Niobium was added on the basis of titanium microalloying. The strength was promoted to be in the range of 600–750 MPa.

(3) Development of thin gauge high strength and high toughness magnetic yoke steel (2013–2016). Molybdenum was added on the basis of titanium and niobium microalloying. The problem of balancing the strength and toughness of titanium-bearing high strength magnetic yoke steel with the thickness of no more than 5 mm was solved. The magnetic yoke steel with the longitudinal and transverse yield strength in the range of 650–750 MPa and good toughness was developed.

(4) Development of high precision magnetic yoke steel (2010–2016). The requirement for the shape of the steel plate is getting higher and the roughness is requested to be below 1 mm/m because of the increase in the unit capacity of the generator, the increase in the speed and diameter of the rotor, and the transition of the processing mode from stamping to laser cutting. By the low internal

Table 6.52 Performance indexes of magnetic yoke steel prescribed in the standard EN 10265

Grade	Mechanical properties			Magnetic flux intensity (T)	
	R_{eL} (MPa)	R_m (MPa)	A (%)	B_{50}	B_{150}
250TG178	≥250	≥350	≥26	≥1.60	≥1.80
300TG178	≥300	≥400	≥24	≥1.60	≥1.80
350TG178	≥350	≥450	≥22	≥1.55	≥1.79
400TG178	≥400	≥500	≥19	≥1.55	≥1.79
450TG178	≥450	≥550	≥17	≥1.54	≥1.79
500TG178	≥500	≥600	≥14	≥1.53	≥1.79
550TG178	≥550	≥650	≥14	≥1.52	≥1.78
600TG178	≥600	≥700	≥12	≥1.50	≥1.78
650TG178	≥650	≥750	≥12	≥1.48	≥1.78
700TG178	≥700	≥800	≥12	≥1.56	≥1.78

stress control technology and plate shape control technology, the magnetic yoke steel plate with the roughness below 1 mm/m was developed.

6.5.1 Standards and Performance Requirements

The research and development of magnetic yoke steel was started earlier in foreign countries, and mature standards have been established, such as EN 10265 and JISC 2555. The maximum strength prescribed in these two standards is 700 MPa. The performance indexes are shown in Tables 6.52 and 6.53. Table 6.54 shows the performance indexes of the magnetic yoke steel developed by Wuhan Iron and Steel. The highest longitudinal and transverse yield strength prescribed is 750 MPa. The magnetic yoke steel with high strength, high toughness, high precision and high magnetic flux intensity is being developed in order to meet the requirements for large hydro-generators.

6.5.2 Composition and Process Design

The ideal microstructure of the magnetic yoke steel is composed of uniform fine polygonal ferrite and dispersive fine secondary phase particles. The factors, such as internal stress, large secondary phase particles, lattice defects and grain boundaries, reduce the magnetic induction property by blocking the displacement of the magnetic domain wall. The martensite structure has high internal stress and defects, which seriously reduce the magnetic induction property by affecting the displacement of the magnetic domain wall during the magnetization process. If the microstructure is

6 Design, Development and Application of Titanium …

Table 6.53 Performance indexes of magnetic yoke steel prescribed in the standard JISC 2555

Grade	Mechanical properties			Magnetic flux intensity (T)	
	R_{eL} (MPa)	R_m (MPa)	A (%)	B_{50}	B_{100}
PCYH250	≥250	≥350	≥24	≥1.60	≥1.69
PCYH300	≥300	≥400	≥22	≥1.60	≥1.69
PCYH350	≥350	≥450	≥20	≥1.55	≥1.66
PCYH400	≥400	≥500	≥19	≥1.55	≥1.66
PCYH450	≥450	≥550	≥17	≥1.54	≥1.65
PCYH500	≥500	≥600	≥15	≥1.53	≥1.64
PCYH550	≥550	≥650	≥14	≥1.52	≥1.63
PCYH600	≥600	≥700	≥13	≥1.50	≥1.62
PCYH650	≥650	≥750	≥13	≥1.48	≥1.60
PCYH700	≥700	≥800	≥13	≥1.46	≥1.60

Table 6.54 Performance indexes of magnetic yoke steel developed by Wuhan Iron and Steel

Grade	Mechanical properties				Magnetic flux intensity (T)
	R_{eL} (MPa)	R_m (MPa)	A (%)	Impact energy at −20 °C KV_2 (J)	B_{50}
WDER235	≥235	≥350	≥22		≥1.60
WDER345	≥345	≥450	≥20		≥1.59
WDER450	≥450	≥550	≥17		≥1.54
WDER490	≥490	≥590	≥15		≥1.54
WDER550	≥550	≥650	≥14		≥1.54
WDER600	≥600	≥700	≥13		≥1.50
WDER650	≥650	≥700	≥12		≥1.50
WDER700	≥700	≥800	≥13		≥1.46
WDER750	≥750	≥800	≥11		≥1.48
WDER650MC	≥650	725–880	≥12	60 (40)	
WDER700MC	≥700	780–950	≥12	60 (40)	
WDER750MC	≥750	≥800	≥12	60 (40)	

composed of ferrite and pearlite, it is difficult to obtain high strength and toughness. Meanwhile, the phase interface has significantly influence on the displacement of the magnetic domain wall. In contrast, the internal stress of the structure consisting of uniform and fine polygonal ferrite is small, which is beneficial for achieving high magnetic properties, high strength and high toughness. Dispersive fine secondary phase particles significantly improve the strength, and their resistance on the displacement of the magnetic domain is small. Therefore, in order to develop magnetic yoke steel with high strength, high toughness and high magnetic properties, the ideal

Table 6.55 Chemical composition of the magnetic yoke steel (wt%)

C	Si	Mn	Nb+Ti	Mo
≤0.10	≤0.05	≤2.10	≤0.20	Appropriate amount

microstructure is composed of polygonal ferrite and dispersive fine secondary phase particles.

The chemical composition of the magnetic yoke steel is designed to be low in carbon, high in manganese and microalloyed with titanium, niobium and molybdenum so as to obtain the ideal structure. A part of carbon atoms enter the iron lattice as interstitial atoms, generating large internal stress and reducing the magnetic induction property. Manganese and molybdenum exist in the form of carbides and in the iron lattice in the form of substitutional solid solution atoms. They have little effect on the magnetic induction property and promote the refinement and homogenization of the austenite microstructure. Thus, the effects of manganese and molybdenum are beneficial to the strength and toughness. Titanium and niobium produce significant strengthening effect by formations of fine Nb(C, N) and Ti(C, N) precipitates, whose resistance to the displacement of the magnetic domain wall is small. A small addition of titanium can generate great strengthening effect because of the significant strengthening effect of titanium. The addition of titanium also raises the heating and rolling temperatures, which reduces the rolling load effectively and is beneficial to the control of the plate shape. In summary, the main chemical composition of the yoke steel with high strength and toughness is shown in Table 6.55.

6.5.3 *Microstructure and Properties*

The metallographic microstructure of the magnetic yoke steel WDER750 is shown in Fig. 6.60. It mainly consists of fine polygonal ferrite. The SEM image of WDER750 with the thickness of 4 mm is shown in Fig. 6.61. The microstructure consists of polygonal ferrite. By this kind of microstructure strengthening, the strength of the steel is greatly enhanced. It is the important strengthening mechanism for the magnetic yoke steel. The secondary phases in the WDER750 sample were investigated by TEM equipped with EDS, and the results are shown in Fig. 6.62. A large amount of particles precipitate and they are uniformly distributed. The shape of the precipitates is not regular. The size of most precipitates is in the range of 10–25 nm, and these precipitates mainly contain titanium and niobium. The size of a few precipitates is in the range of 130–500 nm, and these precipitates mainly contain iron and manganese.

The precipitates in steel prevent the growth of austenite grains during the heating process. Moreover, they inhibit the austenite recrystallization during the rolling process. As a result, high density dislocations and deformation bands exist in the deformed austenite, which are favorable for refining the microstructure. Meanwhile, the formation of fine dispersive precipitates provide significant precipitation strengthening effect. Because of grain refinement strengthening and precipitation strengthening, WDER750 steel plate has high strength.

The typical mechanical and magnetic properties of WDER750 are shown in Table 6.56.

6.5.4 Service Performance

Addition of titanium reduces the addition of other alloying elements into the magnetic yoke steel under the premise that the magnetic yoke steel meets the strength requirements, which is favorable for the magnetic properties. Table 6.57 shows the magnetic induction properties of the magnetic yoke steel of different strength grades. The magnetic induction properties of the longitudinal and transverse samples from different locations of WDER750 steel plates are shown in Fig. 6.63. The magnetic induction properties of WDER750 steel plate are uniform and the fluctuation is within 0.05T. Meanwhile, the greater the intensity of the magnetic field, the greater the magnetic flux density.

6.5.5 Applications

As shown in Fig. 6.64, this series of magnetic yoke steel products have been successfully used in many projects, such as Three Gorges Hydroelectric Power Station, Longtan Hydropower Station, Xiluodu Hydropower Station and Xiangjiaba Hydropower Station and so on.

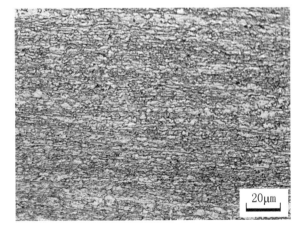

Fig. 6.60 Typical metallographic microstructure of magnetic yoke steel

Fig. 6.61 Microstructure of WDER750 sample with the thickness of 4 mm

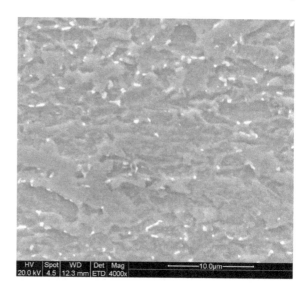

Table 6.56 Mechanical properties of the yoke steel along the longitudinal (L) and transverse (T) directions

Strength grade (MPa)	R_{eL} (MPa)		R_m (MPa)		A (%)		$-20\,°C\,KV_2$ (J)	
	L	T	L	T	L	T	L	T
600	654	710	718	765	20	22	80	72
650	744	804	823	842	18	18	84	72
700	776	817	837	864	23	20	84	68
750	820	847	852	885	20	19	80	68

Table 6.57 Magnetic induction properties of the magnetic yoke steel

Strength grade (MPa)	B_{50} (T)	B_{100} (T)	B_{200} (T)	B_{300} (T)
600	1.63	1.79	1.96	2.07
650	1.65	1.79	1.94	2.02
700	1.64	1.78	1.94	2.03
750	1.66	1.80	1.90	2.02

6.6 Hot Rolled Enamel Steel

Enamel steel is the raw material for manufacturing enameled tanks for water storage. It is widely used in the production of water heater tank, water supply equipment, water tank, water treatment tank, enameled heat sink and other products. The earliest enameled tanks in China was made of cold rolled enamel steel, most of which relied on import. With the development of economy, the demand of enameled tanks continues to grow. However, the production of enamel steel by cold rolling cannot meet the requirements of mass and low-cost production of enameled tanks because

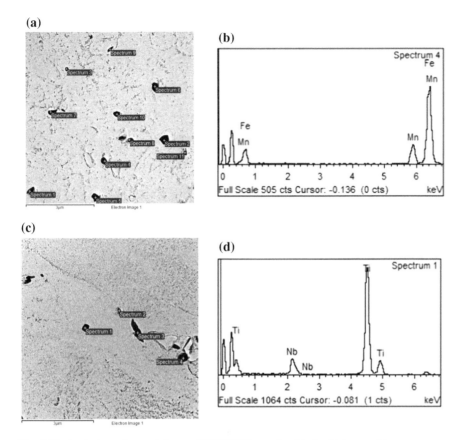

Fig. 6.62 Analysis on the precipitates of WDER750 sample with the thickness of 4 mm

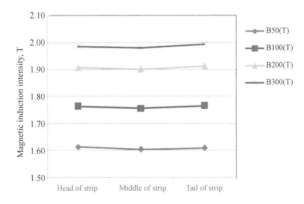

Fig. 6.63 Magnetic property of different locations of WDER50 strip with the thickness of 4 mm

Fig. 6.64 Application of magnetic yoke steel in hydropower projects: **a** Three Gorges hydroelectric power station, **b** Longtan hydropower station, **c** Xiluodu hydropower station, **d** Xiangjiaba hydropower station

of the complexity of production process, low production efficiency and high manufacturing cost. In order to replace cold rolled enamel steel imported with hot rolled for manufacturing enameled tanks, the development of hot rolled enamel steel has become increasingly important.

6.6.1 Standards and Performance Requirements

The technical specifications of hot rolled enamel steel plates include strength, formability, enameling performance and weldability and so on. Because enameled tanks are used as the water storage structure, it is required that the steel plates are able to bear high pressure. Hot rolled enamel steel can only be used for manufacturing enameled tanks under the premises that the steel passes several rigorous tests, such as fish scaling resistance test, acid-proof test, alkali-proof test, adherence test, impact resistance test, current protection test, reliability test, hot and cold water cycling test. At present, there are no unified technical standards for hot rolled enamel steel. The major manufacturers of hot rolled enamel steel are Arcelor Mittal and Nippon Steel. The main grades of Arcelor Mittal are S240EK AM FCE and S300EK AM FCE,

Table 6.58 Chemical composition of the hot rolled enamel steel developed by Arcelor Mittal (wt%)

Grade	C	Mn	P	S	Ti	B
S240EK AM FCE	0.030–0.100	0.30–0.60	≤0.020	≤0.015	≤0.040	≤0.0040
S300EK AM FCE	0.030–0.100	0.30–0.60	≤0.020	≤0.015	≤0.040	≤0.0040

Table 6.59 Mechanical properties of the hot rolled enamel steel developed by Arcelor Mittal

Grade	Sampling direction	Thickness (mm)	R_{eL}(MPa)	R_m(MPa)	A(%)
S240EK AM FCE	Transverse	1.5–3.0	≥240	≥360	≥27
S300EK AM FCE	Transverse	1.5–3.0	260–420	≥360	≥25

Table 6.60 Chemical composition of the hot rolled enamel steel developed by Nippon Steel (wt%)

Grade	C	Si	Mn	P	S	Microalloying element
NHH270	≤0.08	–	≤0.50	≤0.030	≤0.020	Ti, Nb
NHH400	≤0.25	≤0.55	≤1.50	≤0.030	≤0.020	Ti, Nb

Table 6.61 Mechanical properties of the hot rolled enamel steel developed by Nippon Steel

Grade	Sampling direction	Thickness (mm)	R_{eL} (MPa)	R_m (MPa)	A (%)
NHH270	Transverse	1.0–3.0	≥185	≥270	≥30
NHH400	Transverse	1.0–3.0	≥245	≥400	≥22

as shown in Tables 6.58 and 6.59. Nippon Steel's main grades are NHH270 and NHH400, as shown in Tables 6.60 and 6.61. Baosteel has also developed a series of hot rolled enamel steel. The main grades are BTC245R, BTC330R and BTC360R, as shown in Tables 6.62 and 6.63.

6.6.2 Composition and Process Design

The hot rolled enamel steel is generally developed by adding an appropriate amount of microalloying element into low C–Mn steel composition system. Reducing free

Table 6.62 Chemical composition of the hot rolled enamel steel developed by Baosteel (wt%)

Grade	C	Si	Mn	P	S	Alt	Microalloying element
BTC245R	≤0.12	≤0.05	≤0.70	≤0.035	≤0.035	≤0.010	Ti, Nb
BTC330R	≤0.16	≤0.05	≤0.90	≤0.035	≤0.035	≤0.010	Ti, Nb
BTC360R	≤0.16	≤0.05	≤0.90	≤0.035	≤0.035	≤0.010	Ti, Nb

Table 6.63 Mechanical properties of the hot rolled enamel steel developed by Baosteel

Grade	Sampling direction	R_{eL} (MPa)	R_m (MPa)	A (%)
BTC245R	Transverse	≥ 245	≥ 340	≥ 26
BTC330R	Transverse	≥ 330	≥ 400	≥ 22
BTC360R	Transverse	≥ 360	≥ 400	≥ 22

carbon atoms in steel reduces the formation of CO bubbles and prevents fish scaling after enameling, so the carbon content is appropriately decreased under the premises that the strength is maintained after enameling. An appropriate content of manganese ensures that the yield strength of steel meets the requirement. Adding a small amount of titanium and niobium improves the strength by precipitation strengthening. In addition, titanium and niobium-bearing secondary phase precipitates or boron-bearing inclusions increase the number of reversible hydrogen traps in steel, which improves the hydrogen storage performance, thereby enhancing the fish scaling resistance.

A series of hot rolled enamel steel of Wuhan Iron and Steel are developed by adding an appropriate amount of titanium into the low C–Mn steel composition system. The steel has appropriate strength and plasticity, good weldability and good fish scaling resistance. The content of carbon is in the range of 0.06–0.08% and is adjusted according to the strength levels. Silicon affects the surface quality of steel, resulting in the decrease in the adherence of enamel layer. Therefore, silicon is generally controlled to be not more than 0.030%. In order to compensate the strength loss due to the lack of carbon and silicon, an appropriate amount of manganese is added to ensure that the yield strength after enameling meets the requirements for the enameled tank. The content of manganese is generally in the range of 0.50–0.70%. Titanium combines with carbon and nitrogen to form very stable Ti(C, N) secondary phase precipitates. On one hand, the precipitates behave as the hydrogen trap to improve the fish scaling resistance. On the other hand, titanium improves the weldability by controlling the grain size in the heat affected zone and inhibiting the growth of abnormal grains during the welding process. The content of titanium in hot rolled enamel steel is generally controlled to be in the range of 0.030–0.050%.

The plasticity of hot rolled enamel steel is required to be high because the steel needs to go through the coiling, stamping and drawing processes for manufacturing enamel tanks. Therefore, the matrix microstructure is designed to be composed of polygonal ferrite and a small amount of pearlite. In addition, the microstructure should have fine grains and fine secondary phase precipitates to ensure that the strength meets the requirement for pressure resistance. Moreover, the mechanical properties should be isotropical because there is no definite direction for cutting the steel plates during the process of manufacturing tanks. Therefore, the matrix consisting of equiaxed grains is favorable. And the microstructure of equiaxed grains is beneficial to the adherence of the enamel layer on the surface of steel plates.

6.6.3 Microstructure and Property

The RST360 (equivalent to BTC360R) produced by Wuhan Iron and Steel is used to illustrate the microstructure and property of enamel steel. The typical metallographic microstructure is shown in Fig. 6.65. The microstructure mainly consists of ferrite and a small amount of pearlite, and the grain size of ferrite is about 12.

The precipitates were investigated by TEM equipped with EDS, and the results are shown in Fig. 6.66. The titanium-bearing secondary phase particles are fine and uniformly distributed. The size is in the range of 10–340 nm. The precipitation strengthening effect not only improves the strength of hot rolled enamel steel plates, but also helps to improve the fish scaling resistance.

The tensile tests were carried out on RST360 steel samples along the transverse, 45° and longitudinal directions. The results are shown in Table 6.64. The transverse strength is the highest, while the strength along the 45° direction is the lowest. The strength difference among different directions is small and within 20 MPa, indicating that the anisotropy of the strength is not obvious.

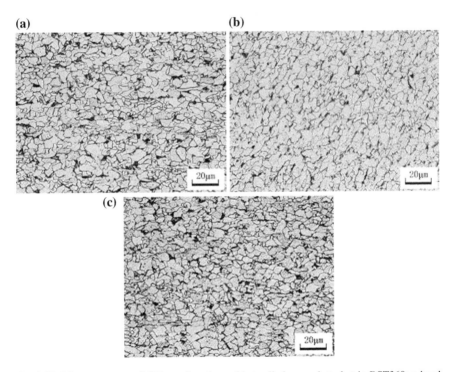

Fig. 6.65 Microstructures of different locations of hot rolled enamel steel strip RST360: **a** head, **b** middle, **c** tail

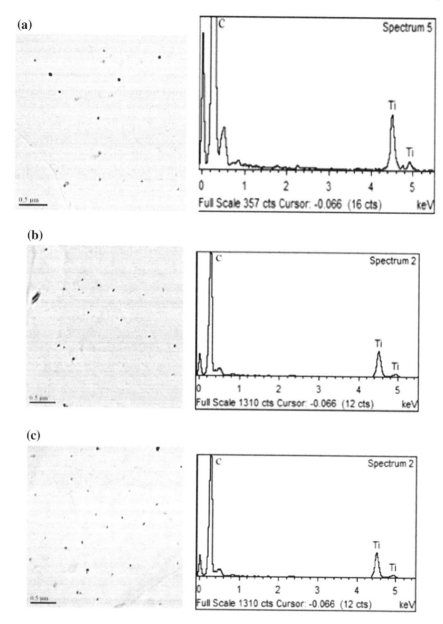

Fig. 6.66 Precipitates of different locations of hot rolled enamel steel strip RST360: **a** head, **b** middle, **c** tail

The mechanical properties were measured along the transverse and longitudinal directions of the head, middle region and tail so as to study the stability of the mechanical properties of RST360. The results shown in Table 6.65 indicate that the difference of the mechanical properties is small and less than 30 MPa.

Table 6.64 Mechanical properties of RST360 along different directions

Sampling direction	Sample number	R_{eL} (MPa)	R_m (MPa)	A (%)
Transverse	1	437	509	31.5
	2	431	512	29.5
45°	1	423	498	29.0
	2	420	501	31.0
Longitudinal	1	434	510	29.0
	2	424	497	30.5

Table 6.65 Mechanical properties of RST360 at different locations

Specification (mm)	Sampling location	Sampling direction	R_{eL} (MPa)	R_m (MPa)	A (%)
2.5	Head	Transverse	432	495	31.5
		Longitudinal	428	491	30.5
	Middle	Transverse	425	496	29
		Longitudinal	427	492	31.5
	Tail	Transverse	444	512	29.5
		Longitudinal	439	510	30.5

6.6.4 Service Performance

The fish scaling resistance is one of the most important properties of hot rolled enamel steel. The secondary phase precipitates have an important effect on the fish scaling resistance. Three samples of RT360 steel were used for surface enameling and accelerated fish scaling test. The sample was put in the refrigerator at −20 °C for 72 h to accelerate the fish scaling of its enamel layer. Then, the morphology and the hydrogen permeation curves were analyzed. The results are shown in Figs. 6.67 and 6.68.

According to Fig. 6.67, after the accelerated fish scaling test, fish scaling points with the diameter of about 0.3–0.5 mm formed on the surface of enamel layer of sample A. Fish scaling points with the diameter of about 0.1 mm were observed on sample C. In contrast, no fish scaling points formed on sample B. The results indicate that the fish scaling degree of sample A is the highest while that of sample B is the lowest. This difference is related to the fish scaling resistance of hot rolled enamel steel RT360. The hydrogen permeability of the steel plate plays a decisive role in the fish scaling of the enamel layer, so the fish scaling resistance of the steel plate is usually characterized by hydrogen permeation performance. The hydrogen permeation curves obtained at room temperature for samples A, B and C are shown in Fig. 6.68. The hydrogen permeation coefficient of sample A is the highest while that of B is the lowest, which is consistent with the observed results of the fish scaling.

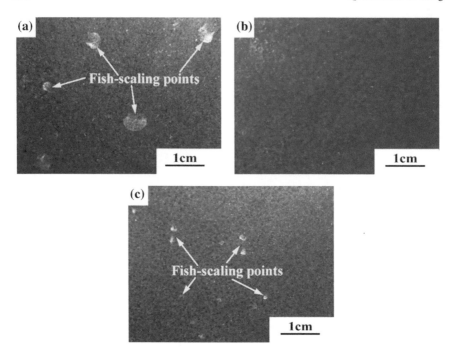

Fig. 6.67 Macroscopic surface morphology of enameled layer after accelerated fish scaling test: **a** sample A, **b** sample B, **c** sample C

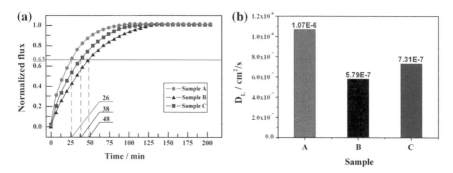

Fig. 6.68 Result of hydrogen permeation test

Hydrogen traps in alloys play an important role in the diffusion and dissolution of hydrogen atoms. All the defects in microstructures such as vacancy, dissolved atoms, dislocations, grain boundaries, micro-pores, and internal stress fields can act as hydrogen traps in alloys. However, these hydrogen traps are reversible hydrogen traps, which do not have a strong force on hydrogen atoms. Hydrogen atoms may escape in a reversible hydrogen trap in a short period of time. Reversible hydrogen traps are usually not very stable. The secondary phase precipitates in the alloy are

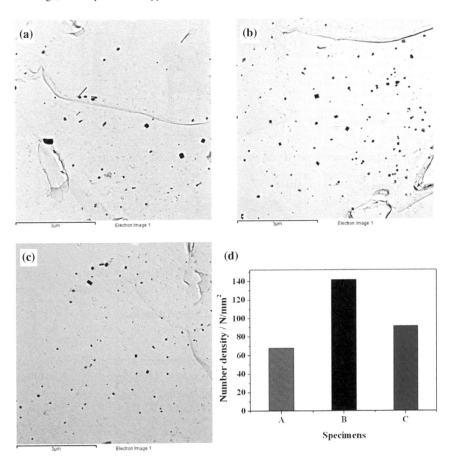

Fig. 6.69 Distribution and density of precipitates of different samples: **a** sample A, **b** sample B, **c** sample C, **d** density

usually regarded as irreversible hydrogen traps. The interaction between irreversible hydrogen traps and hydrogen atoms is strong. Therefore, hydrogen atoms cannot escape from irreversible hydrogen traps. The irreversible hydrogen trap can inhibit the diffusion of hydrogen atoms and reduce the diffusion coefficient of hydrogen, thus improving the fish scaling resistance. The samples were investigated by TEM and the results are shown in Fig. 6.69.

According to Fig. 6.69, a large amount of Ti (C, N) particles are distributed in the matrix. These secondary phase particles adsorb a large amount of hydrogen atoms in the interface between Ti (C, N) particles and matrix, thereby suppressing the free diffusion of hydrogen atoms. Further analysis show that the density of precipitates in sample B is the highest while that of sample A is the lowest, which is consistent

Fig. 6.70 **a** Electric water heater, **b** solar water heater, **c** air-source water heater

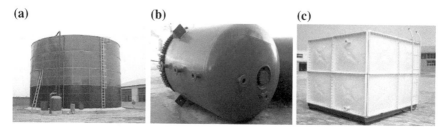

Fig. 6.71 **a** Oil storage tank, **b** chemical reaction tank, **c** large water tank

with the fish scaling resistance. Generally, the precipitation of titanium-bearing secondary phases is enhanced by adding a certain amount of titanium and proper rolling technology, so as to improve the fish scaling resistance.

6.6.5 Applications

Hot rolled enamel steel has been widely used in electric water heaters, solar water heaters, air can heaters, oil storage tanks, enameled reaction tanks, large water tanks and other products, as shown in Figs. 6.70 and 6.71. The thickness of the steel plate is generally required to be in the range of 1.2–6.0 mm. Most of the plates with the thickness in the range of 1.2–3.0 mm are cold rolled previously. Currently, the hot rolled enamel steel produced by TSCR process has been developed to replace the cold rolled steel with the thickness in the range of 1.2–3.0 mm.

6.7 Low Alloyed High Strength Structure Steel

At present, low alloyed high strength structure steel is the main steel material. It includes several grades such as Q345, Q390, Q420, Q460, Q500, Q550, Q620, Q690. Its quality level is graded into five levels, i.e. A, B, C, D and E. The annual consumption is more than 25 million tons and it is widely used in building structures, construction machinery, automobiles, coal mining, maritime works, military machinery and equipment.

Microalloying technology is an effective way to improve the comprehensive performance of low alloyed high strength structure steel mainly through the addition of small amounts of strong carbide forming elements niobium or titanium into conventional low alloyed steel. The price of niobium is expensive, in contrast, the price of titanium is close to that of manganese. By adding an appropriate amount of titanium to provide the grain refinement strengthening and precipitation strengthening, the amount of manganese is greatly reduced. As a result, the manufacturing cost is reduced and the market competitiveness of the steel product is improved.

Meanwhile, if the traditional high manganese steel composition system is used to develop the low alloyed high strength structure steel requiring high cold formability, several problems will occur. Segregation in the slab center is likely to happen because the control of superheat of molten steel in tundish is unstable. After rolling, the problems of inclusions MnS in the center of the plate and the pearlite banded structure are serious, making the steel plates prone to bending cracking. By decreasing manganese and adding titanium, not only the content of manganese is reduced, but also the precipitation of titanium consumes a part of sulfur and carbon to form granular compound inclusions. Therefore, the number of MnS inclusions is effectively reduced and the problem of pearlite banded structure is relieved, which greatly improves the cold formability of low alloyed high strength structure steel.

6.7.1 Standards and Performance Requirements

Currently, the composition system based on addition of titanium and replacement of manganese is mainly used for the low alloyed high strength structure steel with the strength below 460 MPa and the quality levels of A, B and C. The thickness specification is in the range of 2–16 mm. The national standard GB/T 1591-2008 is for low alloyed high strength structure steel. The chemical composition and mechanical properties of the steel with the strength below 460 MPa and the quality levels of A, B and C are shown in Tables 6.66 and 6.67.

Table 6.66 Chemical composition of the low alloyed high strength steel with the strength below 460 MPa (wt%)

Grade	Quality level	C	Si	Mn	P	S	Nb	Ti	N	Als
Q345	A/B	≤0.20	≤0.50	≤1.70	≤0.035	≤0.035	≤0.07	≤0.20	≤0.012	–
	C				≤0.030	≤0.030				≤0.015
Q390/Q420	A/B	≤0.20	≤0.50	≤1.70	≤0.035	≤0.035	≤0.07	≤0.20	≤0.015	–
	C				≤0.030	≤0.030				≤0.015

Table 6.67 Mechanical properties of the low alloyed high strength steel with the strength below 460 MPa

Grade	Yield Strength (MPa)	Strength (MPa)	Elongation (%)	Cold bending test (180 °C)
Q345A/B/C	≥345	470–630	≥20	$d=2a$
Q390A/B/C	≥390	490–650	≥20	$d=2a$
Q420A/B/C	≥420	520–680	≥19	$d=2a$

Table 6.68 Chemical composition of Q345A/B (wt%)

Grade	C	Si	Mn	P	S	N	Als
Q345	0.15–0.20	≤0.50	1.20–1.60	≤0.035	≤0.025	≤0.012	≥0.015

Table 6.69 Chemical Composition of Q345A/B (Ti) (wt%)

Grade	C	Si	Mn	P	S	Ti	N	Als
Q345(Ti)	0.12–0.20	≤0.50	0.40–0.60	≤0.025	≤0.015	0.03–0.05	≤0.008	≥0.015

6.7.2 Composition and Process Design

The purpose of reducing the amount of manganese by adding titanium into the low alloyed high strength structural steel is to reduce the cost and improve the competitiveness of the steel. Therefore, the composition and process design should focus on this purpose, and take the improvement of the properties and quality into account. The following part mainly introduces the development of typical Q345B steel.

The conventional Q345B is based on C–Mn steel composition system, and the manganese content is between 1.2 and 1.6 wt%. By addition of titanium to reduce the content of manganese, Q345A/B (Ti) is developed based on the C–Mn–Ti steel composition system, in which the manganese content is about 0.4–0.6 wt% and the titanium content is 0.03–0.05 wt%. The compositions of the conventional and new Q345B steel are shown in Tables 6.68 and 6.69.

Titanium is liable to react with impurity elements such as oxygen, nitrogen and sulfur to form large Ti_2O_3, TiN and $Ti_4C_2S_2$ particles, which reduces the grain refinement strengthening and precipitation strengthening. As a result, it is necessary to control the contents of oxygen, nitrogen and sulfur.

Titanium forms highly dispersive carbonitride particles, which pin the boundaries of austenite grains and prevent grain growth. Therefore, the austenite grains in Q345B (Ti) steel maintain fine at about 1250 °C, and high temperature heating also ensures that the steel has high strength. The process is designed as follows. The heating temperature for Q345B(Ti) is set in the range of 1250–1300 °C. Two stages of rolling consisting of recrystallization region rolling and non-recrystallization region rolling are used. Large reduction is applied to facilitate the strain induced precipitation of TiC and Ti (C, N) particles. After rolling, the steel strip is cooled by laminar cooling, and is coiled at about 600 °C.

6 Design, Development and Application of Titanium …

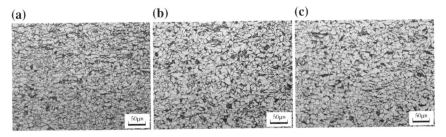

Fig. 6.72 Metallographic microstructure of Q345B (Ti): **a** upper surface, **b** core, **c** lower surface

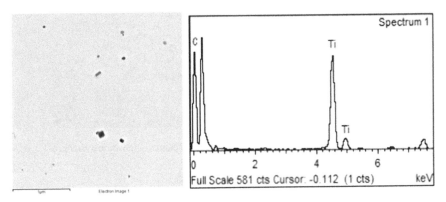

Fig. 6.73 Morphology and composition of precipitates

6.7.3 Microstructure and Properties

The microstructure of Q345B (Ti) with the thickness of 16 mm was investigated and the results are shown in Fig. 6.72. The microstructures of the upper surface, core and lower surface are composed of ferrite and pearlite. In addition, the difference of the grain size is small.

The precipitates were investigated by TEM and the results are shown in Fig. 6.73. A large number of precipitates are observed. The number of the precipitates with the size in the range of 30–50 nm is the most, and the shape is mainly ellipsoid. Most of the precipitates with the size in the range of 50–100 nm are square, while the size of a small amount of square precipitates reaches 200 nm. The EDS analysis shows that most precipitates are TiC. These fine TiC precipitates hinder the growth of recrystallized austenite grains and refine the grains. Adversely, the large rectangular TiN particles with a small amount are disadvantages to the toughness of steel.

Table 6.70 shows the mechanical properties of Q345B(Ti) of different thickness specifications.

Table 6.70 Mechanical properties of Q345B(Ti) of different thickness specifications

Thickness (mm)	$R_{p0.2}$ (MPa)	R_m (MPa)	$R_{p0.2}$ (R_m)	A_{50} (%)
4	439	546	0.80	28
8	439	523	0.84	26
12	394	503	0.78	25

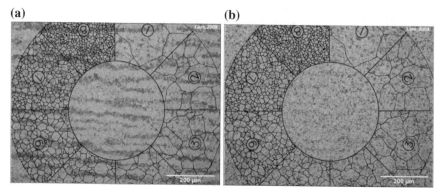

Fig. 6.74 Metallographic microstructure of Q345B of different composition system: **a** 1.5 wt% Mn, **b** 0.5 wt% Mn −0.04 wt% Ti

6.7.4 Service Performance

Formability

Low alloyed high strength structure steel is used for structural parts of vehicles, such as rear beam, body frame and so on. The manufacturing of these parts include large deformation processing, which requires that the steel has good cold bending property. Bending cracking problems may occurr due to the use of conventional high manganese steel Q345B for the rear beam. After Q345B (Ti) was used to replace Q345B, no cracking phenomenon occurred. The microstructures of Q345B (Ti) and Q345B are shown in Fig. 6.74. Compared with Q345B, the grain size of ferrite in Q345B(Ti) is in the grade of 9–10, and the grade of the banded structure decreases from 2–3 to 0.5 or less, indicating the pearlite banded structure almost disappears. The grade of inclusions is improved from class A (2.5–3.0) level to class B (below 1.0), as shown in Fig. 6.75.

A series of cold bending tests were carried out on the samples with the thicknesses of 6, 12 and 16 mm. The diameter of the bending core is $2a$ and the bending angle is 180°. The tests were carried out with the WES-1000G material testing machine according to the national standard GB/T 232-1999. The results show that all the samples are qualified. The cold bending test samples are shown in Fig. 6.76.

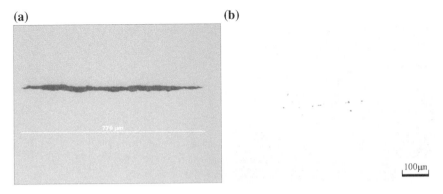

Fig. 6.75 Inclusion grades of Q345B of different composition systems: **a** 1.5 wt% Mn, **b** 0.5 wt% Mn–0.04 wt% Ti

Fig. 6.76 Samples for cold bending test

Low Temperature Toughness

The Charpy V-notch impact tests were carried out on the transverse samples with the thicknesses of 6 and 12 mm to study the toughness of titanium-bearing Q345B steel The dimensions of the samples were $5 \times 10 \times 55$ mm and $10 \times 10 \times 55$ mm. For each thickness specification and each temperature, a group of three tests were applied. The impact test was carried out with the JB-30B impact testing machine according to the national standard GB/T 299-1994. The test temperature was in the range of -80–20 °C, and the results are shown in Fig. 6.77.

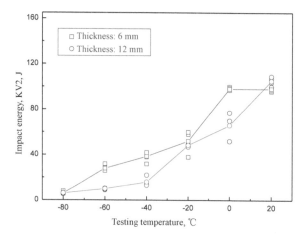

Fig. 6.77 Impact energy of Q345B (Ti) at different testing temperatures

Table 6.71 Welding parameters

Welding materials		Current (A)	Voltage (V)	Speed (cm/min)	Energy input (KJ/cm)	Gas flow rate (L/min)
Grade	Specification (mm)					
WH50C6	Φ1.2	260–280	24–28	30–33	11.3–15.7	14–16

The results show that Q345B (Ti) steel has no obvious low temperature transition temperature, and it has a high impact energy from room temperature to −20 °C. The impact performance is poor when the temperature is below −20 °C. The thicker the plates, the worse the impact toughness. Therefore, for thick steel plates and requirements for high impact toughness at low temperatures, it is not appropriate to choose composition system based on adding titanium and decreasing manganese.

6.7.4.1 Weldability

According to the national standards GB/T 2651-2008 and GB/T 2653-2008, the LincolnPower Wave 455 welder was used to test the weldability of Q345B (Ti) with the thickness of 10 mm. The shielding gas was CO_2. The welding power was the DC reverse. The steel plates were welded without preheating. Multi-layer and multi-channel welding was applied and the interlayer temperature was controlled to be below 150 °C. After the front side was welded, the reverse side was grinded to clear the root, and then was cover welded. The environmental temperature was 25 °C and the humidity was 87%. The welding parameters are shown in Table 6.71.

Figure 6.78 shows the microstructures of the different locations of the samples. The microstructures of the weld, transitional zone and base metal consist of pre-eutectoid ferrite and acicular ferrite, non-uniform ferrite and pearlite, ferrite and

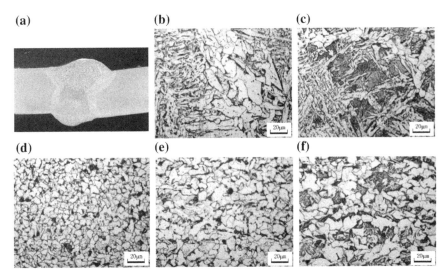

Fig. 6.78 Metallographic microstructure of the weld joint of Q345B(Ti): **a** weld joint, **b** weld, **c** coarse grain zone, **d** fine grain zone, **e** critical heat affected zone, **f** base metal

Table 6.72 Results of the tensile test and bending test

Tensile test		Bending test		
Tensile strength (MPa)	Fraction location	Diameter of the bending core	Bending angle	Evaluation
498	Weld	$D=2a$	180°	Qualified
520	Weld			Qualified

pearlite, respectively. The microstructure of the coarse grain zone of heat affected zone is composed of coarse ferrite and pearlite, while that of the fine grain zone consists of ferrite and pearlite.

The tensile test and 180 °C bending test were carried out on the samples welded, and the results are shown in Table 6.72. The fracture location of the tensile sample was at the weld. The tensile strength of the welds are 498 MPa and 520 MPa. The results of the 180° bending tests are qualified. The results indicate that the steel has good weldability, but the strength of the weld is low.

6.7.5 Applications

Titanium microalloyed low alloyed high strength structure steel with the strength below 460 MPa is widely used in the structural parts of engineering machinery, rectangular tube of vehicle body, as shown in Fig. 6.79. In order to reduce the cost

Fig. 6.79 Application of titanium microalloyed low alloyed high strength structure steel: **a** rear beam of truck body, **b** excavator arm

and improve the efficiency, the steel manufacturers in China, such as Wuhan Iron and Steel, Taiyuan Iron and Steel, Han Steel have widely used titanium composition system. By adding 0.03%–0.05 wt% titanium, the content of manganese can be reduced by 0.8%–1.1wt%, which significantly reduces the cost.

References

1. Mao X P, ChenQ L, ZhuD Y. Development and application of new generation of container steel, compiled in Frontiers of Modern Chemical, Metallurgy and Materials Technology [C]. Metallurgy and Materials Engineering of Chinese Academy of Engineering. Proceedings of the 7th Academic Conference of the Department of Chemical. Beijing: Chemical Industry Press, 2009: 809–813.
2. Chen X W, Li L J, Zhuang H Z, et al. Research and design of the composition of titanium microalloyed high strength weathering steel [C]. Proceedings of the Smelting and Continuous Casting Conference of Pan Pearl River Delta Region, 2007: 296–300.
3. Zhu D Y, Gao J X, Chen Q L, et al. Development and application of 700 MPa ultra-high strength steel [J]. Metallurgical Collections, 2011, 3:8–10.
4. Lu J X. Research and development of a high strength micro-alloyed steel with yield strength of 700 MP [D]. Shengyang: Northeastern University, 2005.
5. Qu P. Development of 700 MPa grade high strength container steel in FTSC of Bensteel [J]. Physics Examination and Testing, 2009, 27(3): 10–13, 28.
6. Wang Y T. Experiment and development of TISCO TH800 hot rolled coil sheet with high strength [J]. Shanxi Metallurgy, 2009, 32(1): 17–19.
7. Chen L, Zhang C, Zhu S, et al. Rolling process and microstructure and properties of thin gauge ultra-high strength strip manufactured by CSP line [J]. Iron and Steel, 2014, 49(1): 57–60.
8. Kang Y L. Theory and technology of processing and forming for advanced automobile steel sheets [M]. Beijing: Metallurgy Industry Press, 2009.
9. Ma M T. Advanced automobile steel [M]. Beijing: Chemical Industry Press, 2007.
10. Zrnik J, Mamuzic I, Dobatkin S V. Recent process in high strength low carbon steels [J]. Metabk, 2006, 45(4):323.
11. Kang Y L. The technology and application process of ultra-thin hot rolled strip produced by thin slab casting and rolling [J]. Steel Rolling, 2015, 32(1): 7–11.

12. Tao W Z, Liang W, Liu Z Y. Analysis on the cold bending cracking mechanism of 700 MPa thin gauge high strength steel [J]. Wugang Technology, 2013, 3: 12–15.
13. Han B, Shi X G, Dong Y. Development and production of ultra-high strength steel for automotive heavy beam [J]. Angang Technology, 2012, 5: 10–13.
14. Yu Y, Meng X T, Wang L, et al. Analysis on stamping cracking mechanism of Shougang high strength beam steel of 750 MPa level [J]. Shougang Science and Technology, 2013, 4: 16–20.
15. Zhao P L, Lu F, Wang J J. Research and development of 700 MPa grade super high strength automobile frame steel [J]. Steel Rolling, 2011, 28(2): 12–15.